Lecture Notes in Mathematics

Edited by A. Dold and B. Eckmann

423

S. S. Abhyankar
A. M. Sathaye

Geometric Theory of Algebraic Space Curves

Springer-Verlag
Berlin · Heidelberg · New York 1974

Prof. Dr. Shreeram Shankar Abhyankar
Purdue University
Division of Mathematical Sciences
West Lafayette, IN 47907/USA

Prof. Dr. Avinash Madhav Sathaye
University of Kentucky
Department of Mathematics
Lexington, KY 40506/USA

Library of Congress Cataloging in Publication Data

Abhyankar, Shreeram Shankar.
Geometric theory of algebraic space curves.

(Lecture notes in mathematics ; 423)
Includes bibliographical references and indexes.
1. Curves, Algebraic. 2. Algebraic varieties.
I. Sathaye, Avinash Madhav, 1948- joint author.
II. Title. III. Series: Lecture notes in mathematics (Berlin) ; 423.
QA3.L28 no. 423 [QA567] 510'.8s [516'.35] 74-20717

AMS Subject Classifications (1970): 14-01, 14H99, 14M10

ISBN 3-540-06969-0 Springer-Verlag Berlin · Heidelberg · New York
ISBN 0-387-06969-0 Springer-Verlag New York · Heidelberg · Berlin

Offsetdruck: Julius Beltz, Hemsbach/Bergstr.

PREFACE

The original main part of this book was just a sequel to the Montreal Notes [3]. The main part was the proof to the Theorem (36.9), namely that "All irreducible nonsingular space curves of degree at most five and genus at most one over an algebraically closed ground field are complete intersections." As such, the Theorem was completely proved in 1971. Two versions of the proof have been written and circulated, but none published. We intended to give a completely self-contained treatment of the Theorem, and in the process, the size of the proof, or rather, the preparatory material, enlarged; while the proof continued to become clearer and somewhat sharper. The present version was finally started in June 1973, and we finally decided that it had to be a book.

During October 1973, however, Murthy rendered our main Theorem obsolete by proving that "All irreducible nonsingular space curves of genus at most one over an algebraically closed ground field are complete intersections. [12]. In fact he proved the well known "Serre's Conjecture" that all projective modules over $k[X,Y,Z]$ are free (when k is algebraically closed)." However he also illustrated in his proof that, concrete detailed proofs can sometimes be more useful than the theorem itself, by using a concrete description of a basis of three elements for the ideal of an irreducible nonsingular space curve given in the Montreal Notes [3] as one of the main steps in his proof.

Another important feature of the book is a new treatment of the genus and differentials of a separably generated one-dimensional function field over an arbitrary ground field. The treatment was developed by Abhyankar during the Purdue Seminar in Summer and Fall 1973. One virtue of the present treatment is, that it does not need any artificial devices such as repartitions or derivations of the ground field. (See Chapter III for the treatment and §40 for the comparison with other treatments.)

Returning to the proof of the complete intersection theorem, we note that it has basically two parts: one part is to obtain a "nice" projection of the space curve with a "nice" adjoint (our Theorem (26.12)) and the second part is to construct a basis of two elements for the ideal of the curve (our Theorem (36.7)).

In the present version of the proof, the second part that is needed was already developed in [3], but we include a proof for the sake of completeness. In an older version, a generalization was needed namely, the treatment of the so-called chains of euclidean domains, and is presented in §39, mainly because it is of interest in itself.

The treatment of the first part about projections is essentially different from its counterpart in [3]. In [3], we write down explicit equations of the curve and carry out the proof by complicated calculations; here a similar method was first tried and proved to be too cumbersome. What we present here is a geometric argument in which we never even need a coordinate system. However, it might be difficult to convince anybody, that this is geometry, for we have deliberately avoided the use of geometric terms, so that the proof may stay rigorours, self-contained and still reasonably short.

Thus we have taken the useful geometric concepts, translated them into precise algebraic terms and almost never gone back to the geometric terms. For a geometric minded reader, however, we have provided a dictionary in §43 so that he may be able to read the underlying geometric argument very easily. The name of this book owes its 'geometry' to this arrangement.

The proof of existence of the "nice projection" may be appropriately described as repeated applications of "Bezout's Theorem." We need basically the special (but most well known) case, termed as "Bezout's Little Theorem" in the present book, namely, the case of the intersection of a hypersurface with a curve. This case is treated in §(23.9). The general case of intersection of two hypersurfaces is presented in §38; mainly because a readily accessible proof of this case is not available in the literature.

Bezout's Theorem is, however, a projective theorem. This made it necessary to embed the given affine curve in a projective space, and study its projective completion. To avoid clashes in affine and projective terminology we first concentrate only on projective curves in Chapter II, get a projective counterpart of the theorem about existence of nice projections (§26) and then return to affine curves in Chapter IV.

Bezout's theorem also needs a precise theory of intersection multiplicity; for projective as well as affine curves, in our case. We develop the theory for projective curves in Chapter II §23,24 and for affine curves in Chapter I §5,6; and put it together in Chapter IV. The only use of Chapter III, so far as complete intersections are concerned is to provide a proof of the well known genus formula for plane curves.

We have taken the viewpoint that varieties are represented by their coordinate rings, if they are abstract, and by ideals in the coordinate rings of ambient spaces in case they are embedded. Several sections appear to be almost in duplicate (§5,6; §23,24 etc.) because they describe the same results for these two types of varieties.

Here is a general summary of various chapters.

Chapter I gives the theory of intersection multiplicity of two curves at a point, essentially for affine or local curves. It also contains some general terminology.

Chapter II gives a treatment of "homogeneous domains" which represent abstract irreducible projective varieties. It develops the concepts of intersection multiplicity, projection, tangential spaces etc. for projective curves.

Chapter III gives a treatment of differentials in separably generated function fields. It also has various genus formulas for abstract and embedded projective curves. This chapter is almost self-contained except for some use of Chapter I and some alternative proofs using Chapter II.

Chapter IV studies affine irreducible curves with an equivalence class of affine coordinate systems preassigned; translated algebraically, such curves have coordinate rings which are "filtered domains". We also study the concepts of taking a projective completion and taking an affine piece; algebraically, homogenization and dehomogenization. Then we go on to finish the proof of the main Theorem (36.9).

Chapter V is a supplement. It deals with generalizations of some concepts of the first four chapters and has several statements whose proof are only sketched or referred to other sources.

An elementary knowledge of general algebra is assumed to be available to the reader (for example, results like 'Nakayama's Lemma' , "Krull's Intersection Theorem', '$\sum e_i f_i = n$ formula' etc.). There is only one "official exercise" (in §15), but several properties stated in Chapter I, II and IV may very well be treated as exercises with varying degree of difficulty.

The contents are intended to give brief descriptions in geometric words of what is being treated in the relevant sections. A list of interdependences of sections follows the contents.

Shreeram S. Abhyankar

Avinash Sathaye

CONTENTS

<u>Interdependence of sections</u>.

In the following, $\S b \leftarrow \S a_1, \ldots, \S a_r$ means $\S a_1, \ldots, \S_r$ are directly referred to in §b. Except for such references, the only other prerequisites for §b are the notations and definitions from previous sections and they may be located from the index.

§1, §2, §3, §4, §5, basic.

§6 ← §5.

§7 ← §6.

§8,§9, independent.

§10 ← §5.

§11, §12, §13, §14, §15, §16, §17, basic.

§18 ← §15.

§19 ← §18.

§20 ← §18,§19.

§21 ← §18,§20.

§22 ← §21.

§23 ← §4,§5,§15,§17,§18.

§24 ← §5,§10,§15,§23.

§25 ← §8,§9,§15,§23.

§26 ← §18,§21,§23,§25.

§27, basic.

§28 ← §27 (§3,§5,§23 and §25 optional use).

§29 ← §4,§15,§17,§24,§25,§28.

§30 ← §25,§26,§29.

§31,§32, basic.

§33 ← §32.

§34 ← §15,§25,§32,§33.

§35 ← §19,§26,§30,§33,§34.

§36 ← §10,§26,§30.

§37 ← §10.

§38 ← §23,§25.

§39 ← §36.

§40, independent.

§41, related to §28.

§42 ← §29,§30 (also related to §10).

§43,§44,§45, index.

CHAPTER I: LOCAL GEOMETRY OR LENGTH

§1. General terminology.

We shall assume the terminology introduced in [1: (1.1),(1.6), (1.7),(1.8),(1.9)]. This includes general algebraic terms, definitions and elementary properties of models and some well known and useful results about local rings. We shall also use the following notation.

By card we denote cardinal number.

For subsets $J_1, J_2, \ldots, J_n$ of an additive group E (for instance $E =$ a module over a ring), where n is a positive integer, by $J_1 + J_2 + \ldots + J_n$ we denote the set $\{x_1 + x_2 + \ldots + x_n : x_i \in J_i\}$.

For subsets $J_1, J_2, \ldots, J_n$ of a ring A, where $n > 1$ is a positive integer, by $J_1 J_2 \ldots J_n$ we denote the additive subgroup of A generated by the set $\{x_1 x_2 \ldots x_n : x_i \in J_i\}$. For a subset J of A and a positive integer n, by J^n we denote the additive subgroup of A generated by the set $\{x_1 x_2 \ldots x_n : x_i \in J\}$. In case J is an ideal in A, we also put $J^0 = A$.

For a module E over a ring A, by $[E:A]$ we denote the <u>length</u> of E as an A-module, i.e.,

$$[E : A] = \sup\left\{n : \begin{array}{l}\text{there exists a sequence } E_0 \subset E_1 \subset E_2 \subset \ldots \subset E_n \\ \text{of } A\text{-submodules of } E \text{ with} \\ E_0 \neq E_1 \neq E_2 \neq \ldots \neq E_n\end{array}\right\}$$

Note that then $[E : A]$ is either a nonnegative integer or ∞. Also note that, if A is a field, then $[E : A]$ is simply the vector-space-dimension of E over A.

§2. Principal ideals and prime ideals.

For a ring A we <u>define</u>:

$F^*(A) =$ the set of all nonzero principal ideals in A,

$$\mathfrak{P}(A) = \text{the set of all prime ideals in } A ,$$

$$\mathfrak{P}_i(A) = \{P \in \mathfrak{P}(A) : \dim A/P = i\}$$

$$\left.\begin{aligned} \mathfrak{P}(A,x) &= \{P \in \mathfrak{P}(A) : x \in P\} \\ \mathfrak{P}_i(A,x) &= \mathfrak{P}(A,x) \cap \mathfrak{P}_i(A) \end{aligned}\right\} \text{ for any } x \in A ,$$

$$\left.\begin{aligned} \mathfrak{P}(A,I) &= \{P \in \mathfrak{P}(A) : I \subset P\} \\ \mathfrak{P}_i(A,I) &= \mathfrak{P}(A,I) \cap \mathfrak{P}_i(A) \end{aligned}\right\} \text{ for any } I \subset A ,$$

$$\left.\begin{aligned} \mathfrak{P}([A,I]) &= \mathfrak{P}(A,I) \\ \mathfrak{P}_i([A,I]) &= \mathfrak{P}_i(A,I) \end{aligned}\right\} \text{ for any } I \in A \text{ or } I \subset A ,$$

and

$$\left.\begin{aligned} \mathfrak{P}([A,I],J) &= \mathfrak{P}(A,I) \cap \mathfrak{P}(A,J) \\ \mathfrak{P}_i([A,I],J) &= \mathfrak{P}([A,I],J) \cap \mathfrak{P}_i(A) \end{aligned}\right\} \begin{aligned} &\text{for any } I \in A \text{ or } I \subset A \\ &\text{and any } J \in A \text{ or } J \subset A . \end{aligned}$$

We note that then

$$\mathfrak{P}(A) = \mathfrak{P}(A,0) ,$$

$\mathfrak{P}(A,IA) = \mathfrak{P}([A,0],I) = \mathfrak{P}([A,I],0) = \mathfrak{P}([A,I]) = \mathfrak{P}(A,I)$ for any $I \in A$ or $I \subset A$, and for any ideals I and J in A we have:

$$\mathrm{rad}_A I = \bigcap_{P \in \mathfrak{P}([A,I])} P$$

$$\mathfrak{P}(A,I) = \mathfrak{P}(A) \Leftrightarrow I \subset \mathrm{rad}_A\{0\}$$

$$\mathfrak{P}(A,I) = \emptyset \Leftrightarrow I = A$$

$$\mathfrak{P}(A,I) = \mathfrak{P}(A,J) \Leftrightarrow \mathrm{rad}_A I = \mathrm{rad}_A J$$

$$\mathfrak{P}(A,I) \cup \mathfrak{P}(A,J) = \mathfrak{P}(A,I \cap J) = \mathfrak{P}(A,IJ)$$

and

$$\mathfrak{P}(A,I) \cap \mathfrak{P}(A,J) = \mathfrak{P}(A,I + J) .$$

§3. Total quotient ring and conductor.

For a nonnull ring R we <u>define</u>

$\mathfrak{J}(R)$ = the total quotient ring of R.

and we <u>define</u>

$\mathfrak{C}(R)$ = the conductor of R in the integral closure of R in $\mathfrak{J}(R)$, and we note that, upon letting R^* to be the integral closure of R in $\mathfrak{J}(R)$, we then have

$$\begin{aligned}\mathfrak{C}(R) &= \text{the largest ideal in } R \text{ which remains an ideal in } R^* \\ &= \{x \in R: xR^* \subset R\} \\ &= \{x \in R^*: xR^* \subset R\} \\ &= \{x \in \mathfrak{J}(R): xR^* \subset R\}.\end{aligned}$$

For a nonunit ideal C in a ring A we <u>define</u>

$$\mathfrak{J}([A,C]) = \mathfrak{J}(A/C)$$

and, upon letting $f: A \to A/C$ to be the canonical epimorphism, we <u>define</u>

$$\mathfrak{C}([A,C]) = f^{-1}(\mathfrak{C}(A/C)) .$$

We take note of the fact that the conductor localizes properly, i.e.:

(3.1) <u>If</u> R <u>is a domain such that the integral closure of</u> R <u>in</u> $\mathfrak{J}(R)$ <u>is a finite model, and</u> S <u>is the quotient ring of</u> R <u>with respect to some multiplicative system in</u> R, <u>then</u> $\mathfrak{C}(R)S = (S)$.

§4. Normal model.

For a domain R and a subring k of R we <u>define</u>

$\mathfrak{X}(R,k)$ = the set of all subrings V of $\mathfrak{J}(R)$ with $k \subset V$ such that $\mathfrak{J}(V) = \mathfrak{J}(R)$ and V is a one-dimensional regular local ring.

For a domain R we <u>define</u>

$$\mathfrak{D}(R) = \mathfrak{X}(R,R)$$

and for any $Q \in R$ or $Q \subset R$ we <u>define</u>

$$\mathfrak{Y}(R,Q) = \{V \in \mathfrak{Y}(R): \text{ord}_V Q > 0\}.$$

For a ring A and any $C \in \mathfrak{P}(A)$ we <u>define</u>

$$\mathfrak{Y}([A,C]) = \mathfrak{Y}(A/C).$$

For a ring A, $C \in \mathfrak{P}(A)$, and $Q \in A$ or $Q \subset A$, upon letting $f: A \to A/C$ to be the canonical epimorphism, we <u>define</u>

$$\text{ord}_{[A,C],V} Q = \text{ord}_V f(Q) \qquad \text{for any } V \in \mathfrak{Y}([A,C])$$

and we <u>define</u>

$$\mathfrak{Y}([A,C],Q) = \left\{V \in \mathfrak{Y}([A,C]): \text{ord}_{[A,C],V} Q > 0\right\}.$$

We observe that then

$$\mathfrak{Y}([A,C],Q) = \mathfrak{Y}(f(A), f(Q)).$$

We note that, if K is a function field (i.e. a finitely generated field extension) over a field k with $\text{trdeg}_k K = 1$, then $\mathfrak{X}(K,k)$ is a projective model of K over k and: $\mathfrak{X}(K,k)$ is the unique complete model of K over k all whose members are normal. We also take note of the well-known fact that "the degree of the divisor of a function is zero", i.e.:

(4.1) <u>If</u> K <u>is a function field over a field</u> k <u>with</u> $\text{trdeg}_k K = 1$, <u>then for every</u> $x \in K$ <u>we have</u>,

$$\sum_{V \in \mathfrak{X}(K,k)} (\text{ord}_V x)[V/M(V): k] = \begin{cases} 0, & \text{if } x \neq 0 \\ \infty, & \text{if } x = 0. \end{cases}$$

We remark that by <u>convention</u>

positive real number times ∞ = ∞ times positive real number
= ∞ times ∞
= ∞.

Further, for any family $a(u)_{u\in U}$ where $a(u)$ is an integer or ∞, upon letting

$$U' = \{u \in U: a(u) \neq 0\},\ U^* = \{u \in U: a(u) = \infty\},$$
$$U^{**} = \{u \in U: a(u) < 0\}.$$

by convention we have

$$\sum_{u\in U} a(u) = \begin{cases} 0 & \text{if } U' = \emptyset \\ \sum_{u\in U'} a(u) & \text{if } U' \text{ is a nonempty finite set and } U^* = \emptyset \\ \infty & \text{if } U' \text{ is a nonempty finite set and } U^* \neq \emptyset \\ \infty & \text{if } U' \text{ is an infinite set and } U^{**} \text{ is a finite set.} \end{cases}$$

We recall that (4.1) follows immediately from (4.2) which in turn follows from " $\sum e_i f_i = n$ " formula, i.e., from the well-known fact (4.3) about extensions of Dedekind domains.

(4.2) If K is a function field over a field k with $\mathrm{trdeg}_k K = 1$, then, upon letting k^* to be the (relative) algebraic closure of k in K for every $x \in K$ we have

$$\sum_{V\in\mathfrak{X}(K,k[x])} (\mathrm{ord}_V x)[V/M(V): k] = \begin{cases} [K: k(x)] & \text{if } x \notin k^* \\ 0 & \text{if } 0 \neq x \in k^* \\ \infty & \text{if } x = 0. \end{cases}$$

(4.3) Let S be a Dedekind domain with quotient field L and let R be the integral closure of S in a finite algebraic field extension K of L. Assume that R is a finite S-module. (Note that this assumption is automatically satisfied, if S is an affine ring over a field, and also if S is the quotient ring, with respect to some multiplicative system, of an affine ring over a field.) Then for every $Q \in \mathfrak{P}_0(S)$ we have

$$\sum_{P\in\mathfrak{P}_0(R)} (\operatorname{ord}_{R_P} Q)[R_P/PR_P : S/Q] = [K : L]$$

§5. Length in a one-dimensional noetherian domain.

Let R be a noetherian domain with $\dim R = 1$. Let k be a subring of R such that

(*) for every $P \in \mathfrak{P}_0(R)$ we have $k \cap P \in \mathfrak{P}_0(k)$

and $[R/P : k/(k \cap P)] < \infty$.

(Note that (*) is satisfied for $R = k$; it is also satisfied, if k is a subfield of R such that $[R/P : k] < \infty$, for all $P \in \mathfrak{P}_0(R)$.)

Let R^* be the integral closure of R in $\mathfrak{J}(R)$. Assume that

(**) R^* is a finite R-module.

For any $I \in R$ or $I \subset R$ and any $Q \in R$ or $Q \subset R$ we <u>define</u>

$$\lambda^k(R,I,Q) = \sum_{V\in\mathfrak{D}(R,Q)} (\operatorname{ord}_V I)[V/M(V) : k/(k \cap M(V))]$$

and

$$\lambda(R,I,Q) = \lambda^R(R,I,Q),$$

and for any $I \in R$ or $I \subset R$ we <u>define</u>

$$\lambda^k(R,I) = \lambda^k(R,I,0) \quad \text{and} \quad \lambda(R,I) = \lambda(R,I,0).$$

(We note that, if k is a field then: in the above defining equation of λ^k and in assertion (5.3) below, we have $[V/M(V) : k(k \cap M(V))] = [V/M(V) : k]$. We also note that, if k is an algebraically closed field, then: for any $I \in R$ or $I \subset R$ or $Q \subset R$ we have $\lambda^k(R,I,Q) = \lambda(R,I,Q)$, and for any $I \in R$ or $I \subset R$ we have $\lambda^k(R,I) = \lambda(R,I)$.)

We note that then clearly we have (5.1), (5.2) and (5.3):

(5.1) For any $I \in R$ or $I \subset R$ and any $P \in \mathfrak{P}_0(R)$ we have:

$$\lambda(R,I,P) = \lambda(R_P,I) = \begin{cases} 0 & \text{, if } IR \not\subset P, \\ \text{a positive integer, } & \text{if } \{0\} \neq IR \subset P, \\ \infty & \text{, if } IR = \{0\}, \end{cases}$$

and

$$\lambda^k(R,I,P) = \lambda^k(R_P,I) = \lambda(R,I,P)[R/P\colon k/(k \cap P)]$$
$$= \begin{cases} 0 & \text{, if } IR \not\subset P, \\ \text{a positive integer, } & \text{if } \{0\} \neq IR \subset P, \\ \infty & \text{, if } IR = \{0\}. \end{cases}$$

(Note that, if k is a field, then $[R/P\colon k/(k \cap P)] = [R/P\colon k]$.)

(5.2) For any $I \in R$ or $I \subset R$ and any $Q \in R$ or $Q \subset R$ we have:

$$\lambda(R,I,Q) = \lambda(R,IR,\mathrm{rad}_R(IR + QR)) = \sum_{P \in \mathfrak{P}_0(R,Q)} \lambda(R,I,P)$$

$$= \text{a nonnegative integer or } \infty,$$

$$\lambda^k(R,I,Q) = \lambda^k(R,IR,\mathrm{rad}_R(IR + QR)) = \sum_{P \in \mathfrak{P}_0(R,Q)} \lambda^k(R,I,P)$$

$$= \text{a nonnegative integer or } \infty,$$

$$\lambda(R,I,Q) = \infty \Leftrightarrow \lambda^k(R,I,Q) = \infty \Leftrightarrow IR = \{0\} \text{ and } QR \neq R,$$

$$\lambda(R,I,Q) = 0 \Leftrightarrow \lambda^k(R,I,Q) = 0 \Leftrightarrow IR + QR = R,$$

$$\lambda(R,I) = \lambda(R,I,Q) \Leftrightarrow \lambda^k(R,I) = \lambda^k(R,I,Q) \Leftrightarrow QR \subset \mathrm{rad}_R IR,$$

and for any $J \in R$ or $J \subset R$ with $JR \subset IR$ we have

$$\lambda^k(R,J,Q) \geq \lambda(R,I,Q) \quad \text{and} \quad \lambda^k(R,J,Q) = \lambda^k(R,J,Q).$$

Also

$$\lambda(R,J,Q) = \lambda(R,I,Q) \Leftrightarrow \lambda^k(R,J,Q) = \lambda^k(R,I,Q)$$
$$\Leftrightarrow J(R_P)^* = I(R_P)^* \quad , \quad \text{for all} \quad P \in \mathfrak{P}_0(R,Q) \ ;$$

where $(R_P)^*$ is the integral closure of R_P in $\mathfrak{Z}(R_P)$.

(5.3) For any $I \in R$ or $I \subset R$ we have:

$$\lambda(R,I) = \lambda(R,IR) = \sum_{V \in \mathfrak{D}(R)} (\mathrm{ord}_V I)[V/M(V) : R/(R \cap M(V))]$$

$$= \sum_{P \in \mathfrak{P}_0(R)} \lambda(R,P)$$

$$= \text{a nonnegative integer or } \infty ,$$

$$\lambda^k(R,I) = \lambda^k(R,IR) = \sum_{V \in \mathfrak{D}(R)} (\mathrm{ord}_V I)[V/M(V) : k/(k \cap M(V))]$$

$$= \sum_{P \in \mathfrak{P}_0(R)} \lambda^k(R,P)$$

$$= \text{a nonnegative integer or } \infty ,$$

$$\lambda(R,I) = \infty \Leftrightarrow \lambda^k(R,I) = \infty \Leftrightarrow IR = \{0\}$$

$$\lambda(R,I) = 0 \Leftrightarrow \lambda^k(R,I) = 0 \Leftrightarrow IR = R ,$$

and for any $J \in R$ or $J \subset R$ with $JR \subset IR$ we have

$$\lambda(R,J) \geq \lambda(R,I) \quad \text{and} \quad \lambda^k(R,J) \geq \lambda^k(R,I) \ .$$

Also

$$\lambda(R,J) = \lambda(R,I) \Leftrightarrow \lambda^k(R,J) = \lambda^k(R,I) \Leftrightarrow JR^* = IR^* \ .$$

The following six lemmas provide alternative definitions of $\lambda(R,I)$ and $\lambda^k(R,I)$.

(5.4) LEMMA. Without assuming condition (**), for any ideal I in R we have

$$\sum_{P \in \mathfrak{P}_0(R)} [R_P/IR_P : R_P][R/P : k/(k \cap P)] = [R/I : k].$$

(Note that if k is a field then $[R/P : k/(k \cap P)] = [R/P : k]$.)

PROOF. We shall prove our assertion by induction on $[R/I : R]$. Clearly

$[R/I : R] = \infty \Leftrightarrow I = \{0\} \Rightarrow$ both sides of the above equation are ∞,

and

$[R/I : R] = 0 \Leftrightarrow I = R \Rightarrow$ both sides of the above equation are 0.

So let $0 < [R/I : R] < \infty$ and assume that the assertion is true for all values of $[R/I : R]$ smaller than the given one. Now $\mathfrak{P}_0(R,I)$ is a nonempty finite set. We can fix $Q \in \mathfrak{P}_0(R,I)$ and then take an ideal J in R_Q with $IR_Q \subset J$ such that there is no ideal in R_Q between IR_Q and J. By Nakayama's lemma we then have

(1) $$M(R_Q)J \subset IR_Q ;$$

moreover, upon letting

(2) $$I' = \bigcap_{Q \neq P \in \mathfrak{P}_0(R)} IR_P \cap J ,$$

we have that I' is an ideal in R with $I \subset I'$ such that

(3) $$[I'/I : R] = 1 ,$$

(4) $$[R/I' : R] = [R/I : R] - 1 ,$$

(5) $$\begin{cases} \displaystyle\sum_{P\epsilon\mathfrak{P}_0(R)} [R_P/IR_P : R_P][R/P : k(k \cap P)] \\ = [R/Q : k/(k \cap Q)] + \displaystyle\sum_{P\epsilon\mathfrak{P}_0(R)} [R_P/I' : R_P][R/P : k/(k\cap P)] , \end{cases}$$

and

(6) $[R/I : k] = [I'/I : k] + [R/I' : k].$

In view of (4), by the induction hypothesis we get

(7) $\sum_{P\epsilon\mathfrak{P}_0(R)} [R_P/I' : R_P][R/P : k/(k \cap P))] = [R/I' : k].$

By (1) and (2) we see that

(8) $QI' \subset I$.

In view of (8), the R-module-structure of I'/I induces a (R/Q)-module-structure on it and we have

(9) $[I'/I : R/Q] = [I'/I : R]$

In view of (8), the k-module-structure of I'/I induces a $(k/(k \cap Q))$-module-structure on it and we have

(10) $[I'/I : k/(k \cap Q)] = [I'/I : k].$

Now the above said $(k/k \cap Q))$-module-structure of I'/I coincides with the $(k/(k \cap Q))$-module-structure induced on it by its above said (R/Q-module-structure, and hence by vector space theory we have

(11) $[I'/I : k/(k \cap Q)] = [R/Q : k/(k \cap Q)][I'/I : R/Q].$

By (3), (5), (6), (7), (9), (10) and (11) it follows that

$$\sum_{P\in\mathfrak{P}_0(R)} [R_P/IR_P : R_P][R/P : k/(k \cap P)] = [R/I : k].$$

(5.5) LEMMA. <u>For any</u> $I \in R$ <u>or</u> $I \subset R$ <u>we have</u>

$$\lambda(R,I) = [R^*/IR^* : R] \quad \underline{\text{and}} \quad \lambda^k(R,I) = [R^*/IR^* : k].$$

PROOF. The first equation follows from the second equation by taking $k = R$. For the second equation we have

$$\begin{aligned} \lambda^k(R,I) &= \sum_{V\in\mathfrak{J}(R)} (\text{ord}_V I)[V/M(V) : k/(k \cap M(V))] \\ &= \sum_{P\in\mathfrak{P}_0(R^*)} [R^*_P/IR^*_P : R^*_P][R^*/P : k/(k \cap P)] \\ &= [R^*/IR^* : k] \qquad \text{by (5.4).} \end{aligned}$$

(5.6) LEMMA. <u>Let</u> $I \in R$ <u>or</u> $I \subset R$. <u>Let</u> $P \in \mathfrak{P}_0$ <u>be such that</u> IR_P <u>is principal</u>. <u>Then</u>

$$\lambda(R,I,P) = [R_P/IR_P : R_P] \quad \text{and} \quad \lambda^k(R,I,P) = [R_P/IR_P : k].$$

PROOF. Let $S = R_P$. By assumption there exists $x \in S$ such that $xS = IS$. Now

(1') $\lambda(R,I,P) = \lambda(S,x)$ and $[R_P/IR_P : R_P] = [S/xS : S]$,

(2') $\begin{cases} \lambda^k(R,I,P) = \lambda(R,I,P)[S/M(S) : k/(k \cap M(S))] \text{ and} \\ [R_P/IR_P : k] = [S/xS : k] , \end{cases}$

and by (5.4) we have

(3') $[S/xS : S][S/M(S) : k/(k \cap M(S))] = [S/xS : k]$.

Upon letting S^* to be the integral closure of S in $\mathfrak{J}(S)$, by (5.5) we also get

(4') $\lambda(S,x) = [S^*/xS^* : S]$.

In view of (1'), (2'), (3') and (4'), our assertion is reduced to proving that

$$(1^*) \qquad [S^*/xS^*: S] = [S/xS: S].$$

We now proceed to prove (1^*). Since S and S^* are one-dimensional noetherian domains, we see that if $x = 0$ then both sides of (1^*) are ∞. So henceforth assume that $x \neq 0$. Now

$$xS \subset xS^* \subset S^* \text{ and } xS \subset S \subset S^*$$

and hence

$$(1) \quad [S^*/xS^*: S] + [xS^*/xS: S] = [S^*/xS: S] = [S^*/S: S] + [S/xS: S].$$

Since $0 \neq x \in S$, $y \to xy$ gives an S-isomorphism of S^* onto xS^* and under this isomorphism the image of S is xS ; consequently S^*/S is S-isomorphic to xS^*/xS and hence

$$(2) \qquad [xS^*/xS: S] = [S^*/S: S].$$

Now if we prove that $[S/xS: S]$ and $[S^*/S: S]$ are both finite, then (1) and (2) prove (1^*).
Since x is a nonzero element in the one-dimensional local domain S, we get

$$(3) \qquad [S/xS: S] < \infty .$$

Now S^* is a finite S-module and hence $\mathfrak{C}(S)$ is a nonzero ideal in S. Consequently

$$(4) \qquad \lambda(S,\mathfrak{C}(S)) < \infty .$$

Upon taking S and $\mathfrak{C}(S)$ for R and I in (5.5) we also get

$$(5) \qquad \lambda(S,\mathfrak{C}(S)) = [S^*/\mathfrak{C}(S)S^*: S] .$$

Clearly

$$\mathfrak{C}(S)S^* = \mathfrak{C}(S) \subset S \subset S^* ,$$

and hence by (4) and (5) we see that

$$(6) \qquad [S^*/S : S] < \infty .$$

By (1), (2), (3) and (6) we get (1^*).

(5.7) LEMMA. <u>Let</u> $I \in R$ <u>or</u> $I \subset R$ <u>be such that</u> IR_P <u>is principal for all</u> $P \in \mathfrak{P}_0(R)$. <u>Then</u>

$$\lambda(R,I) = [R/IR : R] \quad \text{and} \quad \lambda^k(R,I) = [R/IR : k].$$

PROOF. The second equality gives the first by putting $k = R$. For the second equality we have,

$$\lambda^k(R,I) = \sum_{P\in\mathfrak{P}_0(R)} \lambda(R,I,P)[R/P : k/(k \cap P)]$$

$$= \sum_{P\in\mathfrak{P}_0(R)} [R_P/IR_P : R_P][R/P : k/(k \cap P)] \qquad \text{by (5.6)}$$

$$= [R/IR : k] \qquad \text{by (5.4)}$$

(5.8) LEMMA. <u>Let</u> $P \in \mathfrak{P}_0(R)$ <u>be such that</u> R_P <u>is regular</u>. <u>Then for any</u> $I \subset R$ <u>or</u> $I \subset R$ <u>we have</u>

$$\lambda(R,I,P) = [R_P/IR_P : R_P] \quad \text{and} \quad \lambda^k(R,I,P) = [R_P/IR_P : k].$$

PROOF. Now R_P is a principal ideal domain and hence our assertion follows from (5.6). Alternatively we can note that now $\mathfrak{D}(R,P) = \{R_P\}$ and hence

$$\lambda(R,I,P) = (\text{ord}_{R_P} I)[R_P/M(R_P) : R/(R \cap M(R_P))] \qquad \text{(by definition)}$$

$$= [R_P/IR_P : R_P] \qquad \text{(obviously)}$$

and more generally,

$$\lambda^k(R,I,P) = (\mathrm{ord}_{R_P} I)[R_P/M(R_P) : k/(k \cap M(R_P))] \quad \text{(by definition)}$$

$$= [R_P/IR_P : R_P][R_P/M(R_P) : (k/(k \cap M(R_P))] \quad \text{(obviously)}$$

$$= [R_P/IR_P : k]. \quad \text{(by (5.4))}.$$

(5.9) LEMMA. <u>If</u> R <u>is normal</u>, <u>then for any</u> $I \in R$ <u>or</u> $I \subset R$ <u>we have</u>

$$\lambda(R,I) = [R/IR : R] \text{ and } \lambda^k(R,I) = [R/IR : k] .$$

PROOF. If R is normal then R_P is a principal ideal domain for all $P \in \mathfrak{P}_0(R)$, and hence our assertion follows from (5.7). Alternatively we can note that if R is normal then $P \to R_P$ gives a bijection of $\mathfrak{P}_0(R)$ onto $\mathfrak{Y}(R)$ and hence

$$\lambda(R,I) = \sum_{V \in \mathfrak{Y}(R)} (\mathrm{ord}_V I)[V/M(V) : R/(R \cap M(V))] \quad \text{(by definition)}$$

$$= \sum_{P \in \mathfrak{P}_0(R)} [R_P/IR_P : R_P] \quad \text{(obviously)}$$

$$= [R/IR : R]. \quad \text{(by 5.4))}.$$

and more generally,

$$\lambda^k(R,I) = \sum_{V \in \mathfrak{Z}(R)} (\mathrm{ord}_V I)[V/M(V) : k/(k \cap M(V))] \quad \text{(by definition)}$$

$$= \sum_{P \in \mathfrak{P}_0(R)} [R_P/IR_P : R_P][R/P : k(k \cap P)] \quad \text{(obviously)}$$

$$= [R/IR : k] \quad \text{(by (5.4))}.$$

(5.10). DEFINITION. In case R is local, we <u>define</u>

$$\lambda(R) = \lambda(R,M(R)) \quad \text{and} \quad \lambda^k(R) = \lambda^k(R,M(R)).$$

We observe that then:

$1 \le \mathrm{card}\, \mathfrak{Y}(R) \le \lambda(R) =$ a positive integer,

$$\lambda^k(R) = \lambda(R)[R/M(R) : k/(k \cap M(R))] = \text{a positive integer},$$

and by Nakayama's Lemma we have

$$\lambda(R) = 1 \Leftrightarrow R \text{ is regular.}$$

(We note that, if k is a field, then $[R/M(R) : k/(k \cap M(R))] = [R/M(R) : k]$. We also note that, if k is an algebraically closed field, then $\lambda^k(R) = \lambda(R)$.)

We also observe that, without assuming R to be local, for any $P \in \mathfrak{P}_0(R)$ we now have

$$\lambda(R,P) = \lambda(R_P),\ \lambda^k(R,P) = \lambda^k(R_P),\text{ and } \mathfrak{D}(R,P) = \mathfrak{D}(R_P).$$

(5.11) DEFINITION. For any $Q \in R$ or $Q \subset R$ we <u>define</u>

$$\lambda_{\mathfrak{C}}(R,Q) = \lambda(R,\mathfrak{C}(R),Q) ,$$

$$\lambda^k_{\mathfrak{C}}(R,Q) = \lambda^k(R,\mathfrak{C}(R),Q) ,$$

$$\text{adj}(R,Q) = \{\Phi \in F^*(R): \Phi R_P \subset \mathfrak{C}(R_P) \text{ for all } P \in \mathfrak{P}_0(R,Q)\},$$

and

$$\text{tradj}(R,Q) = \{\Phi \in F^*(R): \Phi(R_P)^* = \mathfrak{C}(R_P) \text{ for all } P \in \mathfrak{P}_0(R,Q)\},$$

where R_P^* denotes the integral closure of R_P in $\mathfrak{F}(R_P)$.

We note that then:

$$\lambda_{\mathfrak{C}}(R,Q) = \sum_{P\in\mathfrak{P}_0(R,Q)} \lambda_{\mathfrak{C}}(R,P) = \text{a nonnegative integer},$$

$$\lambda^k_{\mathfrak{C}}(R,Q) = \sum_{P\in\mathfrak{P}_0(R,Q)} \lambda^k_{\mathfrak{C}}(R,P) = \text{a nonnegative integer},$$

$$\lambda_{\mathfrak{C}}(R,Q) = 0 \Leftrightarrow \lambda^k_{\mathfrak{C}}(R,Q) = 0 \Leftrightarrow R_P \text{ is regular for all } P \in \mathfrak{P}_0(R,Q) ,$$

$$\lambda(R,\Phi,Q) \geq \lambda_{\mathfrak{C}}(R,Q) \text{ and } \lambda^k_{\mathfrak{C}}(R,\Phi,Q) \geq \lambda^k_{\mathfrak{C}}(R,Q) \text{ for all}$$

$$\Phi \in \text{adj}(R,Q),$$

$$\mathrm{adj}(R,Q) = \bigcap_{P\in\mathfrak{P}_0(R,Q)} \mathrm{adj}(R,P) ,$$

and

$$\mathrm{tradj}(R,Q) = \bigcap_{P\in\mathfrak{P}_0(R,Q)} \mathrm{tradj}(R,P) = \{\Phi \in \mathrm{adj}(R,Q): \lambda(R,\Phi,Q) = \lambda_{\mathfrak{C}}(R,Q)\}$$
$$= \{\Phi \in \mathrm{adj}(R,Q): \lambda_{\mathfrak{C}}^{k}(R,\Phi,Q) = \lambda^{k}(R,Q)\}$$

By an <u>adjoint</u> in R <u>at</u> Q we mean a member of $\mathrm{adj}(R,Q)$, and by a <u>true adjoint</u> in R at Q we mean a member of $\mathrm{tradj}(R,Q)$; from these two we may drop " in R " when the reference to R is clear from the context.

We also <u>define</u>

$\lambda_{\mathfrak{C}}(R) = \lambda(R,\mathfrak{C}(R))$

$\lambda_{\mathfrak{C}}^{k}(R) = \lambda^{k}(R,\mathfrak{C}(R))$

$\mathrm{adj}(R) = \{\Phi \in \mathfrak{J}^{*}(R): \Phi \subset \mathfrak{C}(R)\}$

and

$\mathrm{tradj}(R) = \{\Phi \in \mathfrak{J}^{*}(R): \Phi R^{*} = \mathfrak{C}(R)\}$, where R^{*} is the integral closure of R in $\mathfrak{J}(R)$.

We note that then

$$\lambda_{\mathfrak{C}}(R) = \lambda_{\mathfrak{C}}(R,0) = \sum_{P\in\mathfrak{P}_0(R)} \lambda_{\mathfrak{C}}(R,P) = \text{a nonnegative integer,}$$

$$\lambda_{\mathfrak{C}}^{k}(R) = \lambda_{\mathfrak{C}}^{k}(R,0) = \sum_{P\in\mathfrak{P}_0(R)} \lambda_{\mathfrak{C}}^{k}(R,P) = \text{a nonnegative integer,}$$

$\lambda_{\mathfrak{C}}(R) = 0 \Leftrightarrow \lambda_{\mathfrak{C}}^{k}(R) = 0 \Leftrightarrow R$ is normal $\Leftrightarrow R_P$ is regular for all $P \in \mathfrak{P}_0(R)$,

$\lambda_{\mathfrak{C}}(R,\Phi) \geq \lambda_{\mathfrak{C}}(R)$ and $\lambda_{\mathfrak{C}}^{k}(R,\Phi) \geq \lambda_{\mathfrak{C}}^{k}(R)$ for all $\Phi \in \mathrm{adj}(R)$,

$$\mathrm{adj}(R) = \mathrm{adj}(R,0) = \bigcap_{P\in\mathfrak{P}_0(R)} \mathrm{adj}(R,P) ,$$

$$\mathrm{tradj}(R) = \mathrm{tradj}(R,0) = \bigcap_{P \in \mathfrak{P}_0(R)} \mathrm{tradj}(R,P)$$

$$= \{\Phi \in \mathrm{adj}(R): \lambda(R,\Phi) = \lambda_{\mathfrak{C}}(R)\}$$

$$= \{\Phi \in \mathrm{adj}(R): \lambda^k(R,\Phi) = \lambda^k_{\mathfrak{C}}(R)\}$$

and

$$\mathrm{tradj}(R) \neq \emptyset \Leftrightarrow \mathfrak{C}(R) \text{ is principal in } R^*.$$

By an _adjoint in_ R we mean a member of $\mathrm{adj}(R)$, and by a _true adjoint in_ R we mean a member of $\mathrm{tradj}(R)$.

We observe that for any $P \in \mathfrak{P}_0(R)$ we have:

$\lambda_{\mathfrak{C}}(R,P) = \lambda_{\mathfrak{C}}(R_P) =$ a nonnegative integer,

$\lambda^k_{\mathfrak{C}}(R,P) = \lambda^k_{\mathfrak{C}}(R_P) = \lambda_{\mathfrak{C}}(R,P)[R/P: k/(k \cap P)] =$ a nonnegative integer,

and

$$\begin{aligned} \lambda_{\mathfrak{C}}(R,P) = 0 &\Leftrightarrow \lambda^k_{\mathfrak{C}}(R,P) = 0 \\ &\Leftrightarrow R_P \text{ is normal} \\ &\Leftrightarrow R_P \text{ is regular} \\ &\Leftrightarrow \lambda(R,I,P) = \mathrm{ord}_{R_P} I \text{ for every } I \in R \text{ or } I \subset R. \end{aligned}$$

We note that, if k is a field, then $[R/P : k/(k \cap P)] = [R/P : k]$. Finally we observe that, if k is an algebraically closed field, then: $\lambda^k_{\mathfrak{C}}(R,Q) = \lambda_{\mathfrak{C}}(R,Q)$ for any $Q \in R$ or $Q \subset A$, and in particular, $\lambda^k_{\mathfrak{C}}(R) = \lambda_{\mathfrak{C}}(R)$.

(5.12) LEMMA ON OVERADJOINTS. _Let_ $V \in \mathfrak{I}(R)$ _be residually rational over_ k. _Let_ e _be any nonnegative integer, and let_

$$I = \mathfrak{C}(R)(R^* \cap M(V))^e.$$

Then I _is an ideal in_ R _with_ $I \subset \mathfrak{C}(R)$ _such that,_

$$[R/I : k] = e + [R/\mathfrak{C}(R) : k] \text{ and } \lambda^k(R,I) = e + \lambda^k_{\mathfrak{C}}(R).$$

PROOF. Now $\mathfrak{C}(R)$ is a nonzero ideal in R^*, and upon letting

$$Q = R^* \cap M(V)$$

we have that Q is a nonzero prime ideal in R^*. So

(1) $$I = \mathfrak{C}(R)Q^e \subset \mathfrak{C}(R)Q^{e-1} \subset \ldots \subset (R)Q^0 = \mathfrak{C}(R)$$

are ideals in R^* with

(2) $$\mathfrak{C}(R)Q^i \neq (R)Q^j \quad \text{whenever} \quad i \neq j .$$

Since $\mathfrak{C}(R) \subset R$, we also see that,

(3) $$\mathfrak{C}(R)Q^i \text{ is an ideal in } R \text{ for all } i \geq 0.$$

We note that for every $i \geq 0$ we clearly have,

(4) $$\mathfrak{C}(R)Q^i = \{y \in R^* : \mathrm{ord}_V y \geq i + \mathrm{ord}_V \mathfrak{C}(R) \text{ and } \mathrm{ord}_W \geq \mathrm{ord}_W \mathfrak{C}(R) \text{ whenever } V \neq W \in \mathfrak{D}(R)\}$$

We cleaim that,

for any $d \geq 0$ and any $x \in \mathfrak{C}(R)Q^d \setminus \mathfrak{C}(R)Q^{d+1}$, we have $\mathfrak{C}(R)Q^{d+1} + xR = \mathfrak{C}(R)Q^d$.

In view of (4) we have,

(5_1) $$\mathrm{ord}_V x = d + \mathrm{ord}_V \mathfrak{C}(R)$$

and

(5_2) $$\mathrm{ord}_W x \geq \mathrm{ord}_W \mathfrak{C}(R), \text{ whenever } V \neq W \in \mathfrak{D}(R).$$

Given any $z \in \mathfrak{C}(R)Q^d$, by (4) we get

(5_3) $$\mathrm{ord}_V z \geq d + \mathrm{ord}_V \mathfrak{C}(R)$$

and

(5_4) $\qquad \mathrm{ord}_W z \geq \mathrm{ord}_W\,(R)$ whenever $V \neq W \in \mathfrak{Y}(R)$.

Now V is residually rational over R (because it is assumed to be residually rational over k), and hence in view of (5_1) and (5_3) we can find $\alpha \in R$ such that

(5_5) $\qquad \mathrm{ord}_V(z-\alpha x) \geq d + 1 + \mathrm{ord}_V \mathfrak{C}(R)$.

By (5_2) and (5_4) we have,

(5_6) $\qquad \mathrm{ord}_W(z-\alpha x) \geq \mathrm{ord}_W \mathfrak{C}(R)$, whenever $V \neq W \in \mathfrak{Y}(R)$.

In view of (5_4) and (5_6), by (4) we get that

$$z - \alpha x \in \mathfrak{C}(R)Q^{d+1}$$

and hence $z \in \mathfrak{C}(R)Q^{d+1} + xR$. This completes the proof of (5).

By (5) we see that,

$$(6) \begin{cases} \text{for any } d \geq 0 \text{ and any ideal } J \text{ in } R \text{ with} \\ \mathfrak{C}(R)Q^{d+1} \subset J \subset \mathfrak{C}(R)Q^{d}, \\ \text{we must have either } J = \mathfrak{C}(R)Q^{d+1} \text{ or } J = \mathfrak{C}(R)Q^{d}. \end{cases}$$

By (1), (2), (3) and (6) we get

$$(7) \qquad [R/I : R] = e + [R/\mathfrak{C}(R) : R].$$

Let

$$(8) \qquad P_0 = R \cap M(V).$$

Then $P_0 \in \mathfrak{P}_0(R)$ and we clearly have

(9) $\qquad IR_P = \mathfrak{C}(R)R_P$ whenever $P_0 \neq P \in \mathfrak{P}_0(R)$.

Now V is assumed to be residually rational over k and hence, in view of (7), (8) and (9), by (5.4) we conclude that

$$[R/I : k] = e + [R/\mathfrak{C}(R) : k]$$

Since V is residually rational over k, we obviously have,

$$\lambda^k(R,I) = e + \lambda^k_{\mathfrak{C}}(R).$$

§6. Length in a one-dimensional noetherian homomorphic image.

Let A be a domain and let $C \in \mathfrak{P}_1(A)$ be such that A/C is noetherian. Let $f: A \to A/C$ be the canonical epimorphism. Let k be a subring of A such that

(*) for every $P \in \mathfrak{P}_0([A,C])$ we have $k \cap P \in \mathfrak{P}_0$ and $[A/P : k/(k \cap P] < \infty$.

(Note that (*) is satisfied for $A = k$; it is also satisfied if k is a subfield of A such that $[A/P : k] < \infty$ for every $P \in \mathfrak{P}_0([A,C])$.).

Assume that

(**) the integral closure of A/C in $\mathfrak{F}(A/C)$ is a finite (A/C)-module.

For any $I \in A$ or $I \subset A$ and any $Q \in A$ or $Q \subset A$, we <u>define</u>

$$\lambda([A,C],I,Q) = \lambda(f(A),f(I),f(Q)) \quad \text{and} \quad \lambda^k([A,C],I,Q) = \lambda^{f(k)}(f(A),f(I),f(Q)) \ ;$$

and for $I \in A$ or $I \subset A$, we <u>define</u>

$$\lambda([A,C],I) = \lambda(f(A),f(I)) \text{ and } \lambda^k([A,C],I) = \lambda^{f(k)}(f(A),f(I)).$$

For any $Q \in A$ or $Q \subset A$, we <u>define</u>

$$\lambda_{\mathfrak{C}}([A,C],Q) = \lambda_{\mathfrak{C}}(f(A),f(Q)) \text{ and } \lambda^k_{\mathfrak{C}}([A,C],Q) = \lambda^{f(k)}_{\mathfrak{C}}(f(A),f(Q)),$$

$$\mathrm{adj}([A,C]) = \{\Phi \in F^*(A) : f(\Phi) \in \mathrm{adj}(f(A))\},$$

and

$$\mathrm{tradj}([A,C]) = \{\Phi \in F^*(A) : f(\Phi) \in \mathrm{tradj}(f(A))\}.$$

By an <u>adjoint of</u> C in A we mean a member of $\mathrm{adj}([A,C])$, and by a <u>true adjoint of</u> C in A we mean a member of $\mathrm{tradj}([A,C])$, from these two phrases we may drop "in A " when the reference to A is clear from the context.

We note that then, in view of (5.1), (5.2), (5.3), (5.10) and (5.11) we clearly get (6.1), (6.2), (6.3), (6.4) and (6.5):

(6.1) For any $I \in A$ or $I \subset A$ and any $P \in \mathfrak{P}_0([A,C])$ we have:

$$\lambda([A,C],I,P) = \lambda(f(A)_{f(P)}, f(I)) = \lambda([A_P, CA_P], I)$$

$$= \sum_{V \in \mathfrak{Y}([A,C],P)} (\mathrm{ord}_{[A,C],V} I)[V/M(V) : A/P]$$

$$= \begin{cases} 0 & \text{, if } IA \not\subset P, \\ \text{a positive integer,} & \text{if } IA \subset P \text{ and } IA \not\subset C, \\ \infty & \text{, if } IA \subset C, \end{cases}$$

and

$$\lambda^k([A,C],I,P) = \lambda([A,C],I,P)[A/P : k/(k \cap P)]$$

$$= \lambda^{f(k)}(f(A)_{f(P)}, f(I))$$

$$= \lambda^k([A_P, CA_P], I)$$

$$= \begin{cases} 0 & \text{, if } IA \not\subset P, \\ \text{a positive integer,} & \text{if } IA \subset P \text{ and } IA \not\subset C, \\ \infty & \text{, if } IA \subset C. \end{cases}$$

(We observe that, if k is a field, then $[A/P: k/(k \cap P)] = [A/P : k]$.)

For any $P \in \mathfrak{P}_0([A,C])$ we also have:

$$1 \leq \operatorname{card}([A,C],P) \leq \lambda([A,C],P) = \lambda(f(A)_{f(P)}) = \text{a positive integer},$$

$$\lambda^k([A,C],P) = \lambda([A,C],P)[A/P : k/(k \cap P)] = \lambda^{f(k)}(f(A)_{f(P)})$$
$$= \text{a positive integer},$$

$$\lambda_{\mathfrak{C}}([A,C],P) = \lambda([A,C],\mathfrak{C}([A,C],P) = \lambda_{\mathfrak{C}}(f(A)_{f(P)})$$
$$= \text{a nonnegative integer}$$

$$\lambda_{\mathfrak{C}}^k([A,C],P) = \lambda_{\mathfrak{C}}([A,C],P)[A/P : k/(k \cap P)] = \lambda^k([A,C],\mathfrak{C}([A,C]),P)$$
$$= \lambda_{\mathfrak{C}}^{f(k)}(f(A)_{f(P)})$$
$$= \text{a nonnegative integer},$$

and

$$\lambda_{\mathfrak{C}}([A,C],P) = 0 \Leftrightarrow \lambda_{\mathfrak{C}}^k([A,C],P) = 0$$
$$\Leftrightarrow f(A)_{f(P)} \text{ is normal}$$
$$\Leftrightarrow f(A)_{f(P)} \text{ is regular}$$
$$\Leftrightarrow \lambda([A,C],P) = 1$$
$$\Leftrightarrow \lambda([A,C],I,P) = \operatorname{ord}_{f(A)_{f(P)}} f(I)$$
$$\text{for every } I \in A \text{ or } I \subset A.$$

(We again observe that, if k is a field, then $[A/P : k(k \cap P)] = [A/P : k]$.)

(6.2) For any $I \in A$ or $I \subset A$ or $Q \subset A$ we have:

$$\lambda([A,C],I,Q) = \lambda([A,C],IA+C,\operatorname{rad}_A(IA+QA+C) = \sum_{P \in \mathfrak{P}_0([A,C],Q)} \lambda([A,C],I,P)$$
$$= \text{a nonnegative integer or } \infty,$$

$$\lambda^k([A,C],I,Q) = \lambda^k([A,C],IA+C,\operatorname{rad}_A(IA+QA+C) = \sum_{P \in \mathfrak{P}_0([A,C],Q)} \lambda^k([A,C],I,P)$$
$$= \text{a nonnegative integer or } \infty,$$

$\lambda([A,C],I,Q) = \infty \Leftrightarrow \lambda^k([A,C],I,Q) = \infty \Leftrightarrow IA \subset C$ and $QA + C \neq A$,

$\lambda([A,C],I,Q) = 0 \Leftrightarrow \lambda^k([A,C],I,Q) = 0 \Leftrightarrow IA + QA + C = A$,

$\lambda([A,C],I) = \lambda([A,C],Q) \Leftrightarrow \lambda^k([A,C]),I) = \lambda^k([A,C],I,Q)$

$$\Leftrightarrow QA \subset \mathrm{rad}_A(IA+C).$$

Also, for any $J \in A$ or $J \subset A$ with $JA \subset IA + C$ we have

$$\lambda([A,C],J,Q) \geq \lambda([A,C],I,Q) \quad \text{and} \quad \lambda^k([A,C],J,Q) \geq \lambda^k([A,C],I,Q).$$

and

$$\lambda([A,C],J,Q) = \lambda([A,C],I,Q) \Leftrightarrow \lambda^k([A,C],J,Q) = \lambda^k([A,C],I,Q)$$

$$\Leftrightarrow f(J)(f(A)_{f(P)})^* = f(I)(f(A)_{f(P)})^* \quad \text{for all} \quad P \in \mathfrak{P}_0([A,C],Q) ,$$

where $(f(A)_{f(P)})^*$ is the integral closure of $f(A)_{f(P)}$ in

$\mathfrak{J}(f(A)_{f(P)})$.

For any $Q \in A$ or $Q \subset A$ we also have:

$$\lambda_{\mathfrak{C}}([A,C],Q) = \lambda([A,C],\mathfrak{C}([A,C]),Q) = \sum_{P \in \mathfrak{P}_0([A,C],Q)} \lambda_{\mathfrak{C}}([A,C],P)$$

$$= \text{a nonnegative integer,}$$

$$\lambda_{\mathfrak{C}}^k([A,C],Q) = \lambda^k([A,C],\mathfrak{C}([A,C]),Q) = \sum_{P \in \mathfrak{P}_0([A,C],Q)} \lambda_{\mathfrak{C}}^k([A,C],P)$$

$$= \text{a nonnegative integer,}$$

$$\lambda_{\mathfrak{C}}([A,C],Q) = 0 \Leftrightarrow \lambda_{\mathfrak{C}}^k([A,C],Q) = 0 \Leftrightarrow f(A)_{f(P)} \quad \text{is regular for all}$$

$$P \in \mathfrak{P}_0([A,C],Q),$$

$$\lambda([A,C],\Phi,Q) \geq \lambda_{\mathfrak{C}}([A,C],Q) \quad \text{and} \quad \lambda_{\mathfrak{C}}^k([A,C],\Phi,Q) \geq \lambda^k([A,C],Q) \quad \text{for all}$$

$$\Phi \in \mathrm{adj}([A,C],Q),$$

$$\mathrm{adj}([A,C],Q) = \sum_{P \in \mathfrak{P}_0([A,C],Q)} \mathrm{adj}([A,C],P),$$

and

$$\mathrm{tradj}([A,C],Q) = \bigcap_{P\in\mathfrak{P}_0([A,C],P)} \mathrm{tradj}([A,C],P)$$
$$= \{\Phi \in \mathrm{adj}([A,C],Q) : \lambda([A,C],\Phi,Q) = \lambda_{\mathfrak{C}}([A,C],Q)\}$$
$$= \{\Phi \in \mathrm{adj}([A,C],Q) : \lambda^k([A,C],\Phi,Q) = \lambda^k_{\mathfrak{C}}([A,C],Q\}.$$

(6.3) For any $I \in A$ or $I \subset A$ we have:

$$\lambda([A,C],I) = \lambda([A,C],I,0) = \lambda([A,C],IA) = \sum_{P\in\mathfrak{P}_0([A,C])} \lambda([A,C],I,P)$$
$$= \text{a nonnegative integer or } \infty,$$

$$\lambda^k([A,C],I) = \lambda^k([A,C],I,0) = \lambda^k([A,C],IA) = \sum_{P\in\mathfrak{P}_0([A,C])} \lambda^k([A,C],I,P)$$
$$= \text{a nonnegative integer or } \infty,$$

$$\lambda([A,C],I) = \infty \Leftrightarrow \lambda^k([A,C],I) = \infty \Leftrightarrow IA \subset C,$$

and

$$\lambda([A,C],I) = 0 \Leftrightarrow \lambda^k([A,C],I) = 0 \Leftrightarrow IA + C = A.$$

Also, for any $J \in A$ or $J \subset A$ with $JA + IA + C$ we have:

$$\lambda([A,C],J) \geq \lambda([A,C],I) \quad \text{and} \quad \lambda^k([A,C],J) \geq \lambda^k([A,C],J),$$

and

$$\lambda([A,C],J) = \lambda([A,C],I) \Leftrightarrow \lambda^k([A,C],J) = \lambda^k([A,C],I)$$
$$\Leftrightarrow f(J)f(A)^* = f(I)f(A)^*,$$

where $f(A)^*$ is the integral closure of $f(A)$ in $\mathfrak{J}(f(A))$.

We also have:

$$\lambda_{\mathfrak{C}}([A,C]) = \lambda_{\mathfrak{C}}([A,C],0) = \lambda([A,C],\mathfrak{C}([A,C])) = \sum_{P\in\mathfrak{P}_0([A,C])} \lambda_{\mathfrak{C}}([A,C],P)$$
$$= \text{a nonnegative integer,}$$

$$\lambda_{\mathfrak{C}}^{k}([A,C]) = \lambda_{\mathfrak{C}}^{k}([A,C],0) = \lambda^{k}([A,C],\mathfrak{C}([A,C]) = \sum_{P\in\mathfrak{P}_0([A,C])} \lambda_{\mathfrak{C}}^{k}([A,C],P)$$

$$= \text{a nonnegative integer}$$

$\lambda_{\mathfrak{C}}([A,C]) = 0 \Leftrightarrow \lambda_{\mathfrak{C}}^{k}([A,C]) = 0 \Leftrightarrow A/C$ is normal

$\Leftrightarrow f(A)_{f(P)}$ is regular for all

$P \in \mathfrak{P}_0([A,C])$,

$\lambda([A,C],\Phi) \geq \lambda_{\mathfrak{C}}([A,C])$ and $\lambda_{\mathfrak{C}}^{k}([A,C],\Phi) \geq \lambda_{\mathfrak{C}}([A,C])$

for all $\Phi \in \text{adj}([A,C])$,

$$\text{adj}([A,C]) = \bigcap_{P\in\mathfrak{P}_0([A,C])} \text{adj}([A,C],P),$$

and

$$\text{tradj}([A,C]) = \bigcap_{P\in\mathfrak{P}_0([A,C])} \text{tradj}([A,C],P)$$

$$= \{\Phi \in \text{adj}([A,C]): \lambda([A,C],\Phi) = \lambda_{\mathfrak{C}}([A,C])\}$$

$$= \{\Phi \in \text{adj}([A,C]): \lambda^{k}([A,C],\Phi) = \lambda_{\mathfrak{C}}^{k}([A,C])\}.$$

(6.4) If k is an algebraically closed field, then:

$\lambda^{k}([A,C],I,Q) = \lambda([A,C],I,Q)$ for any $I \in A$ or $I \subset A$ and any $Q \subset A$,

$\lambda^{k}([A,C],I) = \lambda([A,C],I)$ for any $I \in A$ or $I \subset A$,

$\lambda_{\mathfrak{C}}^{k}([A,C],Q) = \lambda_{\mathfrak{C}}([A,C],Q)$ for any $Q \in A$ or $Q \subset A$,

and

$$\lambda_{\mathfrak{C}}^{k}([A,C]) = \lambda_{\mathfrak{C}}([A,C]).$$

(6.5) If $C = \{0\}$, then:

$$\left.\begin{aligned} \lambda^{k}([A,C],I,Q) &= \lambda^{k}(A,I,Q) \\ \lambda([A,C],I,Q) &= \lambda(A,I,Q) \end{aligned}\right\} \text{ for any } I \in A \text{ or } I \subset A \text{ and any } Q \in A \text{ or } Q \subset A\ ;$$

$$\left.\begin{aligned} \lambda^k([A,C],I) &= \lambda^k(A,I) \\ \lambda([A,C],I) &= \lambda(A,I) \end{aligned}\right\} \text{ for any } I \in A \text{ or } I \subset A\ ;$$

$$\left.\begin{aligned} \lambda^k_{\mathfrak{C}}([A,C],Q) &= \lambda^k_{\mathfrak{C}}(A,Q) \\ \lambda_{\mathfrak{C}}([A,C],Q) &= \lambda_{\mathfrak{C}}(A,Q) \\ \mathrm{adj}([A,C],Q) &= \mathrm{adj}(A,Q) \\ \mathrm{tradj}([A,C],Q) &= \mathrm{tradj}(A,Q) \end{aligned}\right\} \text{ for any } Q \in A \text{ or } Q \subset A\ ;$$

$$\lambda^k_{\mathfrak{C}}([A,C]) = \lambda^k_{\mathfrak{C}}(A) \quad\text{and}\quad \lambda_{\mathfrak{C}}([A,C]) = \lambda_{\mathfrak{C}}(A)\ ;$$

and

$$\mathrm{adj}([A,C]) = \mathrm{adj}(A) \quad\text{and}\quad \mathrm{tradj}([A,C]) = \mathrm{tradj}(A).$$

In view of (5.6) to (5.9) we immediately get (6.6) to (6.9):

(6.6) LEMMA. <u>Let</u> $I \in A$ <u>or</u> $I \subset A$ <u>and</u> $P \in \mathfrak{B}_0([A,C])$ <u>be such that</u> $f(I)f(A)_{f(P)}$ <u>is principal</u>. <u>Then</u>

$$\lambda([A,C],I,P) = [A_P/(IA + C)A_P : A_P]$$

<u>and</u>

$$\lambda^k([A,C],I,P) = [A_P/(IA + C)A_P : k].$$

(6.7) LEMMA. <u>Let</u> $I \in A$ <u>or</u> $I \subset A$ <u>be such that</u> $f(I)f(A)_{f(P)}$ <u>is principal for all</u> $P \in \mathfrak{B}_0([A,C])$. <u>Then</u>

$$\lambda([A,C],I) = [A/(IA+C) : A] \quad \underline{\text{and}} \quad \lambda^k([A,C],I) = [A/(IA+C) : k].$$

(6.8) LEMMA. <u>Let</u> $P \in \mathfrak{B}_0([A,C])$ <u>be such that</u> $f(A)_{f(P)}$ <u>is regular</u>. <u>Then, for any</u> $I \in A$ <u>or</u> $I \subset A$, <u>we have</u>,

$$\lambda([A,C],I,P) = [A_P/(IA+C)A_P : A_P]$$

<u>and</u>

$$\lambda^k([A,C],I,P) = [A_P/(IA+C)A_P : k].$$

(6.9) LEMMA. _If_ A/C _is normal, then, for any_ $I \in A$ _or_ $I \subset A$, _we have_

$$\lambda([A,C],I) = [A/(IA+C):A] \quad \text{and} \quad \lambda^k([A,C],I) = [A/(IA+C):k].$$

§7. A commuting lemma for length.

Let A be a domain. Let $C \in \mathfrak{P}_1(A)$ be such that A/C is noetherian and the integral closure of A/C in $\mathfrak{F}(A/C)$ is a finite (A/C)-module. Let $D \in \mathfrak{P}_1(A)$ be such that A/D is noetherian and the integral closure of A/D in $\mathfrak{F}(A/D)$ is a finite (A/D)-module. Let k be a subring of A such that

$$(*)\quad \begin{cases} \text{for every } P \in \mathfrak{P}_0([A,C]) \cap \mathfrak{P}_0([A,D]) \text{ we have } k \cap P \in \mathfrak{P}_0(k) \\ \text{and } [A/P : k/(k \cap P)] < \infty. \end{cases}$$

(Note that (*) is satisfied for $A = k$. It is also satisfied if k is a subfield of A such that $[A/P : k] < \infty$, for every $P \in \mathfrak{P}_0([A,C]) \cap \mathfrak{P}_0([A,D])$.)

(7.1) LEMMA. _If_ $P \in \mathfrak{P}_0([A,C]) \cap \mathfrak{P}_0([A,D])$ _is such that_ $\lambda([A,C],P) = 1 = \lambda([A,D],P)$, _then_

$$\lambda([A,C],D,P) = \lambda([A,D],C,P) \quad \text{and} \quad \lambda^k([A,C],D,P) = \lambda^k([A,D],C,P).$$

PROOF. It suffices to note that by (6.1) and (6.8) we have that:

$$P \in \mathfrak{P}_0([A,C]) \text{ with } \lambda([A,C],P) = 1 \quad \begin{cases} \lambda([A,C],D,P) = [A_P/(C+D)A_P : A_P] \\ \text{and} \\ \lambda^k([A,C],D,P) = [A_P/(C+D)A_P : k] \end{cases}$$

and

$$P \in \mathfrak{P}_0([A,D]) \text{ with } \lambda([A,D],P) = 1 \Rightarrow \begin{cases} \lambda([A,D],C,P) = [A_P/(C+D)A_P : A_P] \\ \text{and} \\ \lambda^k([A,D],C,P) = [A_P/(C+D)A_P : k] \end{cases}$$

(7.2) LEMMA. <u>If</u> $Q \in A$ <u>or</u> $Q \subset A$ <u>is such that</u> $\lambda([A,C],P) = 1 = \lambda([A,D]),P)$ <u>for all</u> $P \in \mathfrak{P}_0([A,C]) \cap \mathfrak{P}_0([A,D]) \cap \mathfrak{P}_0([A,Q])$, <u>then</u> $\lambda([A,C],D,Q) = \lambda([A,D],C,Q)$ <u>and</u> $\lambda^k([A,C],D,Q) = \lambda^k([A,D],C,Q)$.

PROOF. Follows from (6.2) and (7.1).

(7.3) LEMMA. <u>If</u> $\lambda([A,C],P) = 1 = \lambda([A,D],P)$ <u>for all</u> $P \in \mathfrak{P}_0([A,C]) \cap \mathfrak{P}_0([A,D])$, <u>then</u>

$$\lambda([A,C],D) = \lambda([A,D],C) \text{ and } \lambda^k([A,C],D) = \lambda^k([A,D],C).$$

PROOF. In (7.2) take $Q = 0$.

§8. Length in a two-dimensional regular local domain.

Let A be a domain, let $P \in \mathfrak{P}_0(A)$ be such that A_P is a two-dimensional regular local domain, and let $\Phi \in F^*(A)$ and $\Psi \in F^*(A)$. We <u>define</u>

$$\lambda([A,\Phi,\Psi],P) = [A_P/(\Phi + \Psi)A_P : A_P].$$

We note that then:

$$\lambda([A,\Phi,\Psi] = \lambda([A,\Psi,\Phi],P) = \lambda([A_P,\Phi A_P,\Psi A_P],M(A_P))$$

$$= \text{a nonnegative integer or } \infty,$$

$$\lambda([A,\Phi,\Psi],P) = 0 \Leftrightarrow \Phi + \Psi \not\subset P,$$

and

$$0 \neq \lambda([A,\Phi,\Psi],P) \neq \infty \Leftrightarrow (\Phi+\Psi)A_P \text{ is primary for } M(A_P).$$

The next lemma says that most of the time when $\lambda([A,\Phi],\Psi,P)$ and $\lambda([A,\Phi],\Psi,P)$ are both defined they coincide.

(8.1) LEMMA. Assume that A/Φ is a one-dimensional noetherian domain such that the integral closure of A/Φ in $\mathfrak{J}(A/\Phi)$ is a finite (A/Φ)-module. Also assume that either $\Phi \subset P$ or $P \in \mathfrak{P}_0(A)$. Then $\lambda([A,\Phi,\Psi],P) = \lambda([A,\Phi],\Psi,P)$.

PROOF. Follows from (6.2) and (6.6).

(8.2) LEMMA. *Assume that* A/Φ *is a one-dimensional noetherian domain such that the integral closure of* A/Φ *in* $\mathfrak{J}(A/\Phi)$ *is a finite* (A/Φ)*-module, and* A/Ψ *is a one-dimensional noetherian domain such that the integral closure of* A/Ψ *in* $\mathfrak{J}(A/\Psi)$ *is a finite* (A/Ψ)*-module. Also assume that either* $\Phi + \Psi \subset P$ *or* $P \in \mathfrak{P}_0(A)$. *Then* $\lambda([A,\Phi],\Psi,P) = \lambda([A,\Psi],\Phi,P)$.

PROOF. Follows from (8.1).

(8.3) LEMMA. *Let also* $\Phi' \in F^*(A)$ *be given. Then*

$$\lambda([A,\Phi\Phi',\Psi],P) = \lambda([A,\Phi,\Psi],P) + \lambda([A,\Phi',\Psi],P)$$

PROOF. If $\Psi A_P = A_P$, then both sides of the above equation are zero. So now suppose that $\Psi A_P \neq A_P$. Let $f: A_P \to B = A_P/\Psi A_P$ be the canonical epimorphism. Then B is a one-dimensional local ring such that for any $z \in B$, we have,

(1) $[B/zB : B] = \infty \Rightarrow z$ is a zerodivisor in B.

We can take elements x and x' in B such that $xB = f(\Phi' A_P)$. Now clearly

$$\lambda([A,\Phi,\Psi],P) = [B/xB : B],$$

$$\lambda([A,\Phi',\Psi],P) = [B/x'B : B],$$

$$\lambda([A,\Phi\Phi',\Psi],P) = [B/xx'B : B],$$

and hence we are reduced to proving that

$$(*) \qquad [B/xx'B : B] = [B/xB : B] + [B/x'B : B].$$

If x or x' is a zero divisor in B, then by (1) we see that both sides of (*) are ∞. So now assume that x and x' are nonzero-divisors in B. Then xx' is also nonzerodivisor in B, and so by (1) we know that all the three terms in (*) are noninfinite. Now

$$xx'B \subset xB \subset B$$

and hence

$$(1) \qquad [B/xx'B : B] = [B/xB : B] + [xB/xx'B : B].$$

Since x is a nonzerodivisor in B, $y \to xy$ gives a B-isomorphism of B onto xB and under this isomorphism the image of $x'B$ is $xx'B$. Consequently, $B/x'B$ is B-isomorphic to $xB/xx'B$ and hence

$$(2) \qquad [B/x'B : B] = [xB/xx'B : B].$$

Now by (1) and (2) we get (*).

§9. Multiplicity in a regular local domain.

Let A be a domain, let $P \in \mathfrak{P}(A)$ be such that A_P is a regular local domain, and let $\Phi \in F^*(A)$. We <u>define</u>

$$\lambda'([A,\Phi,],P) = \mathrm{ord}_{A_P} \Phi$$

We note that then:

$$\lambda'([A,\Phi],P) = \lambda'([A_P,\Phi A_P], M(A_P)) = \text{a nonnegative integer},$$

$$\lambda'([A,\Phi],P) = 0 \Leftrightarrow \Phi \not\subset P,$$

and

$$\lambda'([A,\Phi],P) = 1 \Leftrightarrow A_P/\Phi A_P \text{ is a regular local domain.}$$

The next lemma says that most of the time when $\lambda'([A,\Phi],P)$ and $\lambda([A,\Phi],P)$ are both defined they coincide.

(9.1) LEMMA. <u>Assume that</u> $P \in \mathfrak{P}_0(A)$ <u>and</u> A/Φ <u>is a one-dimensional noetherian domain such that the integral closure of</u> A/Φ <u>in</u> $\mathfrak{J}(A/\Phi)$ <u>is a finite</u> (A/Φ)-<u>module</u>. <u>Then</u> $\lambda'([A,\Phi],P) = \lambda([A,\Phi],P)$.

PROOF. If $\Phi \not\subset P$, then $\lambda'([A,\Phi],P) = 0 = \lambda([A,\Phi],P)$. If $\Phi \subset P$ then upon taking $S = A_P$ and $\varphi \in \Phi$ with $\varphi A = \Phi$, the assertion follows from (9.2):

(9.2) LEMMA. <u>Let</u> S <u>be a two-dimensional regular local domain and let</u> $\varphi \in S$ <u>be such that; upon letting</u> $f: S \to R = S/\varphi S$ <u>to be the canonical epimorphism</u>, <u>we have that</u>, R <u>is a one-dimensional local domain and the integral closure of</u> R <u>in</u> $\mathfrak{J}(R)$ <u>is a finite</u> R-<u>module</u>. <u>Then</u> $\lambda(R) = \text{ord}_S\varphi$.

PROOF. First we shall prove the assertion when $S/M(S)$ is infinite and then we shall reduce the general case to that special case.

<u>Case when</u> .$S/M(S)$ <u>is infinite</u>. Let k be a coefficient set for S, let (x,y) be a basis of $M(S)$, and let $e = \text{ord}_S\varphi$. Now e is a positive integer and there exists a unique nonzero homogeneous polynomial $t(X,Y)$, of degree e, in indeterminates X,Y, with coefficients in k, such that $\varphi - t(x,y) \in M(S)^{e+1}$. Let $k_1 = \{b \in k: t(1,b) \in M(S)\}$. Then k_1 is finite, and for any $a \in k\backslash k_1$, upon letting $h: S \to E = S/(y-ax)S$ to be the canonical epimorphism, we have that: E is a one-dimensional regular local domain and $\text{ord}\, h(\varphi) = e$. Consequently

$$[E/h(\varphi)E : E] = e ,$$

and clearly

$$[R/f(y-ax)R : R] = [S/(\varphi,y-ax)S : S] = [E/h(\varphi)E : E]$$

By (5.6) we also have

$$\lambda(R,f(y-ax)) = [R/f(y-ax)R : R].$$

By the above three displayed equations we get that $\lambda(R,f(y-ax)) = e$. Thus

(1) $$\lambda(R,f(y-ax)) = \operatorname{ord}_S\varphi \quad \text{for all} \quad a \in k\backslash k_1.$$

Now for any $V \in \mathfrak{Y}(R)$ there is at most one element b in k such that $\operatorname{ord}_V f(y-bx) \neq \operatorname{ord}_V M(R)$. Since $\mathfrak{Y}(R)$ is finite, upon letting

$$k_2 = \{b \in k: \operatorname{ord}_V f(y-bx) \neq \operatorname{ord}_V M(R) \quad \text{for some} \quad V \in \mathfrak{Y}(R)\},$$

we conclude that k_2 is finite. For any $a \in k\backslash k_2$, we clearly have $\lambda(R,M(R)) = \lambda(R,f(y-ax))$. By definition $\lambda(R) = \lambda(R,M(R))$, and hence we conclude that

(2) $$\lambda(R) = \lambda(R,f(y-ax)) \quad \text{for all} \quad a \in k\backslash k_2.$$

Since k_1 and k_2 are finite, by (1) and (2) it follows that, if $S/M(S)$ is infinite, then $\lambda(R) = \operatorname{ord}_S\varphi$.

(REMARK. By (9.2) and the arguments used in its proof, it can be seen that actually, without assuming $S/M(S)$ to be infinite, we have $k_1 = k_2$.)

<u>General case</u>. Let X be an indeterminate. Now $M(S)S[X]$ is a prime ideal in $S[X]$ and, upon letting $S' = S[X]_{M(S)S[X]}$, we have that S' is a two-dimensional regular local domain and $\operatorname{ord}_{S'}z = \operatorname{ord}_S z$, for all $z \in S$. In particular

(3) $$\operatorname{ord}_{S'}\varphi = \operatorname{ord}_S\varphi.$$

Let Y be an indeterminate. Now $M(R)R[Y]$ is a prime ideal in $R[Y]$; and upon letting $R' = R[Y]_{M(R)R[Y]}$, we have that R' is a one-dimensional local domain and $M(R)R' = M(R')$. Also we have a unique epimorphism $f'\colon S' \to R'$ such that $f'(s) = f(s)$ for all $s \in S$ and $f'(X) = Y$. Clearly $\operatorname{Ker} f' = \varphi S'$.

Let R^* be the integral closure of R in (R). Let R'^* be the quotient ring of $R^*[Y]$ with respect to the multiplicative set $R[Y]\backslash M(R)R[Y]$. The R'^* is the integral closure of R' in $\mathfrak{Q}(R')$. Since R^* is a finite R-module, we also have that R'^* is a finite R'-module.

Let $V_1, V_2, \ldots, V_p$ be all the distinct members at $\mathfrak{Y}(R)$. Let V_i' be the quotient ring of $V_i[Y]$ with respect to the prime ideal $M(V_i)V_i[Y]$. Then $V_1', V_2', \ldots V_p'$ are exactly all the distinct members of $\mathfrak{Y}(R')$. For $1 \le i \le p$, we have:

$$[V_i'/M(V_i') : R'/(R' \cap M(V_i'))] = [V_i/M(V_i) : R/(R \cap M(V_i))]$$

and

$$\operatorname{ord}_{V_i'} z = \operatorname{ord}_{V_i} z \quad \text{for all } z \in V_i \ .$$

Since $M(R)R' = M(R')$, we conclude that

(4) $$\lambda(R') = \lambda(R) \ .$$

Clearly $S'/M(S')$ is infinite and hence by the special case proved above we have

(5) $$\lambda(R') = \operatorname{ord}_{S'} \varphi \ .$$

By (3), (4) and (5) we get that $\lambda(R) = \operatorname{ord}_S \varphi$.

§10. Double points of algebraic curves.

Let R be a one-dimensional local domain such that the integral closure R^* of R in $\mathfrak{F}(R)$ is a finite R-module. Let k be any subring of R over which R is residually finite algebraic.

In this section we shall prove the following theorem.

(10.1) THEOREM. _Assume that_ $\lambda(R) = 2$ _and let_ $d = [R/\mathfrak{C}(R) : R]$. _Then we have the following_.

(10.1.1) Exactly one of the following three cases occurs.

Case (1^*). $\mathfrak{D}(R) = \{V,W\}$, for some $V \neq W$. Then necessarily $V/M(V) = W/M(W) = R/M(R)$ and $\text{ord}_V(M(R)) = \text{ord}_W(M(R)) = 1$.

Case (2^*). $\mathfrak{D}(R) = \{V\}$, for some V such that $V/M(V) = R/M(R)$. Then necessarily $\text{ord}_V M(R) = 2$.

Case (3^*). $\mathfrak{D}(R) = \{V\}$, for some V such that $V/M(V) \neq R/M(R)$. Then necessarily $[V/M(V) : R/M(R)] = 2$ and $\text{ord}_V M(R) = 1$.

(10.1.2) There exists $x \in M(R)$ such that $\lambda(R,x) = 2$. Further we have:

in case (1^*), $\lambda(R,x) = 2 \Leftrightarrow \text{ord}_V x = \text{ord}_W x = 1$;

in case (2^*), $\lambda(R,x) = 2 \Leftrightarrow \text{ord}_V x = 2$;

in case (3^*), $\lambda(R,x) = 2 \Leftrightarrow \text{ord}_V x = 1$.

(10.1.3) There exists $z \in R^*$ such that $R^* = R[z]$.

(10.1.4) There exists $y \in M(R)$ such that:

in case (1^*), $\text{ord}_V y = d < \text{ord}_W y$;

in case (2^*), $\text{ord}_V y = 2d + 1$

in case (3^*), $\text{ord}_V y = d$ and residue of $y/_{x}d$ modulo $M(V)$ does not belong to $R/M(R)$.

(10.1.5) If x and y are as in (10.1.2) and (10.1.4) respectively, then

$$M(R) = \{x,y\}R.$$

(10.1.6) If x is as in (10.1.2), then, upon letting $J_i = \mathfrak{C}(R) + x^i R$ for $i = 0,1,\ldots,d$, we have that $[R/J_i = R] = i$. Further, if $J \supset \mathfrak{C}(R)$ is any ideal in R, then $J = J_i$ for some $0 \leq i \leq d$. In more detail we have:

in case (1^*), $J = J_i$ for some $0 \leq i \leq q$. In more detail we have

<u>in case</u> (2^*), $J = J_i$, <u>where</u> $i = 1/2\ \mathrm{ord}_V J$;

<u>in case</u> (3^*), $J = J_i$, <u>where</u> $i = \mathrm{ord}_V J$.

(10.1.7) <u>We have</u>:

<u>in case</u> (1^*), $\mathfrak{C}(R) = M(V)^d M(W)^d$;

<u>in case</u> (2^*), $\mathfrak{C}(R) = M(V)^{2d}$;

<u>in case</u> (3^*), $\mathfrak{C}(R) = M(V)^d$.

(10.1.8) <u>We have</u>:

<u>in case</u> (1^*), $d = \min\{s: \mathrm{ord}_V y = s < \mathrm{ord}_W y$ for some $y \in R\}$;

<u>in case</u> (2^*), $d = \min\{s: \mathrm{ord}_V y = 2s + 1$ for some $y \in R\}$;

<u>in case</u> (3^*), $d = \min\{s$: for some $y, \theta \in R$, $\mathrm{ord}_V y = \mathrm{ord}_V \theta = s$, and the residue of y/θ modulo $M(V)$ does not belong to $R/M(R)\}$.

<u>The following can be deduced from the above</u>.

(10.1.9) emdim $R = 2$.

(10.1.10) $\lambda\,\mathfrak{C}(R) = 2d$, $\lambda^k_{\mathfrak{C}}(R) = 2d[R/M(R) : k] = 2[R/\mathfrak{C}(R) : k]$.

(10.1.11) $[R^*/\mathfrak{C}(R) : R] = 2d$.

(10.1.12) <u>For</u> $i = 0,1,\ldots,d$, <u>there exist unique ideals</u> $J_i \supset \mathfrak{C}(R)$ <u>in</u> R, <u>such that</u> $[R/J_i : R] = i$.
<u>Further we have</u>:
$\lambda^k(R,J_i) = 2[R/J_i : k] = 2i[R/M(R) : k]$, <u>and in particular</u> $\lambda(R,J_i) = 2i$.

(10.1.13) $\alpha \in \mathfrak{C}(R) \Leftrightarrow \lambda^k(R,\alpha) \geq \lambda^k_{\mathfrak{C}}(R)$,
$\alpha R^* = \mathfrak{C}(R) \Leftrightarrow \lambda^k(R,\alpha) = \lambda^k_{\mathfrak{C}}(R)$,
$\lambda(R,\alpha) < \lambda_{\mathfrak{C}}(R) \Rightarrow \lambda(R,\alpha) \equiv 0(2)$, <u>for every</u> $\alpha \in R$.

(10.1.14) <u>Let</u> A <u>be a two dimensional regular local domain and let</u> $f: A \to R$ <u>be an epimorphism</u>.
<u>Let</u> $z \in R^*$ <u>such that</u> $R^* = R[z]$ <u>and</u>

<u>let</u> $\alpha, \beta \in A$ <u>such that</u> $f(\alpha)z = f(\beta)$ <u>and</u> $f(\alpha)R^* = \mathfrak{C}(R)$. <u>Then</u>

$$\operatorname{Ker} f \subset (\{\alpha, \beta\}A)^2.$$

PROOF. Note that, for $I \subset R$ or $I \in R$ we have:

in case (1^*), $\lambda(R,I) = \operatorname{ord}_V I + \operatorname{ord}_W I$;

in case (2^*), $\lambda(R,I) = \operatorname{ord}_V I$;

in case (3^*), $\lambda(R,I) = 2 \operatorname{ord}_V I$.

In view of this, and (5.4) and (5.5) it is easy to deduce (10.1.10), (10.1.11) and (10.1.12) from (10.1.6) (since we must have $\mathfrak{C}(R) = J_d$); and also (10.1.13) can be easily deduced from (10.1.7), (10.1.8) and (10.1.10). Now we claim that (10.1.9) is implied by (10.1.2) (10.1.4) and (10.1.5).

From (10.1.5) we have emdim $R \leq 2$ (since the said x,y exist by (10.1.2) and (10.1.4)). Also R is not regular since $\lambda(R) = 2$ (using (5.10)) and hence emdim $R > 1$. Thus emdim $R = 2$.

Proof of (10.1.1), (10.1.2) and (10.1.3) is as follows:

Now either (1) card $\mathfrak{D}(R) \neq 1$, or (2) card $\mathfrak{D}(R) = 1$ and the unique member of $\mathfrak{D}(R)$ is residually rational over R, or (3) card $\mathfrak{D}(R) = 1$ and the unique member of $\mathfrak{D}(R)$ is not residually rational over R.

Since $\lambda(R) = 2$, if (1) holds then: card $\mathfrak{D}(R) = 2$, and, upon letting V and W to be the distinct members of $\mathfrak{D}(R)$, we have the V and W are residually rational over R, $\operatorname{ord}_V x_1 = 1$ for some $x_1 \in R$ and $\operatorname{ord}_W x_2 = 1$ for some $x_2 \in R$; now upon letting

$$x_0 = \begin{cases} x_1 & \text{if } \operatorname{ord}_W x_1 = 1 \\ x_2 & \text{if } \operatorname{ord}_W x_1 \neq 1 = \operatorname{ord}_V x_2 \\ x_1 + x_2 & \text{if } \operatorname{ord}_W x_1 \neq 1 \neq \operatorname{ord}_V x_2 \end{cases}$$

we get $x_0 \in R$ with $\mathrm{ord}_V x_0 = 1 = \mathrm{ord}_W x_0$; also for any $x \in R$ we clearly have

$$\mathrm{ord}_V x = 1 = \mathrm{ord}_W x \Leftrightarrow \lambda(R,x) = 2 ;$$

finally, clearly there exists $z \in (R^* \cap M(V))\backslash M(W)$ and for any such z, by Nakayama's lemma, we have $R^* = R[z]$.

Since $\lambda(R) = 2$, if (2) holds, then: upon letting $V = R^*$, we have that $\mathfrak{D}(R) = \{V\}$, V is residually rational over R, $R \cap (M(V)\backslash M(V)^2) = \emptyset$, $R \cap (M(V)^2\backslash M(V)^3) \neq \emptyset$, and for any $x \in R$ we clearly have

$$x \in M(V)^2\backslash M(V)^3 \Leftrightarrow \lambda(R,x) = 2 ;$$

also, clearly there exists $z \in M(V)\backslash M(V)^2$ and for any such z, by Nakayama's lemma, we have $R^* = R[z]$.

Since $\lambda(R) = 2$, if (3) holds then: upon letting $V = R^*$ we have that $\mathfrak{D}(R) = \{V\}$, $[V/M(V) : R/M(R)] = 2$, $R \cap (M(V)\backslash M(V^2)) = \emptyset$, and for any $x \in R$ we clearly have

$$x \in M(V)\backslash M(V)^2 \Leftrightarrow \lambda(R,x) = 2 ;$$

also, clearly there exists $z \in V$ such that $V/M(V) = g(R)(g(z))$ where $g: V \to V/M(V)$ is the canonical epimorphism; for any such z, by Nakayama's lemma, we have $R^* = R[z]$.

Now, (10.1.14) can be deduced from (10.1.5) and (10.1.9). This deduction, as well as proofs of (10.1.4), (10.1.5), (10.1.6), (10.1.7) and (10.1.8) shall be presented in the Lemmas (10.2), (10.3) and (10.4) for the cases (1^*), (2^*) and (3^*) respectively.

REMARK. Geometrically speaking: In case (1^*), R represents (or, R is like the local ring of a singularity of an algebraic curve when the singularity is what may be called a "high node"; the high node is an "ordinary node" if $d = 1$. In case (2^*), R represents what may be called a "high cusp"; the high cusp is an "ordinary cusp" if $d = 1$.

In case (3^*), R represents what may be called a "nonrational high cusp".

(10.2) LEMMA ON HIGH NODES. Assume that $\text{card}\mathfrak{D}(R) = 2$, and let V and W be the distinct members of $\mathfrak{D}(R)$. Assume that V and W are residually rational over R. Assume that there exists $x \in R$ such that $\text{ord}_V x = 1 = \text{ord}_W x$, and fix any such $x \in R$. Let G denote the set of all pairs (p,q) of nonnegative integers, let

$$G^* = \{(p,q) \in G: p = q\}$$

and for every nonnegative integer n let

$$G_n = \{(p,q) \in G: p \geq n \leq q\}.$$

For every $I \subset R^*$ let

$$G(I) = \{(\text{ord}_V r, \text{ord}_W r): 0 \neq r \in I\} ;$$

(Note that then $G(I) \subset G$, and if I is an ideal in R or R^*, then G(I) is a subsemigroup of G, i.e., $(p + p',q + q') \in G(I)$ whenevery $(p,q) \in G(I)$ and $p',q') \in G(I)$.).

Let

$$P = R^* \cap M(V) \quad \text{and} \quad Q = R^* \cap M(W).$$

(Note that then: P and Q are exactly all the distinct maximal ideals in R^*. Further we have

$$R \cap P = R \cap Q = R \cap (PQ) = M(R).$$

For every $(m,n) \in G$, we have

$$M(V)^m \cap M(W)^n = P^m \cap Q^n = P^mQ^n = \{r \in R^*: \text{ord}_V r \geq m \text{ and } \text{ord}_W R \geq n\}$$

and

$$G(P^mQ^n) = \{(p,q) \in G: p \geq m \text{ and } q \geq n\} ;$$

and, in particular, for every nonnegative integer n we have

$$M(V)^n \cap M(W)^n = P^n \cap Q^n = P^nQ^n = \{r \in R^*: (\mathrm{ord}_V r, \mathrm{ord}_W r) \in G_n \cup \{\infty,\infty\}\}$$

and

$$G(P^nQ^n) = G_n.)$$

Let

$$d = [R/\mathfrak{C}(R) : R].$$

Then we have the following:

(10.2.1) _For any_ $y \in R$ _with_ $\mathrm{ord}_V y \neq \mathrm{ord}_W y$, _upon letting_ $a = \min(\mathrm{ord}_V y, \mathrm{ord}_W y)$, _we have_ $P^aQ^a \subset (x,y)R$; _whence in particular,_ $G_a \subset G((x,y)R)$.

(10.2.2) _We have_

$$\mathrm{ord}_V \mathfrak{C}(R) = \mathrm{ord}_W \mathfrak{C}(R) = d = \text{a positive integer}$$

$$\mathfrak{C}(R) = P^dQ^d \quad \text{and} \quad \lambda\mathfrak{C}(R) = 2d$$

and

$$G(R) = G^* \cup G_d \quad \text{and} \quad \mathrm{emdim}\, R = 2.$$

For every integer i _with_ $0 \leq i \leq d$ _we have that there exists a unique ideal_ J_i _in_ R _with_ $\mathfrak{C}(R) \subset J_i$ _such that_ $[R/J_i : R] = i$; _moreover we have_

$$J_i = \mathfrak{C}(R) + x^iR \quad \text{and} \quad \lambda(R,J_i) = 2i \quad \text{for} \quad 0 \leq i \leq d .$$

For every $y \in R$ _with_ $\mathrm{ord}_V y \neq \mathrm{ord}_W y$ _and_ $\min(\mathrm{ord}_V y, \mathrm{ord}_W y) = d$, _we have_ $(x,y)R = M(R)$. _For any_ $\alpha \in R$ _we have_

$$\alpha \in \mathfrak{C}(R) \Leftrightarrow \lambda(R,\alpha) \geq 2d ,$$

$$\alpha R^* = \mathfrak{C}(R) \Leftrightarrow \lambda(R,\alpha) = 2d ,$$

and

$$\lambda(R,\alpha) \le 2d \Rightarrow \lambda(R,\alpha) \equiv 0(2).$$

(10.2.3) Let $f: A \to R$ be an epimorphism where A is a two-dimensional regular local ring. Let $z \in R^*$, $\alpha \in A$ and $\beta \in A$ be any elements such that $R^* = R[z]$, $f(\alpha)R^* = \mathfrak{C}(R)$, and $f(\alpha)z = f(\beta)$. Then $\operatorname{Ker} f \subset ((\alpha,\beta)A)^2$.

PROOF OF (10.2.1). Let any $y \in R$ be given. First we claim that:

$$(1)\quad \begin{cases} \text{if } a = \operatorname{ord}_V y < \operatorname{ord}_W y = b, \text{ then there exists } r \in (x,y)R \\ \text{such that } \operatorname{ord}_V r = a \text{ and } \operatorname{ord}_W r > b. \end{cases}$$

Namely, since W is residually rational over R, there exists $\delta \in R\backslash M(R)$ such that $\operatorname{ord}_W(y - \delta x^b) > b$; it suffices to take $r = y - \delta x^b$.

By induction on m, from (1) we get:

$$(2)\quad \begin{cases} \text{if } a = \operatorname{ord}_V y < \operatorname{ord}_W y \text{ and } m \text{ is any nonnegative integer, then} \\ \text{there exists } r \in (x,y)R \text{ such that } \operatorname{ord}_V r = a \text{ and } \operatorname{ord}_W r > m. \end{cases}$$

Next we claim that:

$$(3)\quad \begin{cases} \text{if } a = \operatorname{ord}_V y < \operatorname{ord}_W y \text{ and } (p,q) \in G \text{ with } a \le p \le q, \text{ then} \\ \text{there exists } s \in (x,y)R \text{ such that } \operatorname{ord}_V s = p \text{ and } \operatorname{ord}_W s = q. \end{cases}$$

Namely, by (2) we can find $r \in (x,y)R$ such that $\operatorname{ord}_V r = a$ and $\operatorname{ord}_W r > q - p + a$; it suffices to take $s = x^p$ in case $p = q$, and $s = x^{p-a}r + x^q$ in case $p < q$.

Now we claim that:

$$(4)\quad \begin{cases} \text{if } a = \mathrm{ord}_V y < \mathrm{ord}_W y, \text{ then there exists } y' \in (x,y)R \\ \text{such that } a = \mathrm{ord}_W y' < \mathrm{ord}_V y' . \end{cases}$$

Namely, since V is residually rational over R, there exists $\delta' \in R\backslash M(R)$ such that $\mathrm{ord}_V(y - \delta' x^a) > a$; it suffices to take $y' = y - \delta' x^a$.

By (3) and (4) we get that:

$$(5)\quad \begin{cases} \text{if } \mathrm{ord}_V y \neq \mathrm{ord}_W y, \text{ then } G_a \subset G((x,y)R) , \\ \text{where } a = \min(\mathrm{ord}_V y , \mathrm{ord}_W y). \end{cases}$$

Now we claim that:

$$(6)\quad \begin{cases} \text{if } \mathrm{ord}_V y \neq \mathrm{ord}_W y \text{ and } 0 \neq \eta \in R^* \text{ with } \min(\mathrm{ord}_V \eta, \mathrm{ord}_W \eta) \geq a = \\ \min(\mathrm{ord}_V y, \mathrm{ord}_W y), \text{ then there exists } \eta' \in R^* \text{ such that } \eta - \eta' \\ \in (x,y)R \text{ and } \min(\mathrm{ord}_V \eta', \mathrm{ord}_W \eta') > \min(\mathrm{ord}_V \eta, \mathrm{ord}_W \eta). \end{cases}$$

Namely, upon letting $e = \min(\mathrm{ord}_W \eta, \mathrm{ord}_W \eta)$, by (5) we can find elements $t_0, t_1 t_2$ in R such that $\mathrm{ord}_V t_0 = e = \mathrm{ord}_V t_0$, $\mathrm{ord}_V t_1 = e < \mathrm{ord}_W t_1$, and $\mathrm{ord}_V t_2 > e = \mathrm{ord}_W t_2$; in case $\mathrm{ord}_V \eta = e = \mathrm{ord}_V \eta$, first (since V is residually rational over R) we can find $\delta_0 \in R\backslash M(R)$ such that $\mathrm{ord}_V(\eta - \delta_0 t_0) > e$ and then (since W is residually rational over R we can find $\rho \in R$ such that $\mathrm{ord}_W(\eta - \delta_0 t_0 - \rho t_1) > e$, and now it suffices to take $\eta' = \eta - \delta_0 t_0 - \rho t_1$; in case $\mathrm{ord}_V \eta = e < \mathrm{ord}_W \eta$, (since V is residually rational over R) we can find $\delta_1 \in R\backslash M(R)$ such that $\mathrm{ord}_V(\eta - \delta_1 t_1) > e$, and now it suffices to take $\eta' = \eta - \delta_1 t_1$; finally, in case $\mathrm{ord}_V \eta > e = \mathrm{ord}_W \eta$, (since W is residually rational over R) we can find $\delta_2 \in R\backslash M(R)$ such that $\mathrm{ord}_W(\eta - \delta_2 t_2) > e$, and now it suffices to take $\eta' = \eta - \delta_2 t_2$.

By induction on i, from (6) we get:

(7) $\begin{cases} \text{if } \operatorname{ord}_V y \neq \operatorname{ord}_W y, \ i \text{ is any nonnegative integer, and } \eta \text{ is} \\ \text{any element in } R^*, \text{ with } \min(\operatorname{ord}_V\eta, \operatorname{ord}_W\eta) \geq \min(\operatorname{ord}_V y, \operatorname{ord}_W y), \\ \text{then there exists } \eta' \in R^* \text{ such that } \eta - \eta' \in (x,y)R \text{ and} \\ \min(\operatorname{ord}_V\eta', \operatorname{ord}_W\eta') > i. \end{cases}$

Now (7) can clearly be reformulated thus:

(8) $\begin{cases} \text{if } \operatorname{ord}_V y \neq \operatorname{ord}_W y \text{ and } a = \min(\operatorname{ord}_V y, \operatorname{ord}_W y), \text{ then} \\ P^aQ^a \subset (x,y)R + P^iQ^i \text{ for every nonnegative integer } i. \end{cases}$

Next we claim that:

(9) $\begin{cases} \text{there exists a positive integer } u \text{ such that} \\ P^{uj}Q^{uj} \subset M(R)^j \text{ for every positive integer } j. \end{cases}$

Namely, since $\mathfrak{C}(R)$ is a nonzero ideal in R^*, we can find a positive integer u such that $P^uQ^u \subset \mathfrak{C}(R)$: since $\mathfrak{C}(R) \subset R$, we then get $P^uQ^u \subset M(R)$; it follows that $P^{uj}Q^{uj} \subset M(R)^j$ for every positive integer j.

Finally we claim that:

(10) $\begin{cases} \text{if } \operatorname{ord}_V y \neq \operatorname{ord}_W y \text{ and } a = \min(\operatorname{ord}_V y, \operatorname{ord}_W y) \text{ then} \\ P^aQ^a \subset (x,y)R. \end{cases}$

Namely, by (8) and (9) we get $P^aQ^a \subset (x,y)R + M(R)^j$ for every positive integer j, and by the Krull intersection theorem we have

$$\bigcap_{j=1}^{\infty} [(x,y)R + M(R)^j] = (x,y)R.$$

This completes the proof of (10.2.1).

PROOF OF (10.2.2). Let Ω be the set of all nonnegative integer w such that for some $\zeta \in R$ we have $\operatorname{ord}_V\zeta \neq \operatorname{ord}_W\zeta$ and $\min(\operatorname{ord}_V\zeta, \operatorname{ord}_W\zeta) = w$.

Since $\mathfrak{C}(R) \subset R$ and $\mathfrak{C}(R)$ is a nonzero ideal in R^*, we see that $\Omega \neq \emptyset$. Upon letting

(1) $$a = \min\{w: w \in \Omega\}$$

we get that a is a positive integer, and for some $y \in R$ we have

(2) $$\mathrm{ord}_V y \neq \mathrm{ord}_W y \quad \text{and} \quad \min(\mathrm{ord}_V y, \mathrm{ord}_W y) = a\ ;$$

henceforth fix any such $y \in R$.

By assumption

(3) $$x \in R \quad \text{with} \quad \mathrm{ord}_V x = 1 = \mathrm{ord}_W x$$

and so $G^*\backslash\{0,0\} = G(\{x,x^2,x^3,\ldots\}) \subset G((x,y)R)$; by (10.2.1) and (2) we have $G_a \subset G((x,y)R)$; consequently, in view of (1) we get that

(4) $$G(M(R)) = G((x,y)R) = (G^*\backslash\{(0,0)\} \cup G_a \quad \text{and} \quad G(R) = G^* \cup G_a.$$

In view of (10.2.1) and (2) we have

(5) $$P^aQ^a \subset (x,y)R$$

and hence in particular $P^aQ^a \subset \mathfrak{C}(R)$; also, for any $r \in R^*\backslash P^aQ^a$ we clearly have

$$\min(\mathrm{ord}_V r, \mathrm{ord}_W r) \leq a$$

and we can find $s \in rR^*$ such that

$$\min(\mathrm{ord}_V r, \mathrm{ord}_W r) = \min(\mathrm{ord}_V s, \mathrm{ord}_W s) < \max(\mathrm{ord}_V s, \mathrm{ord}_W s)$$

and then in view of (4) we see that $s \notin R$ and hence $r \notin \mathfrak{C}(R)$; thus we have proved that

(6) $$\mathfrak{C}(R) = P^aQ^a\ .$$

By (6), we get

(7) $$\mathrm{ord}_V \mathfrak{C}(R) = \mathrm{ord}_W \mathfrak{C}(R) = a$$

and, since V and W are residually rational over R, we also get

(8) $$\lambda\mathfrak{C}(R) = 2a.$$

Since V and W are residually rational over R, by (4) and (6) we see that

(9) for any $\alpha \in R$ we have: $\alpha \in \mathfrak{C}(R) \Leftrightarrow \lambda(R,\alpha) \geq 2a$

(10) for any $\alpha \in R$ we have: $\alpha R^* = \mathfrak{C}(R) \Leftrightarrow \lambda(R,\alpha) = 2a$

and

(11) for any $\alpha \in R$ we have: $\lambda(R,\alpha) \leq 2a \Rightarrow \lambda(R,\alpha) \equiv 0(2)$.

We claim that

(12) $$\begin{cases} \text{for any } \xi \in R \text{ with } \mathrm{ord}_V\xi = \mathrm{ord}_W\xi = i < a \text{ we have} \\ (R \cap (P^{i+1}Q^{i+1})) + \xi R = R \cap (P^iQ^i). \end{cases}$$

Namely, given any $\xi^* \in R \cap (P^iQ^i)$, (since V is residually rational over R) we can find $\rho \in R$ such that $\mathrm{ord}_V(\xi^*-\rho\xi) > i$ and then in view of (4) we must have $\mathrm{ord}_W(\xi^*-\rho\xi) > i$ and hence $\xi^* - \rho\xi \in R \cap (P^{i+1}Q^{i+1})$; thus

$$(R \cap (P^{i+1}Q^{i+1})) + \xi R \supset (P^iQ^i) ,$$

and the reverse inclusion is of course obvious.

By decreasing induction on i, in view of (3) and (12) we get:

(13) $$(R \cap (P^aQ^a)) + x^iR = R \cap (P^iQ^i) \quad \text{for} \quad 0 \leq i \leq a.$$

Next we claim that:

(14) $$[(R \cap (P^iQ^i))/(R \cap (P^{i+1}Q^{i+1})) : R] = 1 \quad \text{for} \quad 0 \leq i < a.$$

Namely, for every

$$\xi \in (R \cap (P^iQ^i))\setminus(R \cap (P^{i+1}Q^{i+1}))$$

by (4) we must have $\operatorname{ord}_V\xi = i = \operatorname{ord}_W\xi$ and hence by (12) we have

$$(R \cap (P^{i+1}Q^{i+1})) + \xi R = R \cap (P^iQ^i) \;;$$

this shows that

$$(14_1) \qquad [(R \cap (P^iQ^i))/(R \cap (P^{i+1}Q^{i+1})) : R] \leq 1 \;;$$

by (3)

$$x^i \in (R \cap (P^iQ^i))\setminus(R \cap (P^{i+1}Q^{i+1}))$$

and hence

$$(14_2) \qquad R \cap (P^iQ^i) \neq R \cap (P^{i+1}Q^{i+1}) \;;$$

now by (14_1) and (14_2) we get

$$[(R \cap (P^iQ^i))/(R \cap (P^{i+1}Q^{i+1}) : R] = 1$$

Now $R \cap (PQ) = M(R)$; hence by applying (13) with $i = 1$, in view of (5) we get

$$(15) \qquad (x,y)R = M(R) \;.$$

By (15) we have emdim $R \leq 2$; now R is not regular because $\mathfrak{D}(R) \neq 1$; therefore we must have

$$(16) \qquad \operatorname{emdim} R = 2 \;.$$

Since $R \cap (P^0Q^0) = R$, upon setting

$$(17) \qquad J_i = \mathfrak{C}(R) + x^iR \;,$$

by (6), (13) and (14) we get

$$(18) \qquad [R/J_i : R] = i \quad \text{for} \quad 0 \leq i \leq a \;.$$

By (3), (6) and (17) we have

(19) $$\lambda(R,J_i) = 2i \quad \text{for} \quad 0 \le i \le a.$$

We claim that:

(20) $$\begin{cases} \text{given any ideal } J \text{ in } R \text{ with } \mathfrak{C}(R) \subset J, \text{ upon letting} \\ i = \mathrm{ord}_V J, \\ \\ \text{we have that } i \text{ is an integer with } 0 \le i \le a, \text{ and } J = J_i. \end{cases}$$

Namely, in view (4) and (6) we see that i is an integer with $0 \le i \le a$, and

(20_1) $$J \subset R \cap (P^i Q^i) \ ;$$

if $i = a$ then (since $\mathfrak{C}(R) \subset J$) by (3), (6), (17) and (20_1) we see that $J = J_a$; so henceforth assume that $i < a$; we can take

(20_2) $$\xi \in J$$

with $\mathrm{ord}_V \xi = i$ and then in view of (2), (3) and (15) we get

(20_3) $$\xi = \delta x^i + \sigma y \quad \text{with} \quad \delta \in R \backslash M(R) \quad \text{and} \quad \sigma \in R$$

by (2) and (6) we have $y \in \mathfrak{C}(R)$, and hence by (20_3) we get

(20_4) $$x^i \in \mathfrak{C}(R) + \xi R \ ;$$

since (R) $\subset J$, by (6), (13), (17), (20_1), (20_2) and (20_4) we conclude that $J = J_i$.

Since $R \cap (P^0 Q^0) = R$, by (6), (13) and (14) we get

$$[R/\mathfrak{C}(R) : R] = a$$

and hence by the definition of d we have

(21) $$a = d.$$

Now in view of (21), the proof of (10.2.2) is complete by (4), (6) to (11), and (15) to (20).

PROOF OF (10.2.3). Since $R^* = R[z]$, we must have $z \notin P \cap Q$, and so, upon relabelling V and W suitably, we may suppose that $z \notin P$. Now, since V is residually rational over R, we can find

(1) $$\delta \in A \backslash M(A)$$

such that

(2) $$z + f(\delta) \in P.$$

Now $R^* = R[z + f(\delta)]$, and hence we must have

(3) $$z + f(\delta) \notin Q .$$

Since $f(\alpha)R^* = \mathfrak{C}(R)$, by (10.2.2) we get that

(4) $$\mathrm{ord}_V f(\alpha) = d = \mathrm{ord}_W f(\alpha) .$$

Upon letting

(5) $$\eta = \beta + \delta\alpha$$

we clearly have

(6) $$(\alpha, \eta)A = (\alpha, \beta)A$$

By assumption $f(\alpha)z = f(\beta)$, and hence by (5) we have

$$f(\eta) = f(\alpha)(z + f(\delta)) ;$$

consequently, by (2), (3) and (4) we get that

(7) $$\mathrm{ord}_W f(\eta) = d < \mathrm{ord}_V f(\eta) .$$

By assumption $\mathrm{ord}_V x = 1 = \mathrm{ord}_W x$, and hence upon fixing any

(8) $$\xi \in A \text{ with } f(\xi) = x$$

we have

(9) $$\mathrm{ord}_V f(\xi) = 1 = \mathrm{ord}_W f(\xi) \ ;$$

Now, in view of (7) and (8), by (10.2.2) we get that

(10) $$(f(\xi), f(\eta))R = M(R) \ .$$

By (10.2.2) we have emdim $R = 2$, and hence in view of (10) we conclude that

(11) $$(\xi, \eta)A = M(A).$$

Since $d > 0$, in view of (4) we get $\alpha \in M(A)$; in view of (4) and (7) we also have $\alpha \notin \eta A$; consequently, in view of (11) we can write

(12) $$\alpha = \delta'\xi^b + \rho\eta \quad \text{where} \quad \delta' \in A\backslash M(A) \quad \text{and} \quad \rho \in A$$

and b is a positive integer; now, by (4), (7), (9) and (12) we get

(13) $$b = d \ .$$

By (12) and (13) we see that $(\xi^d, \eta)A = (\alpha, \eta)A$, and hence by (6) we get

(14) $$(\xi^d, \eta)A = (\alpha, \beta)A \ .$$

Let any $\gamma \in A$ with $f(\gamma) = 0$ be given. We shall show that then $\gamma \in ((\xi^d, \eta)A)^2$, and in view of (14) this will complete the proof. Now, in view of (11), we can write

(15) $$\gamma = r + s\eta + t\eta^2 \quad \text{with} \quad r, s, t \quad \text{in} \quad A$$

such that

(16) $$\begin{cases} \text{either} \quad r = 0 \\ \text{or} \quad r = \delta_0\xi^a \quad \text{with} \quad \delta_0 \in A\backslash M(A) \quad \text{and nonnegative integer} \quad a \end{cases}$$

$$(17)\ \begin{cases} \text{either} \quad s = 0 \\ \text{or} \quad s = \delta_1 \xi^b \quad \text{with} \quad \delta_1 \in A \backslash M(A) \quad \text{and nonnegative integer} \quad b. \end{cases}$$

By (9), (16) and (17) we deduce that

$$(18)\qquad \text{ord}_V f(r) = \text{ord}_W f(r) \quad \text{and} \quad \text{ord}_V f(s) = \text{ord}_W f(s)$$

Since $f(\gamma) = 0$, in view of (7), (15) and (18) we conclude that

$$(19)\qquad \text{ord}_V f(r) = \text{ord}_W f(r) \geq 2d \quad \text{and} \quad \text{ord}_V f(s) = \text{ord}_W f(s) \geq d\ .$$

Now by (15), (16), (17) and (19) it follows that $\gamma \in ((\xi^d, \eta)A)^2$.

(10.3) LEMMA ON HIGH CUSPS. _Let_ $V = R^*$. _Assume that_ $\mathfrak{Y}(R) = \{V\}$ _and that_ V _is residually rational over_ R. _Assume that_ $R \cap (M(V) \backslash M(V)^2) = \emptyset$, _and fix any_ $x \in R \cap (M(V)^2 \backslash M(V)^3)$. _Let_ G b _the set of all nonnegative integers, let_

$$G^* = \{2m : m \in G\}\ ,$$

and for every $n \in G$ _let_

$$G_n = \{2m+1 : m \in G \quad \text{with} \quad m \geq n\}\ .$$

For every $I \subset V$ let

$$G(I) = \{\text{ord}_V r : 0 \neq r \in I\}\ .$$

Let

$$d = [R/\mathfrak{C}(R) : R]\ .$$

Then we have the following,

(10.3.1). d is a positive integer,

$$\text{ord}_V \mathfrak{C}(R) = 2d = \lambda_{\mathfrak{C}}(R) \quad \text{and} \quad \mathfrak{C}(R) = M(V)^{2d}$$

and

$$G(R) = G^* \cup G_d \quad \text{and emdim } R = 2.$$

For every integer i with $0 \le i \le d$ we have that there exists a unique ideal J_i in R with $\mathfrak{C}(R) \supset J_i$ such that $[R/J_i : R] = i$; moreover we have

$$J_i = \mathfrak{C}(R) + x^i R \quad \text{and} \quad \lambda(R,J_i) = 2i \quad \text{for} \quad 0 \le i \le d.$$

For every $y \in R$ with $\operatorname{ord}_V y = 2d + 1$ we have $(x,y)R = M(R)$. For any $\alpha \in R$ we have

$$\alpha \in \mathfrak{C}(R) \Leftrightarrow \lambda(R,\alpha) \ge 2d$$

$$\alpha R^* = \mathfrak{C}(R) \Leftrightarrow \lambda(R,\alpha) = 2d$$

and

$$\lambda(R,\alpha) \le 2d \Rightarrow \lambda(R,\alpha) \equiv 0(2).$$

(10.3.2) Let $f\colon A \to R$ be an epimorphism where A is a two-dimensional regular local ring. Let $z \in R^*$, $\alpha \in A$, and $\beta \in A$ be any elements such that $R^* = R[z]$, $f(\alpha)R^* = \mathfrak{C}(R)$, and $f(\alpha)z = f(\beta)$. Then $\operatorname{Ker} f \subset ((\alpha,\beta)A)^2$.

PROOF OF (10.3.1). Given any $0 \ne \zeta \in V$ we can write $\zeta = \zeta'/\zeta^*$ with $0 \ne \zeta' \in R$ and $0 \ne \zeta^* \in R$, and then we have $\operatorname{ord}_V\zeta = \operatorname{ord}_V\zeta' - \operatorname{ord}_V\zeta^*$; since $\operatorname{ord}_V\zeta = 1$ for some $\zeta \in V$, we conclude that $G(R)$ must contain some odd integer, i.e.,

$$\{w \in G : 2w + 1 \in G(R)\} = \emptyset.$$

Now upon letting

$$a = \min\{w \in G : 2w + 1 \in G(R)\}$$

(since by assumption $R \cap (M(V)\backslash M(V)^2) = \emptyset$) we get that a is a positive integer and for some $y \in R$ we have

(2) $$\operatorname{ord}_V y = 2a + 1 ;$$

henceforth fix any such $y \in R$.

By assumption

(3) $$x \in R \text{ with } \operatorname{ord}_V x = 2$$

and hence by (1) and (2) we see that

(4) $$G(M(R)) = G((x,y)R) = (G^*\setminus\{0\}) \cup G_A \text{ and } G(R) = G^* \cup G_a .$$

We claim that

(5) $$\begin{cases} \text{given any } 0 \neq \eta \in V \text{ with } \operatorname{ord}_V\eta \geq 2a, \text{ there exists } \eta' \in V \\ \text{such that } \eta - \eta' \in (x,y)R \text{ and } \operatorname{ord}_V\eta' > \operatorname{ord}_V\eta. \end{cases}$$

Namely, upon letting $e = \operatorname{ord}_V\eta$, by (4) we can find $t \in (x,y)R$ such that $\operatorname{ord}_V t = e$ and then (since V is residually rational over R) we can find $\delta^* \in R\setminus M(R)$ such that $\operatorname{ord}_V(\eta-\delta^* t) > e$; now it suffices to take $\eta' = \eta - \delta^* t$.

By induction on i, from (5) we get:

(6) $$M(V)^{2a} \subset (x,y)R + M(V)^i \text{ for every nonnegative integer } i.$$

Next we claim that:

(7) $$\begin{cases} \text{there exists a positive integer } u \text{ such that} \\ M(V)^{uj} \subset M(R)^j \text{ for every positive integer } j . \end{cases}$$

Namely, since $\mathfrak{C}(R)$ is a nonzero ideal in V, we can find a positive integer u such that $M(V)^u \subset \mathfrak{C}(R)$; since $\mathfrak{C}(R) \subset R$, we then get $M(V)^u \subset M(R)$; it follows that $M(V)^{uj} \subset M(R)^j$ for every positive integer j.

By (6) and (7) we get

$M(V)^{2a} \subset (x,y)R + M(R)^j$ for every positive integer j .

and by the Krull intersection theorem we have

$\bigcap_{j=1}^{\infty} [(x,y)R + M(R)^j] = (x,y)R$; therefore

(8) $M(V)^{2a} \subset (x,y)R$.

and hence in particular $M(V)^{2a} \subset \mathfrak{C}(R)$; also given any $r \in R^*\backslash M(V)^{2a}$ we can find $s \in rR^*$ with $\mathrm{ord}_V s = 2a - 1$ and then by (4) we have $s \notin R$ and hence $r \notin \mathfrak{C}(R)$; thus we have proved that

(9) $\mathfrak{C}(R) = M(V)^{2a}$ and hence

(10) $\mathrm{ord}_V \mathfrak{C}(R) = 2a$.

Since $\mathfrak{D}(R) = \{V\}$ and V is residually rational over R, by (4) and (1) we see that, for any $\alpha \in R$, we have the following.

(11) $\lambda_{\mathfrak{C}}(R) = 2a$.

(12) $\alpha \in \mathfrak{C}(R) \Leftrightarrow \lambda(R,\alpha) \geq 2a$

(13) $\alpha R^* = \mathfrak{C}(R) \Leftrightarrow \lambda(R,\alpha) = 2a$.

and

(14) $\lambda(R,\alpha) \leq 2a \Rightarrow \lambda(R,\alpha) \equiv 0(2)$.

We claim that

(15) $\begin{cases} \text{for any } \xi \in R \text{ with } \mathrm{ord}_W \xi = 2i \text{ where } i \text{ is an integer} < a , \\ \text{we have } (R \cap M(V)^{2i+2}) + \xi R = R \cap M(V)^{2i} . \end{cases}$

Namely, given any $\xi^* \in R \cap M(V)^{2i}$, (since V is residually rational over R) we can find $\rho \in R$ such that $\mathrm{ord}_V(\xi^* - \rho\xi) > 2i$ and then in view of (4) we must have $\mathrm{ord}_V(\xi^* - \rho\xi) \geq 2i + 2$, and hence $\xi^* - \rho\xi \in R \cap M(V)^{2i+2}$; thus

$$(R \cap M(V)^{2i+2}) + \xi R \supset R \cap M(V)^{2i} ,$$

and the reverse inclusion is of course obvious.

By decreasing induction on i, in view of (3) and (15) we get

(16) $$(R \cap M(V)^{2a}) + x^i R = R \cap M(V)^{2i} \quad \text{for} \quad 0 \leq i < a .$$

Next we claim that

(17) $$[(R \cap M(V)^{2i})/(R \cap M(V)^{2i+2}) : R] \quad \text{for} \quad 0 \leq i < a .$$

Namely, for every

$$\xi \in (R \cap M(V)^{2i}) \setminus (R \cap M(V)^{2i+2})$$

by (4) we must have $\text{ord}_V \xi = 2i$ and hence by (15) we have

$$(R \cap M(V)^{2i+2}) + \xi R = R \cap M(V)^{2i} ;$$

this shows that

(17_1) $$[(R \cap M(V)^{2i})/(R \cap M(V)^{2i+2}) : R] \leq 1 ;$$

by (3) we have

$$x^i \in (R \cap M(V)^{2i} \setminus (R \cap M(V)^{2i+2})$$

and hence

(17_2) $$(R \cap M(V)^{2i}) \setminus (R \cap M(V)^{2i+2}) \neq \emptyset ;$$

now by (17_1) and (17_2) we get

$$[(R \cap M(V)^{2i})/(R \cap M(V)^{2i+2}) : R] = 1 .$$

Now by assumption $R \cap (M(V) \setminus M(V)^2) = \emptyset$, and hence $R \cap M(V)^2 = M(R)$; consequently, by applying (16) with $i = 1$, in view of (8) we get

(18) $$(x,y)R = M(R).$$

By (18 we get $\text{emdim } R \leq 2$; now R is not regular because $R \cap (M(V) \setminus M(V)^2) = \emptyset$; therefore we must have

(19) $$\text{emdim } R = 2 .$$

Since $R \cap M(V)^0 = R$, upon letting

$$J_i = \mathfrak{C}(R) + x^i R , \tag{20}$$

by (9), (16) and (17) we see that

$$[R/J_i : R] = i \quad \text{for} \quad 0 \le i \le a . \tag{21}$$

By (3), (9) and (20) we have

$$\lambda(R, J_i) = 2i \quad \text{for} \quad 0 \le i \le a . \tag{22}$$

We claim that:

(23) $\begin{cases} \text{given any ideal } J \text{ in } R \text{ with } \mathfrak{C}(R) \subset J, \text{ upon letting} \\ i = (1/2)\mathrm{ord}_V J, \\ \\ \text{we have that } i \text{ is an integer with } 0 \le i \le a, \text{ and } J = J_i . \end{cases}$

Namely, in view of (4) and (9) we see that i is an integer with $0 \le i \le a$, and clearly

$$J \subset R \cap M(V)^{2i} ; \tag{23_1}$$

we can take

$$\xi \in J \tag{23_2}$$

with $\mathrm{ord}_V \xi = 2i$ and then in view of (2), (3) and (18) we get

$$\xi = \delta x^i + \sigma y \quad \text{with} \quad \delta \in R \backslash M(R) \quad \text{and} \quad \sigma \in R ; \tag{23_3}$$

by (2) and (9) we have $y \in \mathfrak{C}(R)$, and hence by (23_3) we get

$$x^i \in \mathfrak{C}(R) + \xi R ; \tag{23_4}$$

since $\mathfrak{C}(R) \subset J$, by (9), (16), (20), ($23_1$), ($23_2$) and ($23_4$) we conclude that $J = J_i$.

Since $R \cap M(V)^0 = R$, by (9), (16) and (17) we get

$$[R/\mathfrak{C}(R) : R] = a$$

and hence by the definition of d we have

(24) $$a = d .$$

Now in view of (24), the proof of (10.3.1) is complete by (4), (9) to (14), and (18) to (23).

PROOF OF (10.3.2). By assumption $R^* = R[z]$, V is residually rational over R, and $R \cap (M(V)\backslash M(V)^2 = \emptyset$, therefore we can find $\delta \in A$ such that

(1) $$\text{ord}_V(z+f(\delta)) = 1$$

and

(2) either $\delta \in A\backslash M(A)$ or $\delta = 0$.

Since $f(\alpha)R^* = \mathfrak{C}(R)$, by (10.3.1) we get that

(3) $$\text{ord}_V f(\alpha) = 2d .$$

Upon letting

(4) $$\eta = \beta + \delta\alpha$$

we clearly have

(5) $$(\alpha,\eta)A = (\alpha,\beta)A$$

By assumption $f(\alpha)z = f(\beta)$, and hence by (4) we have

$$f(\eta) = f(\alpha)(z + f(\delta)) ;$$

consequently by (1) and (3) we get that

(6) $$\text{ord}_V f(\eta) = 2d + 1$$

By assumption $\operatorname{ord}_V x = 2$ and hence, upon fixing any

(7) $$\xi \in A \text{ with } f(\xi) = x ,$$

we have

(8) $$\operatorname{ord}_V f(\xi) = 2$$

Now, in view of (6) and (7), by (10.3.1) we get that

(9) $$(f(\xi), f(\eta))R = M(R).$$

By (10.3.1) we have emdim $R = 2$, and hence in view of (9) we conclude that

(10) $$(\xi, \eta)A = M(A) .$$

Since $d > 0$, in view of (3) we get $\alpha \in M(A)$; in view of (3) and (6) we also have $\alpha \notin \eta A$; consequently, in view of (10) we can write

(11) $$\alpha = \delta' \xi^p + \rho\eta \quad \text{where} \quad \delta' \in A \setminus M(A) \quad \text{and} \quad \rho \in A$$

and p is a positive integer; now, by (3), (6), (8) and (11) we get

(12) $$p = d .$$

By (11) and (12) we see that $(\xi^d, \eta)A = (\alpha, \eta)A$, and hence by (5) we get

(13) $$(\xi^d, \eta)A = (\alpha, \beta)A .$$

Let any $\gamma \in A$ with $f(\gamma) = 0$ be given. We shall show that then $\gamma \in ((\xi^d, \eta)A)^2$, and in view of (13) this will complete the proof. Now, in view of (10), we can write

(14) $$\gamma = r + s\eta + t\eta^2 \text{ with } r, s, t \text{ in } A$$

such that

(15) $\begin{cases} \text{either } r = 0 \\ \text{or } r = \delta_0 \xi^a \text{ with } \delta_0 \in A \setminus M(A) \text{ and nonnegative integer } a \end{cases}$

and

(16) $\begin{cases} \text{either } s = 0 \\ \text{or } s = \delta_1 \xi^b \text{ with } \delta_1 \in A \setminus M(A) \text{ and nonnegative integer } b \end{cases}$

Since $f(\gamma) = 0$, by (6), (8), (14), (15) and (16) we deduce that

(17) $$r \in \xi^{2d}A \quad \text{and} \quad s \in \xi^d A .$$

Now by (14) and (17) it follows that $\gamma \in ((\xi^d, \eta)A)^2$.

(10.4) LEMMA ON NONRATIONAL HIGH CUSPS. Let $V = R^*$. _Assume that_ $\mathfrak{D}(R) = \{V\}$. _Assume that_ $[V/M(V) : R/M(R)] = 2$. _Assume that_ $R \cap M(V) \setminus M(V)^2 \neq \emptyset$ _and fix any_ $x \in R \cap (M(V) \setminus M(V)^2)$. _Let_ $g: V \to V/M(V)$ _be the canonical epimorphism_. _Let_

$$d = [R/\mathfrak{C}(R): R] .$$

Then we have the following

(10.4.1) _We have_

$$\text{ord}_V \mathfrak{C}(R) = d = \text{a positive integer}$$

$$\mathfrak{C}(R) = M(V)^d \quad \text{and} \quad \lambda_{\mathfrak{C}}(R) = 2d$$

and

$$\text{emdim } R = 2 .$$

For every integer i _with_ $0 \le i \le d$ _we have that there exists a unique ideal_ J_i _in_ R _with_ $\mathfrak{C}(R) \subset J$ _such that_ $[R/J_i : R] = i$; _moreover we have_

$$J_i = \mathfrak{C}(R) + x^i R \quad \text{and} \quad \lambda(R, J_i) = 2i \quad \text{for} \quad 0 \le i \le d .$$

<u>For any</u> $0 \neq \theta \in R$ <u>we have</u>:

$$\operatorname{ord}_V \theta \geq d \Leftrightarrow \begin{cases} \text{there exists } \theta' \in R \text{ such that} \\ \operatorname{ord}_V \theta' = \operatorname{ord}_V \theta \text{ and } g(\theta'/\theta) \notin g(R). \end{cases}$$

<u>For every</u> $y \in R$ <u>with</u> $\operatorname{ord}_V y = d$ <u>and</u> $g(y/x^d) \notin g(R)$, <u>we have</u> $(x,y)R = M(R)$. <u>For any</u> $\alpha \in R$ <u>we have</u>

$$\alpha R^* = \mathfrak{C}(R) \Leftrightarrow \lambda(R,\alpha) \geq 2d$$
$$\alpha R^* = \mathfrak{C}(R) \Leftrightarrow \lambda(R,\alpha) = 2d$$

<u>and</u>

$$\lambda(R,\alpha) \leq 2d \Rightarrow \lambda(R,\alpha) \equiv 0(2).$$

(10.4.2). <u>Let</u> $f: A \to R$ <u>be an epimorphism where</u> A <u>is a two-dimensional regular local ring</u>. <u>Let</u> $z \in R^*$, $\alpha \in R$ <u>and</u> $\beta \in R$ <u>be any elements such that</u> $R^* = R[z]$, $f(\alpha)R^* = \mathfrak{C}(R)$, <u>and</u> $f(\alpha)z = f(\beta)$. <u>Then</u> $\operatorname{Ker} f \subset ((\alpha,\beta)A)^2$.

PROOF OF (10.4.1). We claim that:

$$(1) \quad \begin{cases} \text{there exists a positive integer } u \text{ such that} \\ M(V)^{uj} \subset M(R)^j \text{ for every positive integer } j. \end{cases}$$

Namely, since $\mathfrak{C}(R)$ is a nonzero ideal in V, we can find a positive integer u such that $M(V)^u \subset \mathfrak{C}(R)$; since $\mathfrak{C}(R) \subset R$, we then get $M(V)^u \subset M(R)$; it follows that $M(V)^{uj} \subset M(R)^j$ for every positive integer j.

Recall that by assumption

$$(2) \qquad x \in R \text{ with } \operatorname{ord}_V x = 1.$$

Let Ω be the set of all nonnegative integers w such that for some $\zeta \in R$ with $\operatorname{ord}_V \zeta = w$ we have $g(\zeta/x^w) \notin g(R)$. By assumption $g(V) \neq g(R)$, and hence in view of (1) we see that $\Omega \neq \emptyset$.

Now upon letting

(3) $$a = \min\{w\colon w \in \Omega\}$$

we get that a is a positive integer and there exists $y \in R$ such that

(4) $$\operatorname{ord}_V y = a \quad \text{and} \quad g(y/x^a) \notin g(R) \ ;$$

henceforth fix any such $y \in R$.

We claim that for any $0 \neq \theta \in R$ we have

(5) $$\operatorname{ord}_V\theta \geq a \Leftrightarrow \begin{cases} \text{there exists } \theta' \in R \text{ such that} \\ \operatorname{ord}_V\theta' = \operatorname{ord}_V\theta \text{ and } g(\theta'/\theta) \notin g(R). \end{cases}$$

Namely, if $\operatorname{ord}_V\theta = b \geq a$ then by (4) we see that $\operatorname{ord}_V yx^{b-a} = \operatorname{ord}_V\theta = \operatorname{ord}_V x^b$ and either $g(yx^{b-a}/\theta) \notin g(R)$ or $g(x^b/\theta) \notin g(R)$; consequently it suffices to take $\theta' = yx^{b-a}$ or $\theta' = x^b$. Conversely, if θ and θ' are elements in R such that $\operatorname{ord}_V\theta' = \operatorname{ord}_V\theta = b \neq \infty$ and $g(\theta'/\theta) \notin g(R)$, then either $g(\theta'/x^b) \notin g(R)$ or $g(\theta/x^b) \notin g(R)$; hence by (3) we must have $b \geq a$.

Next we claim that

(6) $$\begin{cases} \text{given any } 0 \neq \eta \in V \text{ with } \operatorname{ord}_V = e \geq a, \text{ there exists } \eta' \in V \\ \text{such that } \eta - \eta' \in (x,y)R \text{ and } \operatorname{ord}_V\eta' > \operatorname{ord}_V\eta \ . \end{cases}$$

Namely, since $[V/M(V) : R/M(R)] = 2$, in view of (4) we can find elements ρ and ρ^* in R such that $\operatorname{ord}_V(\eta x^{-e} - \rho - \rho^* yx^{-a}) > 0$ and now it suffices to take $\eta' = \eta - \rho x^e - \rho^* yx^{e-a}$.

By induction on i, from (6) we get:

(7) $$M(V)^a \subset (x,y)R + M(V)^i \quad \text{for every nonnegative integer } i.$$

By (1) and (7) we get

$$M(V)^a \subset (x,y)R + M(R)^j \quad \text{for every positive integer } j$$

and by the Krull intersection theorem we have

$$\bigcap_{j=1}^{\infty} [(x,y)R + M(R)^j] = (x,y)R \ ;$$

therefore

(8) $$M(V)^a \subset (x,y)R$$

and hence in particular $M(V)^a \subset \mathfrak{C}(R)$; also, given any $r \in V$ with $\operatorname{ord}_V r = p < a$, upon letting $s = ry/x^a$, by (2) and (4) we get that $s \in V$ with $\operatorname{ord}_V s = p$ and $g(s/r) \notin g(R)$, and then by (5) we see that either $r \notin R$ or $s \notin R$; thus we have proved that

(9) $$\mathfrak{C}(R) = M(V)^a$$

and hence

(10) $$\operatorname{ord}_V \mathfrak{C}(R) = a \ .$$

Since $\mathfrak{D}(R) = \{V\}$ and $[V/M(V) : R/M(R)] = 2$, in view of (10) we see that

(11) $$\lambda_{\mathfrak{C}}(R) = 2a$$

(12) for any $\alpha \in R$ we have: $\alpha \in \mathfrak{C}(R) \Leftrightarrow \lambda(R,\alpha) \geq 2a$

(13) for any $\alpha \in R$ we have: $\alpha R^* = \mathfrak{C}(R) \Leftrightarrow \lambda(R,\alpha) = 2a$

and

(14) for any $\alpha \in R$ we have: $\lambda(R,\alpha) \leq 2a \Rightarrow \lambda(R,\alpha) \equiv 0(2)$.

We claim that

$$(15)\quad \begin{cases} \text{for any } \xi \in R \text{ with } \operatorname{ord}_V \xi = i < a, \text{ we have} \\ (R \cap M(V)^{i+1}) + \xi R = R \cap M(V)^i . \end{cases}$$

Namely, given any $\xi^* \in R \cap M(V)^i$ we must have (if $\operatorname{ord}_V \xi^* > i$, then obviously, and if $\operatorname{ord}_V \xi^* = i$, then by (5)) $g(\xi^*/\xi) \in g(R)$ and hence we can find $\rho \in R$ such that $\xi^* - \rho\xi \in R \cap M(V)^{i+1}$; thus

$$(R \cap M(V)^{i+1}) + \xi R \supset R \cap M(V)^i ,$$

and the reverse inclusion is of course obvious.

By decreasing induction on i, in view of (2) and (15) we get

$$(16)\qquad (R \cap M(V)^a) + x^i R = R \cap M(V)^i \quad \text{for} \quad 0 \le i < a .$$

Next we claim that

$$(17)\qquad [(R \cap M(V)^i/(R \cap M(V)^{i+1}) : R] = 1 \quad \text{for} \quad 0 \le i < a .$$

Namely, for every

$$\xi \in (R \cap M(V)^i)\backslash(R \cap M(V)^{i+1})$$

by (15) we have

$$(R \cap M(V)^{i+1}) + \xi R = R \cap M(V)^i$$

which shows that

$$(17_1)\qquad [(R \cap M(V)^i)/(R \cap M(V)^{i+1}) : R] \le 1 ;$$

by (2) we have

$$x^i \in (R \cap M(V)^i)\backslash(R \cap M(V)^{i+1})$$

and hence

$$(17_2)\qquad (R \cap M(V)^i)\backslash(R \cap M(V)^{i+1}) \ne \emptyset ;$$

now by (17_1) and (17_2) we get

$$[(R \cap M(V)^i)/(R \cap M(V^{i+1}) \; ; R] = 1 \; .$$

Now $R \cap M(V) = M(R)$ and hence by applying (16) with $i = 1$, in view of (8) we get

(18) $$(x,y)R = M(R) \; .$$

BY (18) we get $\operatorname{emdim} R \le 2$; now R is not regular because $g(V) \neq g(R)$; therefore we must have

(19) $$\operatorname{emdim} R = 2 \; .$$

Since $R \cap M(V)^0 = R$, upon letting

(20) $$J_i = \mathfrak{C}(R) + x^i R \; ,$$

by (9), (16) and (17) we see that

(21) $$[R/J_i : R] = i \quad \text{for} \quad 0 \le i \le a \; .$$

Since $[V/M(V) : R/M(R)] = 2$, by (2), (9) and (20) we have

(22) $$\lambda(R, J_i) = 2i \quad \text{for} \quad 0 \le i \le a \; .$$

We claim that:

(23) $\left\{ \begin{array}{l} \text{given any ideal } J \text{ in } R \text{ with } \mathfrak{C}(R) \subset J, \text{ upon letting} \\ i = \operatorname{ord}_V J \; , \\ \text{we have that } i \text{ is an integer with } 0 \le i \le a, \text{ and } J = J_i \; . \end{array} \right.$

Namely, in view of (9) we see that i is an integer with $0 \le i \le a$, and clearly

(23_1) $$J \subset R \cap M(V)^i \; ;$$

if $i = a$ then (since $\mathfrak{C}(R) \subset J$) by (2), (9), (20) and (23_1) we see that $J = J_a$; so henceforth assume that $i < a$; we can take

(23_2) $$\xi \in J$$

with $\text{ord}_V \xi = i$ and then in view of (2), (4) and (18) we get

(23_3) $$\xi = \delta x^i + \sigma y \quad \text{with} \quad \delta \in R \setminus M(R) \quad \text{and} \quad \sigma \in R ;$$

by (4) and (9) we have $y \in \mathfrak{C}(R)$, and hence by (23_3) we get

(23_4) $$x^i \in \mathfrak{C}(R) + \xi R ;$$

since $\mathfrak{C}(R) \subset J$, by (9), (16), (20), (23_1), (23_2) and (23_4) we conclude that $J = J_i$.

Since $R \cap M(V)^0 = R$, by (9), (16) and (17) we get

$$[R/\mathfrak{C}(R) : R] = a$$

and hence by the definition of d we have

(24) $$a = d .$$

Now in view of (24), the proof of (10.4.1) is complete by (5), (9) to (14), and (18) to (23).

PROOF OF (10.4.2). Since $V = R[z]$ and $g(V) \neq g(R)$, we must have $g(z) \notin g(R)$. Since $f(\alpha)R^* = \mathfrak{C}(R)$ and $f(\alpha)z = f(\beta)$, in view of (10.4.1), we get that

(1) $$\text{ord}_V f(\alpha) = d = \text{ord}_V f(\beta) \text{ and } g(f(\beta)/f(\alpha)) \notin g(R).$$

By assumption $\text{ord}_V x = 1$ and hence, upon fixing any

(2) $$\xi \in A \quad \text{with} \quad f(\xi) = x ,$$

we have

(3) $$\text{ord}_V f(\xi) = 1 .$$

In view of (1), there is a permutation (η, ζ) of (α, β) such that $g(f(\eta)/f(\xi^d)) \notin g(R)$. Note that now η and ζ are elements in A such that

(4) $$\mathrm{ord}_V f(\eta) = d$$

(5) $$g(f(\eta)/f(\xi^d)) \notin g(R)$$

(6) $$\mathrm{ord}_V f(\zeta) = d$$

(7) $$g(f(\zeta)/f(\eta)) \notin g(R)$$

and

(8) $$(\zeta, \eta)A = (\alpha, \beta)A \ .$$

In view of (2), (4) and (5), by (10.4.1) we get

(9) $$(f(\xi),\ f(\eta))R = M(R) \ .$$

By (10.4.1) we also have $\mathrm{emdim}\ R = 2$, and hence by (9) we conclude that

(10) $$(\xi, \eta)A = M(A) \ .$$

Since $d > 0$, in view of (6) we get $\zeta \in M(A)$; in view of (4), (6) and (7) we also have $\zeta \notin \eta(A)$; consequently, in view of (10) we can write

(11) $$\zeta = \delta' \xi^p + \rho\eta \quad \text{where} \quad \delta' \in A \backslash M(A) \quad \text{and} \quad \rho \in A$$

and p is a positive integer; now by (3), (4), (6), (7) and (11) we get

(12) $$p = d \ .$$

By (11) and (12) we see that $(\xi^d, \eta)A = (\xi, \eta)A$, and hence by (8) we get

(13) $$(\xi^d,\eta)A = (\alpha,\beta)A \ .$$

Let any $\gamma \in A$ with $f(\gamma) = 0$ be given. We shall show that then $\gamma \in ((\xi^d,\eta)A)^2$, and in view of (13) this will complete the proof. Now, in view of (10), we can write

(14) $$\gamma = r + s\eta + t\eta^2 \text{ with } r, s, t \text{ in } A$$

such that

(15) $$\begin{cases} \text{either } r = 0 \\ \text{or } r = \delta_0\xi^a \text{ with } \delta_0 \in A\backslash M(A) \text{ and nonnegative integer } a \end{cases}$$

and

(16) $$\begin{cases} \text{either } s = 0 \\ \text{or } s = \delta_1\xi^b \text{ with } \delta_1 \in A\backslash M(A) \text{ and nonnegative integer } b \ . \end{cases}$$

Since $f(\gamma) = 0$, by (3), (4), (5), (14), (15) and (16) we deduce that

(17) $$r \in \xi^{2d}A \text{ and } s \in \xi^d A \ .$$

Now by (14) and (17) it follows that $\gamma \in ((\xi^d,\eta)A)^2$.

CHAPTER II. PROJECTIVE GEOMETRY OR HOMOGENEOUS DOMAINS

By a homogeneous domain we mean a domain A together with a family $\{H_n(A)\}_{0\le n<\infty}$ of additive subgroups of A such that: the underlying additive group of A is the direct sum of the family $\{H_n(A)\}_{0\le n<\infty}$; $H_m(A)H_n(A) \subset H_{m+n}(A)$ for all m and n ; $H_0(A)$ is a subfield of A (Note that then $H_n(A)$ becomes an $H_0(A)$-vector-space for all n.); $A = H_0(A)[H_1(A)]$; and $0 < [H_1(A) : H_0(A)] < \infty$. We may refer to $H_0(A)$ as the ground field of A.

Note that we automatically have: $H_m(A)H_n(A) = H_{m+n}(A)$ for all m and n ; $H_1(A)^n = H_n(A)$ for all $n > 0$; $0 < [H_n(A) : H_0(A)] < \infty$ for all n ; A is an affine domain over $H_0(A)$; and A is noetheria

Now assume that B_0 is a subfield of domain B and B_1 is a B_0-vector-subspace of B such that $B = B_0[B_1]$, $0 < [B_1 : B_0] < \infty$, $B_0 \cap (B_1)^n = \{0\}$ for all positive integers n, and $(B_1)^m \cap (B_1)^n = \{0\}$ for all distinct postive integers m and n. Then B becomes a homogeneous domain upon taking $H_0(B) = B_0$ and $H_n(B) = (B_1)^n$ for all $n > 0$.

In the rest of Chapter II, let A be a homogeneous domain.

§11. Function fields and projective models.

We define

$$H(A) = \bigcup_{0\le n<\infty} H_n(A).$$

For any $0 \neq x \in H(A)$ we define

$\mathrm{Deg}_A x$ = the unique nonnegative integer n such that $x \in H_n(A)$

and we also define

$$\mathrm{Deg}_A 0 = -\infty .$$

We define

$$\mathfrak{K}(A) = \bigcup_{0\le n<\infty} \{x/y : x \in H_n(A) \text{ and } 0 \ne y \in H_n(A)\}$$

Note that: $\mathfrak{K}(A)$ is a subfield of $\mathfrak{F}(A)$; for every $n > 0$, $0 \ne y \in H_n(A)$, and $I \subset H_n(A)$ such that I generates $H_n(A)$ as an $H_0(A)$-vector-space, we have $\mathfrak{K}(A) = H_0(A)([x/y : x \in I])$: $\mathfrak{K}(A)$ is a function field over $H_0(A)$; for every $z \in H(A)\backslash H_0(A)$ we have that z is transcendental over $\mathfrak{K}(A)$; and for every $z \in H(A)\backslash H_0(A)$ we have $\mathfrak{F}(A) = \mathfrak{K}(A)(z)$.

$\mathfrak{K}(A)$ may be called the <u>function field</u> of A

We <u>define</u>

$$\text{Dim } A = \text{trdeg}_{H_0(A)} \mathfrak{K}(A)$$

and

$$\text{Emdim } A = [H_1(A) : H_0(A)] - 1$$

and we note that then:

$$\text{Emdim } A \ge \text{Dim } A = [\text{trdeg}_{H_0(A)} \mathfrak{F}(A)] - 1$$

and

Dim A = [dim of the underlying (nonhomogeneous) ring of A] - 1.

We <u>define</u>

$$\mathfrak{W}(A) = \mathfrak{W}(H_0(A), H_1(A)) = \bigcup_{0\ne y\in H_1(A)} \mathfrak{W}\left(H_0(A)\left[\left\{\frac{x}{y} : x \in H_1(A)\right\}\right]\right)$$ where,

for any domain R, $\mathfrak{V}(R)$ denotes $\{R_p : p \in \mathfrak{P}(R)\}$.

Note that $\mathfrak{W}(A)$ is a projective model of $\mathfrak{K}(A)$ over $H_0(A)$.

In case Dim A = 1, we <u>define</u>

$$\mathfrak{Z}(A) = \mathfrak{X}(\mathfrak{K}(A), H_0(A)).$$

In case Dim A = 0, we <u>define</u>

$$\text{Deg } A = [\mathfrak{K}(A) : H_0(A)]$$

and we note that then Deg A is a positive integer.

We observe that

$$\text{Emdim } A = 0 \Leftrightarrow \mathfrak{R}(A) = H_0 \Rightarrow \text{Dim } A = 0 \text{ ,}$$

and

$$\text{Dim } A = 0 \text{ and } H_0(A) \text{ algebraically closed} \Rightarrow \text{Emdim } A = 0.$$

§12. Homogeneous homomorphimsms.

Let A' be also a homogeneous domain and let $f: A \to A'$ be a (ring) homomorphism. f <u>is called homogeneous</u> (of degree zero) if $f(H_n(A)) \subset H_n(A')$ for all n. Note that: f is homogeneous $\Leftrightarrow$ $f(H_n(A)) \subset H_n(A')$ for $n = 0,1$. Also note that, if f is homogeneous, then: f is surjective $\Leftrightarrow f(H_n(A)) = H_n(A')$ for $n = 0, 1$ $\Leftrightarrow f(H_n(A)) = H_n(A')$ for all n.

§13. Homogeneous ideals and hypersurfaces.

An <u>ideal</u> Q in A <u>is said to be homogeneous</u> if the following three mutually equivalent conditions are satisfied:

(1) $\sum_{0 \le n < \infty} x_n \in Q$ with $x_n \in H_n(A)$ for all n (where $x_n = 0$ for all sufficiently large n) $\Rightarrow x_n \in Q$ for all n ;

(2) $Q = (Q \cap H(A))A$;

(3) $Q = IA$ for some $I \subset H(A)$.

Note that $H_1(A)A$ is the only homogeneous maximal ideal in A, and for any homogeneous ideal Q in A we have:

$$\text{rad}_A Q = A \text{ or } H_1(A)A \Leftrightarrow H_n(A) \subset Q \text{ for some } n$$

$$\Leftrightarrow H_n(A) \subset Q \text{ for all sufficiently large } n$$

For any nonmaximal homogeneous prime ideal P in A, we regard

$A\backslash P$ as a homogeneous domain by setting $H_n(A\backslash P) = f(H_n(A))$ for all n, where $f: A \to A/P$ is the canonical epimorphism. Note that, then f becomes homogeneous. Conversely, if f is a homogeneous epimorphism of A onto a homogeneous domain A', then Ker f is a nonmaximal homogeneous prime ideal in A.

For any nonmaximal homogeneous prime ideal C in A we <u>define</u>

$$\mathfrak{K}([A,C]) = \mathfrak{K}(A/C) \quad \text{and} \quad \mathfrak{W}([A,C]) = \mathfrak{W}(A/C)$$

and, in case $\operatorname{Dim} A/C = 1$, we also <u>define</u>

$$\mathfrak{Z}([A,C]) = \mathfrak{Z}(A/C) .$$

For any homogeneous ideals $Q_1, Q_2, \ldots, Q_s$ in A (where s is a nonnegative integer) we <u>define</u>

$$\Delta(A, Q_1, Q_2, \ldots, Q_s) = (Q_1 \cap Q_2 \cap \ldots \cap Q_s \cap H_1(A))A$$

and

$$\operatorname{Emdim}[A, Q_1, Q_2, \ldots, Q_s] = (\operatorname{Emdim} A) - [Q_1 \cap Q_2 \cap \ldots \cap H_1(A) : H_0(A)]$$

and we note that then:

$$\Delta(A) = H_1(A)A \quad \text{and} \quad \operatorname{Emdim}[A] = \operatorname{Emdim}[A, \Delta(A)] = -1 ;$$

$$\Delta(A, Q_1, Q_2, \ldots, Q_s) = \Delta(A, Q_1, Q_2, \ldots, Q_s, \Delta(A)) = \text{a nonunit homogeneous ideal in } A ;$$

$$\begin{aligned}\operatorname{Emdim} A \geq \operatorname{Emdim}&[A, Q_1, Q_2, \ldots, Q_s] \\ &= \operatorname{Emdim}[A, Q_1, Q_2, \ldots, Q_s, \Delta(A)] \\ &= \operatorname{Emdim}[A, \Delta(A, Q_1, Q_2, \ldots, Q_s)] \\ &= \operatorname{Emdim}[A, \Delta(A, Q_1, Q_2, \ldots, Q_s, \Delta(A))] \\ &\geq -1 ;\end{aligned}$$

and

$$\text{Emdim}[A,Q_1,Q_2,\ldots,Q_s] = -1$$
$$\Leftrightarrow \Delta(A,Q_1,Q_2,\ldots,Q_s) = \Delta(A)$$
$$\Leftrightarrow \text{for } 1 \le i \le s \text{ we have } Q_i = A \text{ or } \Delta(A),$$
(depending on i).

We observe that for any nonmaximal homogeneous prime ideal P in A we have

$$\text{Emdim}[A,P] = \text{Emdim } A/P .$$

We <u>define</u>

$H^*(A)$ = the set of all nonzero homogeneous principal ideals in A.

For any $\Phi \in H^*(A)$ we <u>define</u>

$$\text{Deg}_A \Phi = \min\{\text{Deg}_A \varphi : 0 \neq \varphi \in \Phi \cap H(A)\}$$

Note that, then $\text{Deg}_A \Phi$ is a nonnegative integer and for any $\varphi \in \Phi \cap H(A)$, we have

$$\text{Deg}_A \Phi = \text{Deg}_A \varphi \Leftrightarrow \Phi = \varphi A .$$

We <u>define</u>

$$H_n^*(A) = \{\Phi \in H^*(A) : \text{Deg}_A \Phi = n\}$$

and we note that $H_0^*(A) = \{A\}$.

§14. Homogeneous subdomains, flats, projections, birational projections, and cones.

By a <u>homogeneous subdomain</u> of A we mean a homogeneous domain B such that B is a subring of A, $H_0(B) = H_0(A)$, and $H_n(B) = B \cap H_n(A)$ for all $n > 0$.

We note that, if B is a homogeneous subdomain of A and if Q is a homogeneous ideal in A, then $B \cap Q$ is clearly a homogeneous ideal in B.

For a homogeneous subdomain B of A, we may regard $\mathfrak{I}(B)$ to be a subfield of $\mathfrak{I}(A)$, and $\mathfrak{K}(B)$ to be a subfield of $\mathfrak{K}(A)$. We note that clearly

$$\text{Dim } B \le \text{Dim } A ,$$

$$(14.1) \quad \begin{cases} \text{Dim } B = \text{Dim } A \Leftrightarrow [\mathfrak{K}(A) : \mathfrak{K}(B)] < \infty , \\ \text{and} \\ \text{Emdim } B \ge 1 = \text{Dim } A \text{ and } H_0(A) \text{ algebraically closed} \\ \Rightarrow \text{Dim } B = 1 . \end{cases}$$

We observe that, if B is a homogeneous subdomain of A and if C is a homogeneous prime ideal in A such that $H_1(B) \not\subset C$, then, upon letting $D = B \cap C$, and, $f: A \to A/C$ and $g: B \to B/D$ to be the canonical epimorphisms, we have the following: D is a nonmaximal homogeneous prime ideal in B. f(B) may be regarded to be a homogeneous subdomain of f(A) by taking $H_n(f(B)) = f(H_n(B))$ for all n, and so $\mathfrak{K}(f(B))$ may be regarded to be a subfield of $\mathfrak{K}([A,C])$. As homogeneous domains, g(B) and f(B) are naturally isomorphic. There exists a unique isomorphism $h^*: \mathfrak{I}(g(B)) \to \mathfrak{I}(f(B))$ such that $h^*(g(\xi)) = f(\xi)$ for all $\xi \in B$; we have $h^*(\mathfrak{K}([B,D])) = \mathfrak{K}(f(B))$ and so we get a unique isomorphism $h: \mathfrak{K}([B,D]) \to \mathfrak{K}(f(B))$ such that $h(\xi) = h^*(\xi)$ for all $\xi \in \mathfrak{K}([B,D])$; we designate h^* and h to be <u>canonical</u> isomorphisms. Via h, $\mathfrak{K}([A,C])$ becomes a $\mathfrak{K}([B,D])$-vector-space and we have

$$\text{Dim } B/D \le \text{Dim } A/C ,$$

$$(14.2) \quad \begin{cases} \text{Dim } B/D = \text{Dim } A/C \Leftrightarrow [\mathfrak{K}([A,C]) : \mathfrak{K}([B,D])] < \infty , \\ \text{and} \\ \text{Emdim}[B,D] \ge 1 = \text{Dim } A/C \text{ and } H_0(A) \text{ algebraically closed} \\ \Rightarrow \text{Dim } B/D = 1. \end{cases}$$

Finally we note that, without assuming $H_1(B) \not\subset C$, we have

$$(14.3)\quad\begin{cases} 1+\operatorname{Emdim}[B,D] = [H_1(B) : H_0(A)] - [C \cap H_1(B) : H_0(A)] \\ \qquad = \operatorname{Emdim}[A,C,H_1(B)A] - \operatorname{Emdim}[A,H_1(B)A] \\ \text{and} \\ \operatorname{Emdim}[B,D] \geq 0 \Rightarrow D \text{ is a nonmaximal homogeneous prime} \\ \text{in } B. \end{cases}$$

We <u>define</u>

$\mathfrak{M}(A)$ = the set of all $H_0(A)$-vector-subspaces of $H_1(A)$.

$\mathfrak{M}_i(A)$ = the set of all (Emdim A - i)-dimensional vector spaces in $\mathfrak{M}(A)$.

For any $\{0\} \neq L \in \mathfrak{M}(A)$ we <u>define</u>

$$A^L = H_0(A)[L] \text{ and } H_n(A^L) = A^L \cap H_n(A), \text{ for all } n,$$

and we note that A^L is then a homogeneous subdomain of A. We observe that $L \to A^L$ gives a (inclusion preserving) bijection of the set of all nonzero members of $\mathfrak{M}(A)$ onto the set of all homogeneous subdomains of A, and the inverse bijection is given by $B \to H_1(B)$.

Given any $\{0\} \neq L \in \mathfrak{M}(A)$ and any $J \subset A$ we <u>define</u>

$$J^{L,A} = A^L \cap J$$

and we call $J^{L,A}$ the <u>projection of</u> J <u>from</u> L in A. When the reference to A is clear from the context, we may write J^L instead of $J^{L,A}$ and we may simply call it the projection of J from L. We note that

$$A^L = A^{L,A} = \text{the projection of } A \text{ from } L \text{ (in } A).$$

By a <u>flat</u> in A we mean an ideal N in A such that $N = (N \cap H_1(A))A$. We note that, N is then clearly a nonunit homogeneous ideal in A. By an i-<u>flat</u> in A we mean a flat N in A such that Emdim[A,N] = i. We <u>define</u>

$$\mathfrak{M}^*(A) = \text{the set of all flats in } A$$

and

$$\mathfrak{M}_i^*(A) = \text{the set of all i-flats in } A.$$

We observe that $L \to LA$ given a (inclusion preserving)bijection of $\mathfrak{M}(A)$ onto $\mathfrak{M}^*(A)$, and the inverse bijection is given by $N \to N \cap H_1(A)$. We note that

$$\mathfrak{M}_{-1}^*(A) = \{\Delta(A)\} ,$$

and for any homogeneous ideals $Q_1, Q_2, \ldots, Q_s$ in A (where s is a nonnegative integer) we have,

$$\begin{cases} \Delta(A, Q_1, Q_2, \ldots, Q_s) \in \mathfrak{M}_e^*(A), \text{ where } e = \text{Emdim}[A, Q_1, Q_2, \ldots, Q_s] , \\ \Delta(A, Q_1, Q_2, \ldots, Q_s) \subset Q_i , \text{ for } 1 \leq j \leq s , \\ \text{and: } N \in \mathfrak{M}^*(A) \text{ with } N \subset Q_j , \text{ for } i \leq j \leq s \Rightarrow \\ N \subset \Delta(A, Q_1, Q_2, \ldots, Q_s) . \end{cases}$$

In case $s > 0$, we may call $\Delta(A, Q_1, Q_2, \ldots, Q_s)$ <u>the flat</u> (or <u>the e-flat</u> where $e = \text{Emdim}[A, Q_1, Q_2, \ldots, Q_s]$) <u>in</u> A <u>spanned by</u> $Q_1, Q_2, \ldots, Q_s$; from these phrases " in A " may be dropped when the reference to A is clear from the context.

We also note that

$$H_1^*(A) = \mathfrak{M}_{r-1}^*(A) \quad \text{where} \quad r = \text{Emdim } A.$$

For any $\{0\} \neq N \in \mathfrak{M}^*(A)$, we <u>define</u>

$$A^N = A^{N \cap H_1(A)} .$$

Again we observe that $N \to A^N$ gives a (inclusion preserving) bijection of the set of all nonzero members of $\mathfrak{M}^*(A)$ onto the set of all homogeneous subdomains of A, and the inverse bijection is given by $B \to H_1(B)A$.

<u>In the rest of</u> §14, <u>let any</u> $\{0\} \neq N \in \mathfrak{M}^*(A)$ <u>be given</u>.

For any $J \subset A$ we <u>define</u>

$$J^{N,A} = A^N \cap J$$

and we call $J^{N,A}$ the <u>projection</u> of J <u>from</u> N in A. In other words, projection from N is the same thing as projection from the corresponding member $N \cap H_1(A)$ of $\mathfrak{M}(A)$. Again, when the referenc to A is clear from the context, we may write J^N instead of $J^{N,A}$ and we simply call it the projection of J from N. We note that,

$$A^N = A^{N,A} = \text{the projection of } A \text{ from } N \text{ (in A)} ,$$

and

$$\text{Emdim } A^N = \text{Emdim } A - \text{Emdim}[A,N] - 1 .$$

We observe that $J \to J^N$ gives a (inclusion preserving) bijection of $\{J \in \mathfrak{M}_i^*(A) : J \subset N\}$ onto $\mathfrak{M}_{i-e-1}^*(A^N)$ where $e = \text{Emdim}[A,N]$, and the inverse bijection is given by $K \to KA$. We note that then, in particular $\pi \to \pi^N$ gives a (inclusion preserving) $\{\pi \in H_1^*(A) : \pi \subset N\}$ onto $H_1^*(A^N)$, and the inverse bijection is again given by $\Theta \to \Theta A$.

We <u>define</u>

$$H^*(A,N) = \{\Phi \in H^*(A) : \Phi = \Phi^N A\}$$

and

$$H_n^*(A,N) = H^*(A,N) \cap H_n^*(A)$$

and we note that $\Phi \to \Phi^N$ gives a (inclusion preserving) bijection of $H_n^*(A,N)$ onto $H_n^*(A^N)$, and the inverse bijection is given by $\Psi \to \Psi A$. We also note that

$$H^*(A,\Delta(A)) = H^*(A) \quad \text{and} \quad H_n^*(A,\Delta(A)) = H_n^*(A) .$$

We shall say that the <u>projection from</u> N in A is birational to mean that $\mathfrak{R}(A^N) = \mathfrak{R}(A)$. Given any nonmaximal homogeneous prime ideal in C in A, we shall say that the projection of C from N i

A <u>is birational</u> to mean that $N \not\subset C$ and $[\mathfrak{K}([A,C]):\mathfrak{K}([A^N,C^N)] = 1$. Again, from these two phrases we may drop " in A " when the reference to A is clear from the context.

§15. Zeroset and homogeneous localization.

We <u>define</u>

$\mathfrak{Q}(A)$ = the set of all nonmaximal homogeneous prime ideals in A. We also <u>define</u>

$$\mathfrak{Q}^1(A) = \{P \in \mathfrak{Q}(A) : \operatorname{Emdim} A/P = \operatorname{Dim} A/P\}$$

$$\mathfrak{Q}_i(A) = \{P \in \mathfrak{Q}(A) : \operatorname{Dim} A/P = i\}$$

$$\mathfrak{Q}_i^1(A) = \mathfrak{Q}^1(A) \cap \mathfrak{Q}_i(A)$$

and we note that

$$\mathfrak{Q}_i(A) \subset \mathfrak{P}_{i+1}(A) \quad \text{and} \quad \mathfrak{Q}_i^1(A) \subset \mathfrak{M}_i^*(A) \quad \text{for all } i.$$

We observe that for any $P \in \mathfrak{Q}(A)$, via the canonical epimorphism $A \to A/P$, every $H_0(A/P)$-vector-space becomes a $H_0(A)$-vector space. With this understanding we have

$$[\mathfrak{K}([A,P]) : H_0(A)] = [\mathfrak{K}([A,P]) : H_0(A/P)]$$

and

$$\mathfrak{Q}_0(A) = \{P \in \mathfrak{Q}(A) : [\mathfrak{K}([A,P]) : H_0(A)] < \infty\} .$$

We <u>define</u>

$$\operatorname{Deg}[A,P] = [\mathfrak{K}([A,P]) : H_0(A)] \text{ for all } P \in \mathfrak{Q}_0(A)$$

and we note that then

$$\operatorname{Deg}[A,P] = \operatorname{Deg}(A/P) \quad \text{for all } P \in \mathfrak{Q}_0(A) .$$

We also note that

$$\mathfrak{D}_0^1(A) = \{P \in \mathfrak{D}_0(A) : \mathrm{Deg}[A,P] = 1\}$$

and in particular:

$$H_0(A) \quad \text{algebraically closed} = \mathfrak{D}_0(A) = \mathfrak{D}_0^1(A) \ .$$

We <u>define</u>

$$\mathfrak{D}(A,x) = \{P \in \mathfrak{D}(A) : x \in P\}$$
$$\mathfrak{D}^1(A,x) = \mathfrak{D}(A,x) \cap \mathfrak{D}^1(A)$$
$$\mathfrak{D}_i(A,x) = \mathfrak{D}(A,x) \cap \mathfrak{D}_i(A)$$
$$\mathfrak{D}_i^1(A,x) = \mathfrak{D}(A,x) \cap \mathfrak{D}_i^1(A)$$

$\left.\vphantom{\begin{matrix}a\\a\\a\\a\end{matrix}}\right\}$ for any $\ x \in H(A)$

$$\mathfrak{D}(A,I) = \{P \in \mathfrak{D}(A) : I \cap H(A) \subset P\}$$
$$\mathfrak{D}^1(A,I) = \mathfrak{D}(A,I) \cap \mathfrak{D}^1(A)$$
$$\mathfrak{D}_i(A,I) = \mathfrak{D}(A,I) \cap \mathfrak{D}_i(A)$$
$$\mathfrak{D}_i^1(A,I) = \mathfrak{D}(A,I) \cap \mathfrak{D}_i^1(A)$$

$\left.\vphantom{\begin{matrix}a\\a\\a\\a\end{matrix}}\right\}$ for any $\ I \subset A$

$$\mathfrak{D}(A,\backslash I) = \mathfrak{D}(A)\backslash\mathfrak{D}(A,I)$$
$$\mathfrak{D}^1(A,\backslash I) = \mathfrak{D}(A,\backslash I) \cap \mathfrak{D}^1(A)$$
$$\mathfrak{D}_i(A,\backslash I) = \mathfrak{D}(A,\backslash I) \cap \mathfrak{D}_i(A)$$
$$\mathfrak{D}_i^1(A,\backslash I) = \mathfrak{D}(A,\backslash I) \cap \mathfrak{D}_i^1(A)$$

$\left.\vphantom{\begin{matrix}a\\a\\a\\a\end{matrix}}\right\}$ for any $\ I \in H(A) \quad$ or $\quad I \subset A$

$$\mathfrak{D}([A,I]) = \mathfrak{D}(A,I)$$
$$\mathfrak{D}^1([A,I]) = \mathfrak{D}^1(A,I)$$
$$\mathfrak{D}_i([A,I]) = \mathfrak{D}_i(A,I)$$
$$\mathfrak{D}_i^1([A,I]) = \mathfrak{D}_i^1(A,I)$$

$\left.\vphantom{\begin{matrix}a\\a\\a\\a\end{matrix}}\right\}$ for any $\ I \in H(A) \quad$ or $\quad I \subset A$

$$\left.\begin{aligned}
\mathfrak{D}([A,I],J) &= \mathfrak{D}(A,I) \cap \mathfrak{D}(A,J) \\
\mathfrak{D}^1([A,I],J) &= \mathfrak{D}([A,I],J) \cap \mathfrak{D}^1(A) \\
\mathfrak{D}_i([A,I],J) &= \mathfrak{D}([A,I],J) \cap \mathfrak{D}_i(A) \\
\mathfrak{D}_i^1([A,I],J) &= \mathfrak{D}([A,I],J) \cap \mathfrak{D}_i^1(A)
\end{aligned}\right\}\quad \begin{aligned}&\text{for any } I \in H(A) \text{ or } I \subset A \\ &\text{and any } J \in H(A) \text{ or } J \subset A\end{aligned}$$

and

$$\left.\begin{aligned}
\mathfrak{D}([A,I],\backslash J) &= \mathfrak{D}(A,I)\backslash\mathfrak{D}(A,J) \\
\mathfrak{D}^1(A,I],\backslash J) &= \mathfrak{D}^1([A,I],\backslash J) \cap \mathfrak{D}^1(A) \\
\mathfrak{D}_i([A,I],\backslash J) &= \mathfrak{D}_i([A,I],\backslash J) \cap \mathfrak{D}_i(A) \\
\mathfrak{D}_i^1([A,I],\backslash J) &= \mathfrak{D}_i^1([A,J],\backslash J) \cap \mathfrak{D}_i^1(A)
\end{aligned}\right\}\quad \begin{aligned}&\text{for any } I \in H(A) \text{ or } I \subset A \\ &\text{and any } J \in H(A) \text{ or } J \subset A\end{aligned}$$

We note that then

$$\mathfrak{D}(A) = \mathfrak{D}(A,0) \; ,$$

$$\mathfrak{D}(A,xA) = \mathfrak{D}([A,0],x) = \mathfrak{D}([A,x],0) = \mathfrak{D}([A,x]) = \mathfrak{D}(A,x)$$

$$\text{for any } x \in H(A) \; ,$$

$$\mathfrak{D}(A,(I \cap H(A))A) = \mathfrak{D}([A,0],I) = \mathfrak{D}([A,I],0) = \mathfrak{D}([A,I]) = \mathfrak{D}(A,I)$$

$$\text{for any } I \subset A \; ,$$

and for any homogeneous ideals I and J in A we have:

$$\mathfrak{D}(A,I) = \mathfrak{D}(A) \Leftrightarrow \mathfrak{D}_0(A,I) = \mathfrak{D}_0(A) \Leftrightarrow I = \{0\} \; ,$$

$$\mathfrak{D}(A,I) = \emptyset \Leftrightarrow \mathfrak{D}_0(A,I) = \emptyset \Leftrightarrow H_1(A) \subset \operatorname{rad}_A I \; ,$$

$$\mathfrak{D}(A,I) \text{ is a finite set} \Leftrightarrow \mathfrak{D}(A,I) = \mathfrak{D}_0(A,I)$$

$$\begin{aligned}
\mathfrak{D}(A,I) = \mathfrak{D}(A,J) &\Leftrightarrow \mathfrak{D}_0(A,I) = \mathfrak{D}_0(A,J) \\
&\Leftrightarrow (\operatorname{rad}_A I) \cap (H_1(A)A) = (\operatorname{rad}_A J) \cap (H_1(A)A) \; ,
\end{aligned}$$

$$\begin{aligned}
\mathfrak{D}(A,I) \subset \mathfrak{D}(A,J) &\Leftrightarrow \mathfrak{D}_0(A,I) \subset \mathfrak{D}_0(A,J) \\
&\Leftrightarrow (\operatorname{rad}_A I) \cap (H_1(A)A) \supset (\operatorname{rad}_A J) \cap (H_1(A)A) \; ,
\end{aligned}$$

$$\mathfrak{Q}(A,I) \cup \mathfrak{Q}(A,J) = \mathfrak{Q}(A,I \cap J) = \mathfrak{Q}(A,IJ) ,$$

and

$$\mathfrak{Q}(A,I) \cap \mathfrak{Q}(A,J) = \mathfrak{Q}(A,I + J) .$$

We observe that, in particular $P \to \mathfrak{Q}(A,P)$ gives an (inclusion reversing) injection of $\mathfrak{Q}(A)$ into the set of all nonempty subsets of $\mathfrak{Q}(A)$. Finally we note that for any $C \in \mathfrak{Q}(A)$, upon letting $f: A \to A/C$ to be the canonical epimorphism, $P \to f(P)$ gives a (inclusion preserving) bijection of $\mathfrak{Q}([A,C])$ onto $\mathfrak{Q}(A/C)$.

Given any $P \in \mathfrak{Q}(A)$ we <u>define</u>

$$\mathfrak{R}(A,x,P) = \{x/z : z \in H_n(A)\setminus P\} \quad \text{for any} \quad x \in H_n(A) ,$$

$$\mathfrak{R}(A,I,P) = \bigcup_{x \in I \cap H(A)} \mathfrak{R}(A,x,P) \quad \text{for any} \quad I \subset A ,$$

$$\mathfrak{R}(A,P) = \mathfrak{R}(A,A,P) ,$$

and we note that then: $\mathfrak{R}(A,P)$ is a local domain with quotient field $\mathfrak{K}(A)$; $\mathfrak{R}(A,P)$ is a spot over $H_0(A)$; $\dim\mathfrak{R}(A,P) = \dim A - \operatorname{Dim} A/P$; $\mathfrak{R}(A,P)/M(\mathfrak{R}(A,P))$ is naturally isomorphic with $\mathfrak{K}([A,P])$;

$$(15.1) \quad \begin{cases} \text{for every } x \in H_n(A) \text{ we have } \emptyset \neq \mathfrak{R}(A,x,P) \subset \mathfrak{R}(A,P) \text{ and} \\ \mathfrak{R}(A,x,P)\mathfrak{R}(A,P) = (x/z)\mathfrak{R}(A,P) \text{ for every } z \in H_n(A)\setminus P ; \end{cases}$$

$$(15.2) \quad \begin{cases} \text{for every } I \subset A \text{ we have } \mathfrak{R}(A,I,P) \subset \mathfrak{R}(A,P) \text{ and} \\ \mathfrak{R}(A,I,P)\mathfrak{R}(A,P) = \mathfrak{R}(A,(I \cap H(A))A, P) ; \end{cases}$$

$$(15.3) \qquad M(\mathfrak{R}(A,P)) = \mathfrak{R}(A,P,P) ;$$

and

(15.4) given any $z \in H_1(A)\backslash P$, upon letting

$I' = \bigcup_{0\le n<\infty} \{xz^{-n} : x \in I \cap H_n(A)\}$ for every $I \subset A$,

we have that

$A' = H_0(A)[H_1(A)z^{-1}]$, $P' \in \mathfrak{P}(A')$, $\mathfrak{R}(A,P) = A'_{P'}$,

$P \cap H_n(A) = \{x \in H_n(A) : xz^{-n} \in M(\mathfrak{R}(A,P))\}$, and

$I'A' = P'$ for every $I \subset A$ with $(I \cap H(A))A = P$.

We also observe that

(15.5) given any $0 \neq z \in H_1(A)$ and any $R' \in \mathfrak{V}(H_0(A)[H_1(A)z^{-1}])$, upon letting

$P = \{\sum_{0\le n<\infty} r_n : r_n \in H_n(A)$ and $r_n z^{-n} \in M(R')$ for all n, and

$r_n = 0$ for all sufficiently large $n\}$,

We have that $z \notin P \in \mathfrak{D}(A)$ and $\mathfrak{R}(A,P) = R'$.

In view of (15.4) and (15.5) we see that:

(15.6) $P \to \mathfrak{R}(A,P)$ gives a (inclusion reversing) bijection of $\mathfrak{D}(A)$ onto $\mathfrak{W}(A)$.

In particular, given any homogeneous domain A with $\dim A = 1$ and any $V \in \mathfrak{V}(A)$, there exists a unique $P \in \mathfrak{D}(A)$ such that

$\mathfrak{R}(A,P) =$ the center of V on $\mathfrak{W}(A)$, i.e., by definition $\mathfrak{R}(A,P) \subset V$ and $M(\mathfrak{R}(A,P)) = M(V) \cap \mathfrak{R}(A,P)$.

We note that clearly $P \in \mathfrak{D}_0(A)$. We <u>define</u>

$\mathfrak{Z}^*(A,V) =$ <u>the projective center of</u> V in $A = P$.

We observe that via the canonical epimorphism $V \to V/M(V)$, $V/M(V)$ becomes a $H_0(A)$-vector-space as well as $\mathfrak{K}([A,P])$-vector-space and we have

$$[V/M(V) : H_0(A)] < \infty \quad \text{and} \quad [V/M(V) : \Re([A,P])] < \infty$$

and

$$[V/M(V) : H_0(A)] = [V/M(V) : \Re([A,P])]\text{Deg}[A,P] .$$

Finally we note that:

$H_0(A)$ algebraically closed $\Rightarrow [V/M(V) : H_0(A)]$

$$= 1 = [V/M(V) : \Re([A,P])] .$$

For any homogeneous domain A with $\text{Dim } A = 1$ and any $Q \in H(A)$ or $Q \subset A$ we <u>define</u>

$$\mathfrak{Z}(A,Q) = \{V \in \mathfrak{Z}(A) : \mathfrak{Z}^*(A,V) \in \mathfrak{D}(A,Q)\}$$

and

$$\mathfrak{Z}(A,\backslash Q) = \mathfrak{Z}(A)\backslash\mathfrak{Z}(A,Q) ;$$

we note that for any other $Q' \in H(A)$ or $Q' \subset A$ we then have

$$\mathfrak{Z}(A,Q') = \mathfrak{Z}(A,Q) \Leftrightarrow \mathfrak{D}(A,Q') \Leftrightarrow \mathfrak{D}(A,Q) .$$

We observe that, in particular,

$$\mathfrak{Z}(A,Q) = \mathfrak{Z}(A) \Leftrightarrow \mathfrak{Z}(A,Q) \text{ is an infinite set}$$

$$\Leftrightarrow \mathfrak{D}(A,Q) \text{ is an infinite set}$$

$$\Leftrightarrow \mathfrak{D}(A,Q) = \mathfrak{D}(A)$$

$$\Leftrightarrow \{0\} \in \mathfrak{D}(A,Q)$$

$$\Leftrightarrow \begin{cases} Q = 0 & \text{in case } Q \in H(A) \\ \\ Q \cap H(A) \subset \{0\} & \text{in case } Q \subset A , \end{cases}$$

and

$$\mathfrak{Z}(A,Q) = \phi \Leftrightarrow \mathfrak{D}(A,Q) = \phi \Leftrightarrow \begin{cases} 0 \neq Q \in H_0(A) & \text{in case } Q \in H(A) \\ \\ H_1(A) \subset \text{rad}_A(Q \cap H(A))A & \text{in case } Q \subset A ; \end{cases}$$

we also observe that for any $V \in \mathfrak{Z}(A)$ we have

$$\mathfrak{Z}^*(A,V) = \text{the unique } P \in \mathfrak{Q}_0(A) \text{ such that } V \in \mathfrak{Z}(A,P) .$$

Given any $C \in \mathfrak{Q}(A)$ and any $P \in \mathfrak{Q}([A,C])$, upon letting $f: A \to A/C$ to be the canonical epimorphism, <u>we define</u>

$$\mathfrak{R}([A,C],x,P) = \{f(x)/f(z) : z \in H_n(A)\backslash P\} \text{ for any } x \in H_n(A) ,$$

$$\mathfrak{R}([A,C],I,P) = \bigcup_{x \in I \cap H(A)} \mathfrak{R}([A,C],x,P) \text{ for any } I \subset A ,$$

$$\mathfrak{R}([A,C],P) = \mathfrak{R}([A,C],P,A) ,$$

and we note that then $\mathfrak{R}([A,C],P)$ is a local domain with quotient field $\mathfrak{K}([A,C])$; $\mathfrak{K}([A,C],P)$ is a spot over $H_0(A/C)(\approx H_0(A))$; $\dim \mathfrak{K}([A,C],P) = \text{Dim } A/C - \text{Dim } A/P$. Also we have:

(15.7) $\begin{cases} \text{for every } x \in H_n(A) \text{ we have } \emptyset \neq \mathfrak{R}([A,C],x,P) \subset \mathfrak{R}([A,C],P) \\ \text{and} \\ \mathfrak{R}([A,C],x,P)\mathfrak{R}([A,C],P) = (f(x)/f(z))\mathfrak{R}([A,C],P) \\ \text{for every } z \in H_n(A)\backslash P ; \end{cases}$

(15.8) $\begin{cases} \text{for every } I \subset A \text{ we have } \mathfrak{R}([A,C],I,P) \subset \mathfrak{R}([A,C],P) \text{ and} \\ \mathfrak{R}([A,C],I,P)\mathfrak{R}([A,C],P) = \mathfrak{R}([A,C],(I \cap H(A))A,P) \\ \qquad = \mathfrak{R}([A,C],(I \cap H(A))A + C,P) ; \end{cases}$

(15.9) $M(\mathfrak{R}([A,C],P) = \mathfrak{R}([A,C],P,P)$.

Further,

f induces a unique epimorphism $f^*: \mathfrak{R}(A,P) \to \mathfrak{R}([A,C],P)$ such that $f^*(x/z) = f(x)/f(z)$, whenever $x \in H_n(A)$ and $z \in H_n(A)\backslash P$; we have $\text{Ker } f^* = \mathfrak{R}(A,C,P)$; for any $I \in H(A)$ or $I \subset A$ we have $f^*(\mathfrak{R}(A,I,P)) = \mathfrak{R}([A,C],I,P)$; and f induces an isomorphism between $\mathfrak{R}([A,C],P)/M(\mathfrak{R}([A,C],P))$ and $\mathfrak{K}([A,P])$.

We again observe that for any $C \in \mathfrak{Q}(A)$, in view of (15.6), $P \to \mathfrak{R}([A,C],P)$ gives a (inclusion reversing) bijection of $\mathfrak{Q}([A,C])$ onto $\mathfrak{B}([A,C])$.

In particular, given any $C \in \mathfrak{Q}_1(A)$ and $V \in \mathfrak{Z}([A,C])$, there exists a unique $P \in \mathfrak{Q}([A,C])$ such that

$$\mathfrak{R}([A,C],P) = \text{the center of } V \text{ on } \mathfrak{Z}([A,C]).$$

We note that clearly $P \in \mathfrak{Q}_0([A,C])$. We <u>define</u>

$$\mathfrak{Z}^*([A,C],V) = \text{the \underline{projective center of} } V \text{ \underline{on} } C \text{ \underline{in} } A = P.$$

From the phrase "the projective center of V on C in A " we may drop " in A " when the reference to A is clear from the context. We note that, then

$$f(\mathfrak{Z}^*([A,C],V)) = \mathfrak{Z}^*(A/C,V) \quad \text{and} \quad \mathfrak{Z}^*([A,C],V) = f^{-1}(\mathfrak{Z}^*(A/C,V)) ,$$

where $f: A \to A/C$ is the canonical epimorphism. We observe that via the canonical epimorphisms $A \to A/C$ and $V \to V/M(V)$, $V/M(V)$ becomes an $H_0(A)$-vector-space as well as $\mathfrak{K}([A,P])$-vector-space and we have

$$[V/M(V) : H_0(A)] < \infty \quad \text{and} \quad [V/M(V) : \mathfrak{K}([A,P])] < \infty$$

and

$$[V/M(V) : H_0(A)] = [V/M(V) : \mathfrak{K}([A,P])]\text{Deg}[A,P]$$

Finally we note that:

$$H_0(A) \text{ algebraically closed} \Rightarrow [V/M(V) : H_0(A)] = 1 = [V/M(V) : \mathfrak{K}([A,P])] .$$

For any $C \in \mathfrak{Q}_1(A)$ and any $Q \in H(A)$ or $Q \subset A$ we <u>define</u>

$$\mathfrak{Z}([A,C],Q) = \{V \in \mathfrak{Z}([A,C]) : \mathfrak{Z}^*([A,C],V) \in \mathfrak{Q}([A,C],Q)\}$$

and

$$\mathfrak{Z}([A,C],\backslash Q) = \mathfrak{Z}([A,C]) \backslash \mathfrak{Z}([A,C],Q) .$$

We note that, then, for any other $Q' \in H(A)$ or $Q' \subset A$, we have

$$\mathfrak{Z}([A,C],Q') = \mathfrak{Z}([A,C],Q) \Leftrightarrow \mathfrak{Q}([A,C],Q') = \mathfrak{Q}([A,C],Q) .$$

We observe that, in particular, we have

$$\begin{aligned}\mathfrak{Z}([A,C],Q) = \mathfrak{Z}([A,C]) &\Leftrightarrow \mathfrak{Z}([A,C],Q) \text{ is an infinite set}\\ &\Leftrightarrow \mathfrak{Q}([A,C],Q) \text{ is an infinite set}\\ &\Leftrightarrow \mathfrak{Q}([A,C],Q) = \mathfrak{Q}([A,C])\\ &\Leftrightarrow C \in \mathfrak{Q}(A,Q)\\ &\Leftrightarrow \begin{cases} \mathfrak{Q} \in C & \text{in case } Q \in H(A)\\ Q \cap H(A) \subset C & \text{in case } Q \subset A,\end{cases}\end{aligned}$$

and

$$\begin{aligned}\mathfrak{Z}([A,C],Q) = \emptyset &\Leftrightarrow \mathfrak{Q}([A,C],Q) = \emptyset\\ &\Leftrightarrow \begin{cases} 0 \neq Q \in H_0(A) & \text{in case } Q \in H(A)\\ H_1(A) \subset \operatorname{rad}_A((Q \cap H(A))A + C) & \text{in case } Q \subset A.\end{cases}\end{aligned}$$

We also observe that, for any $V \in \mathfrak{Z}([A,C])$, we have

$$\mathfrak{Z}^*([A,C],V) = \text{the unique } P \in \mathfrak{Q}_0([A,C]) \text{ such that } V \in \mathfrak{Z}([A,C],P).$$

We end this section with:

(15.10) LEMMA ON IMPOSED LINEAR CONDITIONS. Let R be a homogeneous domain, and for each $P \in \mathfrak{Q}_0(R)$ let any ideals $I(P) \subset I'(P)$ in $\mathfrak{R}(R,P)$ be given. For every nonnegative integer m, let

$$E_m = \{x \in H_m(R) : \mathfrak{R}(R,x,P) \subset I(P) \text{ for all } P \in \mathfrak{Q}_0(R)\}$$

and

$$E'_m = \{x \in H_m(R) : \mathfrak{R}(R,x,P) \subset I'(P) \text{ for all } P \in \mathfrak{Q}_0(R)\}.$$

Then we have that $E_m \subset E'_m$ are $H_0(R)$-vector-subspaces of $H_m(R)$ and

$$[E_m : H_0(R)] + \sum_{P \in \mathfrak{D}_0(R)} [\mathfrak{R}(R,P)/I(P) : H_0(R)]$$

$$\geq [E'_m : H_0(R)] + \sum_{P \in \mathfrak{D}_0(R)} [\mathfrak{R}(R,P)/I'(P) : H_0(R)]$$

$$\geq [H_m(R) : H_0(R)] .$$

(We observe that, if $\sum_{P \in \mathfrak{D}_0(R)} [\mathfrak{R}(R,P)/I(P) : \mathfrak{R}(R,P)] < \infty$, then it can easily be seen that, for all large m the above two inequalities are actually equalities; however, as we shall have no occasion to use this observation in the present book, we shall not prove it.)

PROOF. Note that, if $I'(P) = \mathfrak{R}(R,P)$ for all $p \in \mathfrak{D}_0(R)$, then, $E'_m = H_m(R)$ and the second inequality follows from the first. Hence we shall only prove the first inequality.

Now, for each $P \in \mathfrak{D}_0(R)$, fix $y(P) \in H_m(R)\backslash P$ and define $g(P) : H_m(R) \to \mathfrak{R}(R,P)$ by $g(P)(x) = x/_{y(P)}$. Clearly $g(P)$ is an $H_0(R)$-isomorphism and we have

(1) $$g(P)^{-1}(I(P)) \subset g(P)^{-1}(I'(P))$$

are $H_0(R)$-vector-subspaces of $H_m(R)$ and

(2) $$[\mathfrak{R}(R,P)/I(P) : H_0(R)] = [H_m(R)/g(P)^{-1}(I(P)) : H_0(R)] .$$

and similarly,

(3) $$[\mathfrak{R}(R,P)/I'(P) : H_0(R)] = [H_m(R)/g(P)^{-1}(I'(P)) : H_0(R)] .$$

First note that, if

$$\sum_{P \in \mathfrak{D}_0(R)} [\mathfrak{R}(R,P)/I(P) = H_0(R)] = \infty ,$$

then the inequality is trivially true. Hence we may assume that there are only finitely many P's, say $P_1,\ldots,P_r$, such that

(4) $$P \in \mathfrak{D}_0(R)\backslash\{P_1,\ldots,P_r\} \Rightarrow I(P) = \mathfrak{R}(R,P)$$ or equivalently

$$g(P)^{-1}(I(P)) = H_m(R).$$

Now put $\Omega_i = g(p_i)^{-1}(I(P_i))$ and $\Omega_i' = g(P_i)^{-1}(I'(P_i))$ for $i = 1,2,\ldots,r$. We clearly have from (1),

(5) $$\Omega_i \subset \Omega_i' \ , \quad E_m = \bigcap_{i=1}^{r} \Omega_i \quad \text{and} \quad E_m' = \bigcap_{i=1}^{r} \Omega_i' \ .$$

The required result is now deduced in the following easy exercise in vector spaces, using (2), (3), (5) and the result of (7) by taking $\Omega = H_m(R)$ and $k = H_0(R)$.

EXERCISE. Let Ω be a finite-dimensional vector space over a field k. For $i = 1,2,\ldots,r$; let k-vector-spaces $\Omega_i \subset \Omega_i' \subset \Omega$ be given. Then

(6) $$[\bigcap_{i=1}^{r} \Omega_i : k] + \sum_{i=1}^{r} [\Omega_i' : k] \geq [\bigcap_{i=1}^{r} \Omega_i' : k] + \sum_{i=1}^{r} [\Omega_i : k]$$

and hence (or similarly)

(7) $$[\bigcap_{i=1}^{r} \Omega_i : k] + \sum_{i=1}^{r} [\Omega/\Omega_i : k] \geq [\bigcap_{i=1}^{r} \Omega_i' : k] + \sum_{i=1}^{r} [\Omega/\Omega_i' : k] .$$

(Hint. Both results are easily deduced by induction on r and the well known fact that,

$$[\Lambda_1 \cap \Lambda_2 : k] + [\Lambda_1 + \Lambda_2 : k] = [\Lambda_1 : k] + [\Lambda_2 : k]$$

for any two subspaces Λ_1 , Λ_2 of Ω.)

§16. Homogeneous coordinate systems.

By a <u>homogeneous coordinate system</u> in A we mean a $H_0(A)$-basis of $H_1(A)$.

Let $(X_0, X_1, \ldots, X_r)$ be a homogeneous coordinate system in A, where $r = \text{Emdim}\, A$. Let $H_0(A)^{(r+1)}$ be the set of all $(r+1)$-tuples $a = (a_0, a_1, \ldots, a_r)$ with $a_0, a_1, \ldots, a_r$ in $H_0(A)$. For any $a \in H_0(A)^{(r+1)} \setminus (0,0,\ldots,0)$, let a^* be the ideal in A generated by the set $\{a_q X_p - a_p X_q : p,q = 0,1,\ldots,r\}$. We note that then: $a \to a^*$ gives a surjective map of $H_0(A)^{(r+1)} \setminus (0,0,\ldots,0)$ onto $\mathfrak{M}_0^*(A)$; for any a and b in $H_0(A)^{(r+1)} \setminus (0,0,\ldots,0)$ we have $a^* = b^* \Leftrightarrow$ there exists $0 \neq \alpha \in H_0(A)$ such that $a_p = \alpha b_p$ for $p = 0,1,\ldots,r$; for any $a \in H_0(A)^{(r+1)} \setminus (0,0,\ldots,0)$, we have that, if $a_q \neq 0$, then a^* is generated by the set $\{a_q X_p - a_p X_q : p = 0,1,\ldots,r\}$.

Similarly, obvious matrix representations can be given for other members of $\mathfrak{M}^*(A)$.

§17. Polynomial rings as homogeneous domains.

A polynomial ring $A^* = k[Y_0,Y_1,\ldots,Y_r]$ in indeterminates $Y_0,Y_1,\ldots,Y_r$ over a field k obviously becomes a homogeneous domain by taking

$$H_n(A^*) = \{0\} \cup (\text{the set of all nonzero homogeneous polynomials of degree } n)$$

and then we have

$$\text{Emdim } A^* = \text{Dim } A^* = r.$$

Moreover, we clearly have (17.1) to (17.5):

(17.1) $\begin{cases} \text{Emdim } A = \text{Dim } A = r \\ \Leftrightarrow A \text{ is isomorphic, as a homogeneous domain, to a polynomial} \\ \quad \text{ring in } r+1 \text{ indeterminates over a field.} \end{cases}$

(17.2) $\left\{\begin{array}{l} \text{Emdim } A = \text{Dim } A \\ \Leftrightarrow \text{some homogeneous coordinate system in } A \text{ consists of} \\ \quad \text{algebraically independent elements over } H_0(A) \\ \Leftrightarrow \text{every homogeneous coordinate system in } A \text{ consists of} \\ \quad \text{algebraically independent elements over } H_0(A) \\ \Leftrightarrow \mathfrak{M}_0^*(A) \subset \mathfrak{Q}(A) \\ \Leftrightarrow \mathfrak{M}_i^*(A) = \mathfrak{Q}_i^1(A) \quad \text{for all } i \\ \Leftrightarrow [H_m(A):H_0(A)] = \begin{pmatrix} m + \text{Emdim } A \\ \text{Emdim } A \end{pmatrix} \text{for all } m . \end{array}\right.$

(17.3) $\left\{\begin{array}{l} \text{Emdim } A = \text{Dim } A = r \\ \Rightarrow \mathfrak{Q}_{r-1}(A) = H^*(A) \cap \mathfrak{Q}(A) \text{ and for any } \Phi \in H_n^*(A) \cap \mathfrak{Q}(A) \\ \quad \text{we have} \\ \quad [H_m(A/\Phi):H_0(A)/\Phi] = \begin{cases} \begin{pmatrix} m + r \\ r \end{pmatrix} & \text{for all } m < n \\ \begin{pmatrix} m + r \\ r \end{pmatrix} - \begin{pmatrix} m-n+r \\ r \end{pmatrix} & \text{for all } m \geq n . \end{cases} \end{array}\right.$

(17.4) Emdim A = Dim A ⇒ the local ring $\mathfrak{R}(A,P)$ is regular for all $P \in \mathfrak{Q}(A)$.

(17.5) Emdim A = the smallest nonnegative integer r such that there exists a homogeneous domain A' with Emdim A' = Dim A' = r and a homogeneous homomorphism $A' \to A$.

§18. Order on an embedded curve and integral projections.

Let $C \in \mathfrak{Q}_1(A)$ and let $f: A \to A/C$ be the canonical epimorphism. Given $V \in \mathfrak{Z}([A,C])$, let $P = \mathfrak{Z}^*([A,C],V)$.

We <u>define</u>

$$\mathrm{ord}([A,C],x,V) = \mathrm{ord}_V \mathfrak{R}([A,C],x,P.) \quad \text{for all} \quad x \in H(A)$$

and

$$\mathrm{ord}([A,C],I,V) = \mathrm{ord}_V \mathfrak{R}([A,C],I,P) \quad \text{for all} \quad I \subset A.$$

We observe that, in view of (15.7), (15.8) and (15.9), we then have (18.1), (18.2) and (18.3):

(18.1) For any $x \in H(A)$ we have

$$\begin{aligned}\mathrm{ord}([A,C],x,V) &= \mathrm{ord}([A,C],xA,V)\\ &= \text{a nonnegative integer or } \infty ,\end{aligned}$$

and

$$\mathrm{ord}([A,C],x,V) = \infty \Leftrightarrow x \in C,$$
$$\mathrm{ord}([A,C],x,V) > 0 \Rightarrow x \in P.$$

(18.2) For any $x \in H_n(A)$ we have

$$\mathrm{ord}([A,C],x,V) = \mathrm{ord}_V f(x)/f(z) \quad \text{for all} \quad z \in H_n(A)\setminus P .$$

(18.3) For any $I \subset A$ we have

$$\begin{aligned}\mathrm{ord}([A,C],I,V) &= \mathrm{ord}([A,C],(I \cap H(A))A,V)\\ &= \mathrm{ord}([A,C],(I \cap H(A))A + C,V)\\ &= \min\{\mathrm{ord}([A,C],x,V) : x \in I \cap H(A)\}\\ &= \text{a nonnegative integer or } \infty ,\end{aligned}$$

and

$$\mathrm{ord}([A,C],I,V) = \infty \Leftrightarrow I \cap H(A) \subset C ,$$
$$\mathrm{ord}([A,C],I,V) > 0 \Leftrightarrow I \cap H(A) \subset P .$$

In view of (18.1) and (18.3) we get (18.4) and (18.5):

(18.4) $\operatorname{ord}([A,C],J,V) \geq \operatorname{ord}([A,C],I,V)$ for all $J \subset I \subset A$.

(18.5) For any $Q \in H(A)$ or $Q \subset A$ we have

$$\mathfrak{Z}([A,C],Q) = \{V' \in \mathfrak{Z}([A,C]) : \operatorname{ord}([A,C],Q,V') > 0\} .$$

In view of (18.1), (18.2) and properties of ord_V we also get (18.6), (18.7), (18.8) and (18.9):

(18.6) $\operatorname{ord}([A,C],a,V) = 0$ for all $0 \neq a \in H_0(A)$.

(18.7) For any $x \in H(A)$ and $0 \neq a \in H_0(A)$ we have

$$\operatorname{ord}([A,C],ax,V) = \operatorname{ord}([A,C],x,V) .$$

(18.8) For any x and y in $H_n(A)$ we have

$$\operatorname{ord}([A,C],x+y,V) \geq \min(\operatorname{ord}([A,C],x,V),\operatorname{ord}([A,C],y,V)) ,$$

where equality holds in case $\operatorname{ord}([A,C],x,V) \neq \operatorname{ord}([A,C],y,V)$,

(18.9) If $[V/M(V) : H_0(A)] = 1$, then given any x and y in $H_n(A)$ such that

$$\operatorname{ord}([A,C],x,V) \geq \operatorname{ord}([A,C],y,V) \neq \infty ,$$

there exists a unique $a \in H_0(A)$ such that

$$\operatorname{ord}([A,C],x - ay,V) > \operatorname{ord}([A,C],y,V) ;$$

moreover: $a = 0 \Leftrightarrow \operatorname{ord}([A,C],x,V) > \operatorname{ord}([A,C],y,V)$.

We shall now prove:

(18.10) LEMMA. <u>For any</u> $x \in H_m(A)\backslash C$ <u>and</u> $y \in H_n(A)\backslash C$ <u>we have</u>

$$\operatorname{ord}([A,C],xy,V) = \operatorname{ord}([A,C],x,V) + \operatorname{ord}([A,C],y,V)$$

and

$$\mathrm{ord}([A,C],x^n,V) - \mathrm{ord}([A,C],y^m,V) = \mathrm{ord}_V f(x^n)/f(y^m) \ .$$

PROOF. We can take $z \in H_1(A)\backslash P$ and then we have

$\mathrm{ord}([A,C],xy,V)$

$= \mathrm{ord}_V f(xy)/f(z^{m+n})$ by (18.2)

$= [\mathrm{ord}_V f(x)/f(z^m) + [\mathrm{ord}_V f(y)/f(z^n)]$

$= \mathrm{ord}([A,C],x,V) + \mathrm{ord}([A,C],y,V)$ by (18.2),

and we also have

$([A,C],x^n,V) - \mathrm{ord}([A,C],y^m,V)$

$= [\mathrm{ord}_V f(x^n)/f(z^{mn})] - [\mathrm{ord}_V f(y^m)/f(z^{mn})]$ by (18.2)

$= \mathrm{ord}_V f(x^n)/f(y^m)$

Now we shall prove:

(18.11) PROJECTION LEMMA. <u>Given</u> $\{0\} \neq N \in \mathfrak{M}(A)$, <u>let</u> $B = A^N$ <u>and</u> $D = C^N$. <u>Assume that</u> $D \in \mathfrak{D}_1(B)$.

(Note that, if $H_0(A)$ is algebraically closed, then

$$D \in \mathfrak{D}_1(B) \Leftrightarrow [N : H_0(A)] - [C \cap N : H_0(A)] \geq 2.)$$

<u>Let</u> $h\colon \mathfrak{R}([B,D]) \to \mathfrak{R}(f(B))$ <u>be the canonical isomorphism, and let</u> $W = h^{-1}(V \cap \mathfrak{R}(f(B)))$.

(Note that now: $\mathrm{ord}([A,C],N,V) \neq \infty$, $H_1(B) = N$, $W \in \mathfrak{Z}([B,D])$, $h(W) = V \cap \mathfrak{R}(f(B))$, and $\mathrm{ord}_V M(h(W))$ is a positive integer.)

<u>Then for all</u> $x \in N\backslash C$ <u>we have</u>

$$\mathrm{ord}([A,C],x,V) - \mathrm{ord}([A,C],N,V) = [\mathrm{ord}_V M(h(W))][\mathrm{ord}([B,D],x,W)] \ ,$$

<u>and for all</u> $J \subset \mathfrak{M}(A)$ <u>with</u> $J \subset N$ <u>and</u> $J \not\subset C$ <u>we have</u>

$$\mathrm{ord}([A,C],J,V) - \mathrm{ord}([A,C],N,V) = [\mathrm{ord}_V M(h(W))][\mathrm{ord}([B,D],J,W) \ .$$

PROOF. Let $g: B \to B/D$ be the canonical epimorphism. Let $Q = \mathfrak{Z}^*([B,D],W)$. We can fix

$$z \in N \backslash Q$$

and then, in view of (18.1) and (18.2), for all $x \in N\backslash D$ we have

$$\mathrm{ord}_W g(x)/g(z) = \mathrm{ord}([B,D],x,W) = \text{a nonnegative integer.}$$

Upon multiplying the above equation by $\mathrm{ord}_V M(h(W))$ we see that

$$(1)\quad \begin{cases} \text{for all } x \in N\backslash D \text{ we have:} \\ \mathrm{ord}_V f(x)/f(z) = [\mathrm{ord}_V M(h(W))][\mathrm{ord}([B,D],x,W)] \\ \qquad = \text{a nonnegative integer.} \end{cases}$$

By (18.10) we have

$$(2)\quad \mathrm{ord}([A,C],x,V) - \mathrm{ord}([A,C],z,V) = \mathrm{ord}_V f(x)/f(z) \quad \text{for all } x \in N\backslash C ,$$

and hence, in view of (18.1) and (1), we conclude that

$$(3)\qquad \mathrm{ord}([A,C],x,V) \geq \mathrm{ord}([A,C],z,V) \quad \text{for all } x \in N.$$

Since $z \in N$, by (18.3) and (3) we get

$$(4)\qquad \mathrm{ord}([A,C],N,V) = \mathrm{ord}([A,C],z,V) .$$

Now by (1), (2) and (4) it follows that

$$(5)\quad \begin{cases} \mathrm{ord}([A,C],x,V) - \mathrm{ord}([A,C],N,V) \\ \quad = [\mathrm{ord}_V M(h(W))][\mathrm{ord}([B,D],x,W)] \quad \text{for all } x \in N\backslash C . \end{cases}$$

In view of (18.3) and (5) we see that for all $J \subset \mathfrak{M}(A)$ with $J \subset N$ and $J \not\subset C$ we have

$$\mathrm{ord}([A,C],J,V) - \mathrm{ord}([A,C],N,V) = [\mathrm{ord}_V M(h(W))][\mathrm{ord}([B,D],J,W)] .$$

In view of (18.3) by (18.11) we get:

(18.12) PROJECTION LEMMA. *Given* $\{0\} \neq N \in \mathfrak{M}^*(A)$, *let* $B = A^N$ *and* $D = C^N$. *Assume that* $D \in \mathfrak{D}_1(B)$.

(Note that, if $H_0(A)$ is algebraically closed, then:

$$D \in \mathfrak{D}_1(B) \Leftrightarrow \text{Emdim}[A,C,N] - \text{Emdim}[A,N] \geq 2.)$$

Let $h\colon \mathfrak{R}([B,D]) \to \mathfrak{R}(f(B))$ *be the canonical isomorphism, and let* $W = h^{-1}(V \cap \mathfrak{R}(f(B)))$.

(Note that now: $\text{ord}([A,C],N,V) \neq \infty$, $W \in \mathfrak{Z}([B,D], h(W)) = V \cap \mathfrak{R}(f(B))$, and $\text{ord}_V M(h(W))$ is a positive integer.)

Then for all $x \in (N \cap H_1(A))\backslash C$, *we have*,

$$\text{ord}([A,C],x,V) - \text{ord}([A,C],N,V) = [\text{ord}_V M(h(W))][\text{ord}([B,D],x,W))] ,$$

and for all $J \in \mathfrak{M}^*(A)$ *with* $J \subset N$ and $J \not\subset C$, *upon letting* $K = J^N$, *we have*

$$\text{ord}([A,C],J,V) - \text{ord}([A,C],N,V) = [\text{ord}_V M(h(W))][\text{ord}([B,D],K,W)] .$$

(18.13) COROLLARY-DEFINITION. Let $\pi \in H_1^*(A)$ and $N \in \mathfrak{M}^*(A)$ be given with $\pi \subset N$. Let $B = A^N$ and $D = C^N$. By (18.12) we see that the following two conditions are equivalent.

(*) $\left\{ \begin{array}{l} D \in \mathfrak{D}_1(B) \text{ and, upon letting } h\colon \mathfrak{R}([B,D]) \to \mathfrak{R}(f(B)) \\ \text{to be the canonical isomorphism, we have:} \\ h^{-1}(V' \cap \mathfrak{R}(f(B))) \in \mathfrak{Z}([B,D],\pi^N) \text{ for all } V' \in \mathfrak{Z}([A,C],\pi). \end{array} \right.$

(**) $\left\{ \begin{array}{l} D \in \mathfrak{D}_1(B) \text{ and:} \\ \text{ord}([A,C],\pi,V') > \text{ord}([A,C],N,V') \text{ for all } V' \in \mathfrak{Z}([A,C],\pi) . \end{array} \right.$

We shall say that the <u>projection of</u> C <u>from</u> N <u>in</u> A is <u>π-integral</u> to mean that one (and hence both) of the two conditions is satisfied. From this phrase we may drop " in A " when the reference to A is clear from the context. In this Chapter we shall use condition (**), whereas in Chapter IV we shall use condition (*). We note that obviously

(18.13.1) $D \in \mathfrak{D}_1(B)$ and $\mathfrak{D}_0([A,C],N) = \emptyset \Rightarrow \begin{cases} \text{the projection of C from N} \\ \text{is } \pi\text{-integral.} \end{cases}$

§19. Order on an abstract curve and integral projections.

(19.1) to (19.12).

Let R be a homogeneous domain with Dim R = 1. Given $V \in \mathfrak{Z}(R)$, let $P = \mathfrak{Z}^*(R,V)$. We <u>define</u>

$$\text{ord}(R,I,V) = \text{ord}([R,\{0\}],I,V) \quad \text{for any} \quad I \in H(A) \quad \text{or} \quad I \subset A.$$

We celarly get assertions (19.1) to (19.10) where, for $1 \le i \le 10$, (19.i) is obtained from (18.i) by replacing: A as well as [A,C] by R; C by {0}; and f by the identity map $R \to R$. We also get assertion (19.11) and (19.12) from (18.11) and (18.12) respectively, where in addition to the above replacements we let $S = R^N$ and then replace: B as well as [B,D] by S; D by {0}; and h by the identity map $\mathfrak{R}(S) \to \mathfrak{R}(S)$.

(19.13) COROLLARY-DEFINITION. Let $\pi \in H_1^*(R)$ and $N \in \mathfrak{M}^*(R)$ be given with $\pi \subset N$. Let $S = R^N$. By (19.12) we see that the following two conditions are equivalent.

(*) Dim S = 1 and: $V' \cap \mathfrak{R}(S) \in \mathfrak{Z}(S,\pi^N)$ for all $V' \in \mathfrak{Z}(R,\pi)$.

(**) Dim S = 1 and: $\text{ord}(R,\pi,V') > \text{ord}(R,N,V)$ for all $V' \in \mathfrak{Z}(R,\pi)$.

We shall say that the <u>projection from</u> N in R <u>is</u> π-<u>integral</u> to mean that one (and hence both) of the conditions (*) and (**) is satisfied. From this phrase we may omit " in R " when the reference to R is clear from the context. In this Chapter we shall use condition (**), whereas in Chapter IV we shall use condition (*).

We note that obviously

(19.13.1) $\text{Dim } S = 1 \text{ and } \mathfrak{D}_0(R,N) = \phi \Rightarrow \begin{cases} \text{the projection from } N \text{ in } R \\ \text{is } \pi\text{-integral} \end{cases}$

and

(19.13.2) $\begin{cases} \text{the projection from } N \text{ in } R \text{ is } \pi\text{-integral} \\ \Leftrightarrow \text{the projection of } \{0\} \text{ from } N \text{ in } R \text{ is } \pi\text{-integral.} \end{cases}$

(19.14) REMARK. Given any $C \in \mathfrak{D}_1(A)$, let $f: A \to A/C$ be the canonical epimorphism.

We note that, then, for any $V^* \in \mathfrak{Z}([A,C])$, we clearly have

$$\text{ord}([A,C],x,V^*) = \text{ord}(f(A),f(x),V^*) \quad \text{for any } x \in H(A)$$

and

$$\text{ord}([A,C],I,V^*) = \text{ord}(f(A),f((I \cap H(A))A),V^*) \quad \text{for any } I \subset A.$$

We also note that, for any $\pi \in H_1^*(A)$ and $N \in \mathfrak{M}^*(A)$, with $\pi \subset N$ and $\pi \not\subset C$, we clearly have that:

the projection of C from N in A is $f(\pi)$-integral

$\Leftrightarrow$ the projection from $f(N)$ in $f(A)$ is $f(\pi)$-integral.

§20. Valued vector spaces.

By a valued vector space we mean a triple (k,Λ,v) where: k is a field, Λ is a k-vector-space with $[\Lambda : k] < \infty$, and

$$v: \Lambda \cup \Omega(\Lambda) \to \mathbb{Z}$$

is a mapping, where $\Omega(\Lambda)$ is the set of all k-vector-subspaces of Λ and $\mathbb{Z}$ is the set of all nonnegative integers together with ∞, such that:

(1^*) $v(0) = \infty$;

(2^*) $v(x+y) \geq \min(v(x),v(y))$ for all x and y in Λ ;

(3^*) $v(x+y) = \min(v(x), v(y))$ for all x and y in Λ with $v(x) \neq v(y)$;

(4^*) $v(zx) = v(x)$ for all $x \in \Lambda$ and $0 \neq a \in k$;

(5^*) given any x and y in Λ with $v(x) \geq v(y) \neq \infty$, there exists

$a \in k$ such that $v(x-ay) > v(y)$; and

(6^*) $v(L) = \min\{v(x) : x \in L\}$ for every $L \in \Omega(\Lambda)$.

In view of (18.1), (18.3), (18.7), (18.8) and (18.9) we see that:

(*) <u>For any</u> $C \in \mathfrak{Q}_1(A)$ <u>and any</u> $V \in \mathfrak{Z}([A,C])$ <u>with</u> $[V/M(V) : H_0(A)] = 1$, <u>we have that</u> $(H_0(A), H_1(A), \mathrm{ord}([A,C],.,V)$ <u>is a valued vector space.</u>

In view of (19.1), (19.3), (19.7), (19.8) and (19.9) we see that:

(**) <u>For any homogeneous domain</u> R <u>with</u> $\mathrm{Dim}\, R = 1$ <u>and any</u> $V \in \mathfrak{Z}(R)$ <u>with</u> $[V/M(V) : H_0(A)] = 1$, <u>we have that</u> $(H_0(R), H_1(R), \mathrm{ord}(R,.,V))$ <u>is a valued vector space</u>.

In §21 and §23 we shall define osculating flats. To prepare the groundwork for defining osculating flats, we shall now prove several lemmas about a valued vector space. To fix the idea, the reader may like to keep in mind situations (*) and (**). We shall deal with the topic of osculating flats much more thoroughly than needed in the rest of this book.

So, <u>in the rest of</u> §20, <u>let</u> (k,Λ,v) <u>be a valued vector space as defined above</u>.

For every $L \in \Omega(\Lambda)$ let:

$\Omega(L) = \{I \in \Omega(\Lambda): I \subset L\}$,

$\Omega_i(L) = \{I \in \Omega(L): [I : k] = i\}$,

$Z(L) = \{v(I): I \in \Omega(L)\}$,

$Z'(L) = \{v(x): x \in L\}$,

$Y(L,j) = \{x \in L: v(x) \geq j\}$,

$Y'(L) = \{I \in \Omega(L): I = Y(L,j)$ for some $j \in \mathbb{Z}\}$,

$Y(L) = \{I \in \Omega(L): I = Y(L,v(I))\}$,

$T^*(L) = \{x \in L: v(x) > v(L)\}$, and

$\Delta^*(L) = \{x \in L: v(x) = \infty\}$.

We note that, in view (1^*), (2^*) and (4^*), we then have $\Delta^*(L) \in \Omega(L)$. Let

$$p(L) = [\Delta^*(L) : k] .$$

we also observe that in view of (18.1) and (19.1)

$$\Delta^*(L) = \begin{cases} C \cap L & \text{in case of } (*) \\ \{0\} & \text{in case of } (**) . \end{cases}$$

Now, in the rest of §20, let $L \in \Omega_d(\Lambda)$ be given and let $p = p(L)$.

We shall first state Lemmas (20.1) to (20.13) and then prove them one by one.

(20.1) LEMMA. $Z(L) = Z'(L)$.

(20.2) LEMMA. card $Z'(L) \leq d - p + 1$.

(20.3) LEMMA. $Y(L) = Y'(L)$.

(20.4) LEMMA. For any $j \in Z$ we have $Y(L,j) \in \Omega(L)$; if moreover $j \in Z(L)$, then $Y(L,j) \in Y(L)$ and $v(Y(L,j)) = j$.

(20.5) LEMMA. card $Y(L) =$ card $Z(L)$.

(20.6) LEMMA. $I \in Y(L) \Rightarrow Y(I) \subset Y(L)$ and $\Delta^*(I) = \Delta^*(L)$.

(20.7) LEMMA. If $d > p$ then $\Delta^*(L) \subset T^*(L) \in Y(L) \cap \Omega_{d-1}(L)$ and $v(T^*(L)) > v(L)$; moreover, if I is any member of $\Omega_{d-1}(L)$ such that $v(I) > v(L)$, then $I = T^*(L)$.

(20.8) LEMMA. There exists a unique sequence $L_p \subset L_{p+1} \subset \ldots \subset L_d = L$ such that $L_i \in \Omega_i(L)$ and $v(L_i) > v(L_{i+1})$ for $p \leq i < d$. Moreover we have $L_i \in Y(L)$ for $p \leq i \leq d$, $L_i = T^*(L_{i+1})$ for $p \leq i < d$, $L_p = \Delta^*(L)$, and $v(L_p) = \infty$.

More generally we can say that given any α with $p \leq \alpha \leq d$, there exists a unique sequence $L_{\alpha,\alpha} \subset L_{\alpha,\alpha+1} \subset \ldots \subset L_{\alpha,d} = L$ such that $L_{\alpha,i} \in \Omega_i(L)$ and $v(L_{\alpha,i}) > v(L_{\alpha,i+1})$ for $\alpha \leq i < d$. Moreover we have $L_{\alpha,i} = L_{\beta,i}$ whenever $p \leq \beta \leq \alpha \leq i \leq d$.

Alternatively, the said sequence $L_p \subset L_{p+1} \subset \ldots \subset L_\alpha = L$ can be a completely characterized by saying the following: there exists a unique sequence $L'_{p'} \subset L'_{p'+1} \subset \ldots \subset L'_d = L$ such that $L'_i \in \Omega_i(L)$ and $v(L'_i) > v(L'_{i+1})$ for $p' \leq i \leq d$, $v(L'_{p'}) = \infty$. Moreover we have $p' = p$ and $L'_i = L_i$ for $p \leq i \leq d$.

(20.9) <u>We have</u>

$$\operatorname{card} Y(L) = \operatorname{card} Y'(L) = \operatorname{card} Z'(L) = \operatorname{card} Z(L) = d - p + 1$$

<u>and</u>

$$\operatorname{card} Y(L) \cap \Omega_i(L) = \operatorname{card} Y'(L) \cap \Omega_i(L) = 1 \quad \underline{\text{for}} \quad p \le i < d.$$

<u>Moreover, with the notation of</u> (20.8) <u>we have that</u>:

$$\begin{cases} \infty = v(L_p) > v(L_{p+1}) > \ldots > v(L_d) \\ \text{is the unique descending labelling of } Z(L), \end{cases}$$

$$Y(L) = Y'(L) = \{L_p, L_{p+1}, \ldots, L_d\},$$

<u>and</u>

$$Y(L) \cap \Omega_i(L) = Y'(L) \cap \Omega_i(L) = \{L_i\} \quad \text{for} \quad p \le i \le d.$$

(20.10) LEMMA. <u>Let</u> $L_p \subset L_{p+1} \subset \ldots \subset L_d = L$ <u>be as in</u> (20.8). <u>Given</u> $J \in \Omega_b(L)$ <u>let</u> $n = p(J)$ <u>and let</u> $J_n \subset J_{n+1} \subset \ldots \subset J_b = J$ <u>be the unique sequence</u> (obtained by applying (20.8) to J) <u>such that</u> $J_i \in \Omega_i(J)$ <u>and</u> $v(J_i) > v(J_{i+1})$ <u>for</u> $n \le i < b$.

<u>In view of the set-theoretic inclusions</u> $L_p \subset L_{p+1} \subset \ldots L_d = L$ <u>and</u> $L \supset J$, <u>we clearly get a unique sequence</u> $p = r(\nu) < r(\nu+1) < \ldots < r(b) \le d$ <u>of integers such that</u>

$$\begin{cases} J \cap L_{r(b)} = J \cap L_s, \text{ for } r(b) \le s \le d, \\ \underline{\text{and}} \\ J \cap L_{r(i)} = J \cap L_s \ne J \cap L_{r(i+1)} \ \underline{\text{for}} \ \nu \le i < b \ \underline{\text{and}} \\ r(i) \le s < r(i+1). \end{cases}$$

[Alternatively, we can instead define the complement of $\{r(\nu), r(\nu+1), \ldots, r(b)\}$ in $\{p, p+1, \ldots, d\}$ as follows. From the facts that $L_p \subset L_{p+1} \subset \ldots \subset L_d = L$, $L_s \in \Omega_s(L)$ for $p \le s \le d$, $J \in \Omega_b(L)$, and $J \cap L_p = J_n$, it follows that: upon letting

$$G_i = \{s \in \{p,p+1,\ldots,d\}: J + L_s \in \Omega_{b+p-n+i}(L)\}$$

we have $G_i \neq \emptyset$ for $1 \le i \le d - b - p + n$ and then upon letting

$$g_i = \min\{s: s \in G_i\} \text{ for } 1 \le i \le d - b - p + n$$

we have that

$$p < g_1 < g_2 < \ldots < g_{d-b-p+n} \le d$$

and

$$\{p,p+1,\ldots,d\} \setminus \{r(\nu),r(\nu+1),\ldots,r(b)\} = \{g_1 g_2,\ldots,g_{d-b-p+n}\} ;$$

whence, in particular, we must have $\nu = n$. We also note that in case of $d - p > b - n$, upon letting

u = the unique integer with $p < u \le d$ such that $L_{u-1} \subset J + L_p$ and $L_u \not\subset J + L_p$

and

$$u^* = \min\{s \in \{p+1,p+2,\ldots,d\}: J \cap L_{s-1} = J \cap L_s\} ,$$

we clearly have

$$u = u^* = g_1 .$$

It follows that if $d - p = b - n + 1$ then: $n < u + n - p \le b + 1$,

$$r(i) = p - n + i \text{ and } L_{p-n+i} \subset J + L_p \text{ for } n \le i < u+n-p,$$

and

$$r(i) = p - n + i + 1 \text{ for } u + n - p \le i \le b.]$$

<u>We claim that</u>, <u>with the sequence</u> $p = r(\nu) < r(\nu+1) < \ldots < r(b) \le d$ <u>as defined above</u>, <u>we have</u> $\nu = n$ <u>and</u>, <u>upon letting</u> $r(b+1) = d + 1$, <u>we have</u>

$$\left.\begin{array}{l} J_i = J \cap L_s \quad \text{for} \quad r(i) \le s < r(i+1) \\[2ex] v(J_i) = v(L_{r(i)}) \end{array}\right\} \text{for} \quad n \le i \le b .$$

In particular, if $d - p = b - n + 1$ then, with u as defined in the above bracketed remark, we have

$$J_i = J \cap L_{p-n+i} \quad \text{and} \quad v(J_i) = v(L_{p-n+i}) \quad \text{for} \quad n \le i < u+n-p$$

and

$$J_i = J \cap L_{p-n+i+1} \quad \text{and} \quad v(J_i) = v(L_{p-n+i+1}) \quad \text{for} \quad u+n-p \le i \le b$$

(we note that in case of $\Delta^*(L) = \{0\}$, we then have $p = 0 = n$ and $J_i = L_i$ for $0 \le i < u$.)

(20.11) LEMMA. Given any sequence $p = r(n) < r(n+1) < \ldots < r(b) \le d$ of integers with $0 \le n \le p$, there exists $J \in \Omega_b(L)$ such that the given sequence is the sequence derived from the pair J and L as in (20.10).

(We observe that, in view of (20.9) and (20.10), this implies that $J \to Z(J)$ gives a surjective map of $\Omega(L)$ onto the set of all non-empty subsets of $Z(L)$.)

(20.12) LEMMA. In the situation (*), let $\{0\} \neq N \in \Omega_e(L)$ be given and let $q = p(N)$; (i.e., $q = [C \cap N : H_0(A)]$). Let $B = A^n$ and $D + C^N$. Assume that $D \in \mathfrak{D}_1(B)$.

(Note that if $H_0(A)$ is algebraically closed then: $D \in \mathfrak{D}_1(B) \Leftrightarrow e - q \ge 2$.)

Let $f: A \to A/C$ be the canonical epimorphism. Let $h: \mathfrak{R}([B,D]) \to \mathfrak{R}(f(B))$ be the canonical isomorphism, and let $W = h^{-1}(V \cap \mathfrak{R}(f(B))$

(Note that now: $\operatorname{ord}([A,C],N,V) \neq \infty$, $H_0(B) = H_1(A)$, $H_1(B) = N$, $W \in \mathfrak{Z}([B,D])$, $h(W) = V \cap \mathfrak{R}(f(B))$, and $\operatorname{ord}_V M(h(W))$ is a positive integer.)

Given any $J \in \Omega_b(N)$, *let* $n = p(J)$ *and let* $J_n \subset J_{n+1} \subset \ldots \subset J_b = J$ *be the unique sequence* (obtained by applying (20.8) to J) *such that* $J_i \in \Omega_i(J)$ *and* $\text{ord}([A,C],J_i,V) > \text{ord}([A,C],J_{i+1},V)$ *for* $n \le i < b$. (Note that now $[J: H_0(B)] = b$ and $[D \cap J: H_0(B)] = n$.) *Also let* $K_n \subset K_{n+1} \subset \ldots \subset K_b = J$ *be the unique sequence* (obtained from (20.8) by replacing A,C,V,L by B,D,W,J respectively) *such that* $K_i \in \Omega_i(L)$ *and* $\text{ord}([B,D],K_i,W) > \text{ord}([B,D],K_{i+1},W)$ *for* $n \le i < b$. *Then we have*

$$K_i = J_i \quad \textit{for} \quad n \le i \le b ,$$

and

$$\left.\begin{aligned} &\text{ord}([B,D]K_i ,W) \\ &= [\text{ord}([A,C],J_i,V) - \text{ord}([A,C],N,V)]/\text{ord}_V M(h(W)) \end{aligned}\right\} \quad \text{for} \quad n < i \le b .$$

(Observe that (20.10) gives recipes for obtaining the sequences $(J_i)_{n \le i \le b}$ and $(\text{ord}([A,C],J_i,V))_{n \le i \le e}$ from the sequences $(L_i)_{p \le i \le d}$ and $(\text{ord}([A,C],L_i,V)_{p \le i \le d}$; this is particularly significant when $L = \Lambda$, i.e., $L = H_1(A)$.)

(20.13) LEMMA. *In the situation* (**), *let* $\{0\} \ne N \in \Omega_e(L)$ *be given. Let* $S = R^n$. *Assume that* $\text{Dim}\, S = 1$.

(Note that, if $H_0(R)$ is algebraically closed, then: $\text{Dim}\, S = 1 \Leftrightarrow e \ge 2$.)

Let $W = V \cap \mathfrak{R}(S)$.

(Note that now: $\text{ord}(R,N,V) \ne \infty$, $H_0(S) = H_0(R)$, $H_1(S) = N$, $W \in \mathfrak{Z}(S)$, and $\text{ord}_V M(W)$ is a positive integer.)

Given any $J \in \Omega_b(N)$, *let* $J_0 \subset J_1 \subset \ldots \subset J_b = J$ *be the unique sequence* (obtained by applying (20.8) to J) *such that* $J_i \in \Omega_i(J)$ *and* $\text{ord}(R,J_i,V) > \text{ord}(R,J_{i+1},V)$ *for* $0 \le i < b$. *Also let* $K_0 \subset K_i \subset \ldots \subset K_b = J$ be the unique sequence (obtained from (20.8) by replacing R,V,L by S,W,J respectively) *such that* $K_i \in \Omega_i(J)$ *and* $\text{ord}(S,K_i,W) > \text{ord}(S,K_{i+1},W)$ *for* $0 \le i < b$. *Then we have*

$$K_i = J_i \quad \underline{\text{for}} \quad 0 \le i \le b ,$$

$$\operatorname{ord}(S,K_i,W) = [\operatorname{ord}(R,J_i,V) - \operatorname{ord}(R,N,V)]/\operatorname{ord}_V M(W) \quad \text{for} \quad 0 < i \le b .$$

(Observe that (20.10) gives recipes for obtaining the sequences $(J_i)_{0 \le i \le b}$ and $(\operatorname{ord}(R,J_i,V))_{0 \le i \le b}$ from the sequences $(L_i)_{0 \le i \le d}$ and $(\operatorname{ord}(R,L_i,V))_{0 \le i \le d}$; this is particularly significant when $L = \Lambda$, i.e., $L = H_1(R)$.)

PROOF OF (20.1). By (6^*) we have $Z(L) \subset Z'(L)$. For any $x \in L$, we have $xk \in \Omega(L)$, and by (1^*), (4^*) and (6^*) we get $v(xk) = v(x)$; consequently $Z'(L) \subset Z(L)$. Therefore $Z(L) = Z'(L)$.

PROOF OF (20.2). Given any elements $x_1,x_2,\dots,x_m$ in L such that $v(x_1), v(x_2),\dots,v(x_m)$ are all distinct and noninfinite, and given any elements $a_1,a_2,\dots,a_m$ in k at least one of which is nonzero, we have

$$v(a_1x_1 + a_2x_2+\dots+a_mx_m) = v(x_i) \quad \text{for some } i \qquad \text{by } (3^*) \text{ and } (4^*)$$
$$\ne \infty$$

and hence

$$a_1x_1 + a_2x_2+\dots+a_mx_m \quad \Delta^*(L) .$$

It follows that

$$\operatorname{card} Z'(L)\backslash\{\infty\} = [L/\Delta^*(L) : k]$$

and hence

$$\operatorname{card} Z'(L) \le [L/\Delta^*(L) : k] + 1 .$$

Clearly $[L/\Delta^*(L) : k] = d - p$, and hence $\operatorname{card} Z'(L) \le d - p + 1$.

PROOF OF (20.3). Obvious in view of (6^*) .

PROOF OF (20.4). By (1^*), (2^*) and (4^*) we see that

$Y(L,j) \in \Omega(L)$, and now by (6^*) we see that, if $j \in Z(L)$, then $Y(L,j) \in Y(L)$ and $v(Y(L,j)) = j$.

PROOF OF (20.5). Clearly $I \to v(I)$ gives an injective map of $Y(L)$ into $Z(L)$, and by (20.4) we see that the said map is surjective. Therefore $\operatorname{card} Y(L) = \operatorname{card} Z(L)$.

PROOF OF (20.6). Obviously: $I \in Y'(L) \Rightarrow Y'(I) \subset Y'(L)$. Therefore, in view of (20.3), we get $I \in Y(L) \Rightarrow Y(I) \subset Y(L)$. In view of (1^*) and (6^*) we also see that $I \in Y(L) \Rightarrow \Delta^*(I) = \Delta^*(L)$.

PROOF OF (20.7). Assume that $d > p$. Then in view of (6^*) we have

(1) $$v(L) < \infty .$$

Consequently, in view of (1^*), (2^*), (4^*) and (6^*) we see that

$$\Delta^*(L) \subset T^*(L) \in Y(L) \quad \text{and} \quad v(T^*(L)) > v(L)$$

and there exists

(2) $y \in L$ such that $v(y) = v(L) = \min\{v(x) : x \in L\}$.

In view of (1), (2) and (5^*) we see that $y \in L \backslash T^*(L)$ and the k-vector-space $L/T^*(L)$ is generated by the image of y; consequently $[L/T^*(L) : k] = 1$ and hence

(3) $$T^*(L) \in \Omega_{d-1}(L) .$$

Let any

(4) $$I \in \Omega_{d-1}(L)$$

be given such that

(5) $$v(I) > v(L) ;$$

then by (5) and (6^*) we get $T^*(L) \subset I$, and hence by (4) and (3) we get $I = T^*(L)$

PROOF OF (20.8). The assertions in the second and the third paragraphs follow from the assertions in the first paragraph. In view of (1^*) and (6^*) we clearly have $L \in Y(L) \cap \Omega_d(L)$ and:

$$v(L) = \infty \Leftrightarrow L = \Delta^*(L) .$$

Therefore the assertions in the first paragraph are obvious in case $d - p = 0$. Now, in view of (20.6) and (20.7), the assertions in the first paragraph in the general case follow by induction on $d - p$.

PROOF OF (20.9). Let the notation be as in (20.8). By (20.8) we have

$$\{L_p, L_{p+1}, \ldots, L_d\} \subset Y(L) \quad \text{and} \quad \text{card}\{L_p, L_{p+1}, \ldots, L_d\} = d - p + 1 ;$$

consequently by (20.1), (20.2), (20.3) and (20.5) we conclude that

$$(1) \quad \begin{cases} Y(L) = Y'(L) = \{L_p, L_{p+1}, \ldots, L_d\} \\ \text{and} \\ \text{card } Y(L) = \text{card } Y'(L) = \text{card } Z'(L) = \text{card } Z(L) = d - p + 1 \end{cases}$$

By (20.8) we have

$$L_i \in Y(L) \cap \Omega_i(L) \quad \text{for} \quad p \le i \le d$$

and clearly

$$\sum_{p \le i \le d} \text{card } Y(L) \cap \Omega_i(L) = \text{card } Y(L) .$$

Consequently, by (1) we conclude that

$$\text{card } Y(L) \cap \Omega_i(L) = \text{card } Y'(L) \cap \Omega_i(L) = 1 \quad \text{for} \quad p \le i \le d$$

and

$$Y(L) \cap \Omega_i(L) = Y'(L) \cap \Omega_i(L) = \{L_i\} \quad \text{for} \quad p \le i \le d .$$

In view of (1), by (20.8) we also see that

$$\begin{cases} \infty = v(L_p) > v(L_{p+1}) > \ldots > v(L_d) \\ \text{is the unique descending labelling of } Z(L) . \end{cases}$$

PROOF OF (20.10). By applying (20.9) to L and J we get that:

(1) $$\begin{cases} \infty = v(L_p) > L(L_{p+1}) > \ldots > v(L_d) \\ \text{is the unique descending labelling of } Z(L) , \end{cases}$$

(2) $$\begin{cases} \infty = v(J_n) > v(J_{n+1}) > \ldots > v(J_b) \\ \text{is the unique descending labelling of } Z(J) , \end{cases}$$

(3) $Y(L) \cap \Omega_i(L) = \{L_i\}$ for $p \le i \le d$,

and

(4) $Y(J) \cap \Omega_i(J) = \{J_i\}$ for $n \le i \le b$.

Since $J \subset \Omega(L)$, we clearly have $Z(J) \subset Z(L)$; consequently, by (1) and (2), there exists a unique sequence $p = t(n) < t(n+1) < \ldots < t(b) \le d$ of integers such that

(5) $v(J_i) = v(L_{t(i)})$ for $n \le i \le b$.

Since $J \subset L$, by (3), (4) and (5) we get that

(6) $J_i = J \cap L_{t(i)}$ for $n \le i \le b$.

Since $J_i \in \Omega_i(J)$ for $n \le i \le b$, we have $J_i \neq J_{i+1}$ for $n \le i \le b$; consequently by (6) we get

(7) $J \cap L_{t(i)} \neq J \cap L_{t(i+1)}$ for $n \le i < b$.

Upon letting

$$\begin{cases} \Gamma_b = \{x \in L : v(J_b) > v(x)\} \\ \text{and} \\ \Gamma_i = \{x \in L : v(J_i) > v(x) > v(J_{i+1})\} \quad \text{for} \quad n \le i < b , \end{cases}$$

in view of (1), (3) and (5) we get

$$(8) \quad \begin{cases} L_s \subset L_{t(b)} \cup \Gamma_b \quad \text{for} \quad t(b) \le s \le d , \\ \text{and} \\ L_s \subset L_{t(i)} \cup \Gamma_i \quad \text{for} \quad n \le i < b \quad \text{and} \quad t(i) \le s < t(i+1) ; \end{cases}$$

now by (20.1) we have $Z'(J) = Z(J)$, and hence in view of (2) we also get

$$(9) \qquad J \cap \Gamma_i = \emptyset \quad \text{for} \quad n \le i \le b .$$

By (7), (8) and (9) it follows that

$$(10) \quad \begin{cases} J \cap L_{t(b)} = J \cap L_s \quad \text{for} \quad t(b) \le s \le d , \\ \text{and} \\ J \cap L_{t(i)} = J \cap L_s \neq J \cap L_{t(i+1)} \quad \text{for} \quad n \le i < b \\ \text{and} \quad t(i) \le s < t(i+1) , \end{cases}$$

and hence we must have $\nu = n$ and $t(i) = r(i)$ for $n \le i \le b$. Consequently, upon letting $r(b+1) = d + 1$, by (5), (6) and (10) we get

$$\left.\begin{array}{l} J_i = J \cap L_s \quad \text{for} \quad r(i) \le s < r(i+1) \\ \\ v(J_i) = v(L_{r(i)}) \end{array}\right\} \text{for} \quad n \le i \le b .$$

The special claim for the case when $d - p = b - n + 1$ now follows from the bracketed remark.

PROOF OF (20.11). Let $L_p \subset L_{p+1} \subset \ldots \subset L_d = L$ be as in (20.8). Then, in view of (20.9), we know that

(1) $$v(L_p) = \infty$$

and

(2) $$\begin{cases} v(L_p) > v(L_{p+1}) > \ldots > (L_d) \\ \text{is the unique descending labelling of } Z(L) \end{cases}$$

Since $0 \le n \le p$, we can take $J^* \in \Omega_n(\Delta^*(L))$. In view of (20.1) and (2) we can take $x_i \in L$ with

(3) $$v(x_i) = v(L_{r(i)}) \quad \text{for} \quad n < i \le b .$$

By (1), (2) and (3) we get

(4) $$\infty > v(x_{n+1}) > \ldots > v(x_b) .$$

Let

$$J = J^* + x_{n+1}k + x_{n+2}k + \ldots + x_b k .$$

Then in view of (4), by (1^*), (3^*) and (4^*) we see that $J \in \Omega_b(L)$, $J^* = \Delta^*(J)$, $n = p(J)$, and

(5) $$Z'(J) = \{\infty, v(x_{n+1}), v(x_{n+2}), \ldots, v(x_b)\} .$$

Let $J_n \subset J_{n+1} \subset \ldots \subset J_n = J$ be the unique sequence (obtained by applying (20.8) to J, such that $J_i \in \Omega_i(J)$ and $v(J_i) \ge v(J_{i+1})$ for $n \le i < b$, and

(6) $$v(J_n) = \infty .$$

By applying (20.1) and (20.9) to J we know that

$$\begin{cases} v(J_n) > v(J_{n+1}) > \ldots > v(J_b) \\ \text{is the unique descending labelling of } Z'(J) \end{cases}$$

and hence, in view of (4) and (5), we must have

(7) $$v(J_i) = v(x_i) \quad \text{for} \quad n < i \le b \; .$$

By (1), (3), (6) and (7) we get

(8) $$v(J_i) = v(L_{r(i)}) \quad \text{for} \quad n \le i \le b \; .$$

In view of (2) and (8), by (20.10) we conclude that the given sequence $p = r(n) < r(n+1) < \ldots < r(b) \le d$ must be the sequence derived from the pair J and L as in (20.10).

PROOF OF (20.12). By (18.11) we have

(1) $$\begin{cases} \operatorname{ord}([B,D],J_i,W) \\ = [\operatorname{ord}([A,C],J_i,V) - \operatorname{ord}([A,C],N,V)]/\operatorname{ord}_V M(h(W)) \quad \text{for} \\ n < i \le b \; . \end{cases}$$

Since

$$\operatorname{ord}([A,C],J_i,V) > \operatorname{ord}([A,C],J_{i+1},V) \quad \text{for} \quad n \le i < b \; ,$$

by (1) we get that

$$\operatorname{ord}([B,D],J_i,W) > \operatorname{ord}([B,D],J_{i+1},W) \quad \text{for} \quad n \le i < b$$

and hence by the uniqueness part of (20.8), applied to B,D,W,J, we conclude that

$$K_i = J_i \quad \text{for} \quad n \le i \le b \; ;$$

therefore, in view of (1), it now follows that

$$\left.\begin{array}{l} \operatorname{ord}([B,D],K_i,W) \\ = [\operatorname{ord}([A,C],J_i,V) - \operatorname{ord}([A,C],N,V)]/\operatorname{ord}_V M(h(W)) \end{array}\right\} \text{for} \quad n < i \le b.$$

PROOF OF (20.13). By (19.11) we have

(1) $\operatorname{ord}(S,J_i,W) = [\operatorname{ord}(R,J_i,V) - \operatorname{ord}(R,N,V)]/\operatorname{ord}_V M(W)$ for $0 < i \le b$.

Since

$$\operatorname{ord}(R,J_i,V) > \operatorname{ord}(R,J_{i+1},V) \quad \text{for} \quad 0 \le i < b ,$$

by (1) we get that

$$\operatorname{ord}(S,J_i,W) > \operatorname{ord}(S,J_{i+1},W) \quad \text{for} \quad 0 \le i < b$$

and hence by the uniqueness part of (20.8), applied to S,W,J, we conclude that

$$K_i = J_i \quad \text{for} \quad 0 \le i \le b ;$$

therefore, in view of (1), it now follows that

$$\operatorname{ord}(S,K_i,W) = [\operatorname{ord}(R,J_i,V) - \operatorname{ord}(R,N,V)]/\operatorname{ord}_V M(W) \quad \text{for} \quad 0 < i \le b .$$

§21. Osculating flats and integral projections of an embedded curve.

Let $C \in \mathfrak{Q}_1(A)$.

(21.1) LEMMA-DEFINITION. Let $V \in \mathfrak{Z}([A,C])$ be such that $[V/M(V) : H_0(A)] = 1$. For any $L \in \mathfrak{M}_d^*(A)$, upon letting $p = \operatorname{Emdim}[A,C,L]$, it is easy to get the following by applying (18.3) and (20.8).

<u>There exists a unique sequence</u> $L = L_d \supset L_{d+1} \supset \dots \supset L_p$ <u>such that</u> $L_i \in \mathfrak{M}_i^*(A)$ <u>and</u> $\operatorname{ord}([A,C],L_{i-1},V) < \operatorname{ord}([A,C],L_i,V)$ <u>for</u> $d < i \le p$. <u>Moreover we have</u>

$$L_i \cap H_1(A) = \{x \in L_{i-1} \cap H_1(A) : \operatorname{ord}([A,C],x,V) > \operatorname{ord}([A,C],L_{i-1},V\} \quad \text{for} \quad d < i \le p ,$$

$L_p = \Delta(A,C,L)$, and $\operatorname{ord}([A,C],L_p,V) = \infty$.

More generally we can say that given any α with $d \le \alpha \le p$, there exists a unique sequence $L = L_{\alpha,d} \supset L_{\alpha,d+1} \supset \ldots \supset L_{\alpha,\alpha}$ such that $L_{\alpha,i} \in \mathfrak{M}_i^*(A)$ and $\operatorname{ord}([A,C],L_{\alpha,i-1},V) < \operatorname{ord}([A,C],L_{\alpha,i},V)$ for $d < i \le \alpha$. Moreover we have $L_{\alpha,i} = L_{\beta,i}$ whenever $d \le i \le \alpha \le \beta \le p$.

Alternatively, the said sequence $L = L_d \supset L_{d+1} \supset \ldots \supset L_p$ can be completely characterized by saying the following: There exists a unique sequence $L = L'_d \supset L'_{d+1} \supset \ldots \supset L'_{p'}$ such that $L'_i \in \mathfrak{M}_i^*(A)$ and $\operatorname{ord}([A,C],L'_{i-1},V) < \operatorname{ord}([A,C],L'_i,V)$ for $d < i \le p'$, and $\operatorname{ord}([A,C],L'_{p'},V) = \infty$. Moreover we have $p' = p$ and $L'_i = L_i$ for $d \le i \le b$.

We define

$$\left.\begin{aligned} T_i([A,C],L,V) &= L_i \\ \tau_i([A,C],L,V) &= \operatorname{ord}([A,C],L_i,V) \end{aligned}\right\} \quad \text{for } d \le i \le p$$

and we note that then: $\tau_p([A,C],L,V) = \infty$, $\tau_i([A,C],L,V)$ is a non-negative integer for $d \le i < p$, and in view of (18.3) we have

$$\tau_d([A,C],L,V) > 0 \Leftrightarrow L \subset \mathfrak{Z}^*([A,C],V) \ .$$

We also define

$$\left.\begin{aligned} T_i([A,C],V) &= T_i([A,C],\Delta(A),V) \ , \\ \tau_i([A,C],V) &= \tau_i([A,C],\Delta(A),V) \ , \end{aligned}\right\} \quad \text{for } -1 \le i \le \operatorname{Emdim}[A,C] \ .$$

We note that in view of (18.3) we have

$$\tau_{-1}([A,C],V) = 0 \quad \text{and} \quad T_0([A,C],V) = \mathfrak{Z}^*([A,C],V) \ .$$

$T_i([A,C],L,V)$ may be called the <u>osculating</u> i-<u>flat of</u> C in A at V relative to L. $T_i([A,C],V)$ may be called the <u>osculating</u> i-<u>flat of</u> C in A <u>at</u> V. From these phrases we may drop " in A ", when the reference to A is clear from the context.

By (18.3) and (20.1) to (20.12) we immediately get Lemmas (21.2), (21.3) and (21.4) as stated below.

(21.2) LEMMA. <u>Let</u> $V \in \mathfrak{Z}([A,C])$ <u>be such that</u> $[V/M(V) : H_0(A)] = 1$. <u>Let</u> $\mathbb{Z}$ <u>denote the set of all nonnegative integers together with</u> ∞. <u>Then for any</u> $L \in \mathfrak{M}_d^*(A)$, <u>upon letting</u>

$$Z(L) = \{\mathrm{ord}([A,C],I,V) : I \in \mathfrak{M}^*(A) \text{ with } I \subset L\}$$

$$Z'(L) = \{\mathrm{ord}([A,C],x,V) : x \in L \cap H_1(A)\}$$

$$Y'(L) = \{I \in \mathfrak{M}^*(A) : I \cap H_1(A) = \{x \in L \cap H_1(A) : \mathrm{ord}([A,C],x,V) \ge j\} \text{ for some } j \in \mathbb{Z}\}$$

$$Y(L) = \{I \in M^*(A) : I \cap H_1(A) = \{x \in L \cap H_1(A) : \mathrm{ord}([A,C],x,V) > \mathrm{ord}([A,C],I,V)\}\}$$

$$p = \mathrm{Emdim}[A,C,L] ;$$

<u>we have the following:</u>

(1) $Y(L) \cap \mathfrak{M}_i^*(A) = Y'(L) \cap \mathfrak{M}_i^*(A) = \{T_i([A,C],L,V)\}$ <u>for</u> $d \le i \le p$.

(2) $\mathrm{card}\, Y(L) \cap \mathfrak{M}_i^*(A) = \mathrm{card}\, Y'(L) \cap \mathfrak{M}_i^*(A) = 1$ <u>for</u> $d \le i \le p$.

(3) $Y(L) = Y'(L) = \{T_i([A,C],L,V) : d \le i \le p\}$.

(4) $Z(L) = Z'(L)$.

(5) $\mathrm{card}\, Y(L) = \mathrm{card}\, Y'(L) = \mathrm{card}\, Z'(L) = \mathrm{card}\, Z(L) = p - d + 1$.

(6) $\tau_d([A,C],L,V) < \tau_{d+1}([A,C],L,V) < \ldots < \tau_p([A,C],L,V) = \infty$ <u>is the unique ascending labelling of</u> $Z(L)$.

$$(7)\quad \begin{cases} J \to \{\mathrm{ord}([A,C],I,V):\ I \in \mathfrak{M}^*(A) \text{ with } I \subset J\} \\ \underline{\text{gives a surjective map of}} \\ \{J \in \mathfrak{M}^*(A):\ J \subset L\} \ \underline{\text{onto the set of all nonempty subsets of}} \\ Z(L)\ . \end{cases}$$

(21.3) LEMMA. Let $V \in \mathfrak{Z}([A,C])$ be such that $[V/M(V):\ H_0(A)] = 1$.

Given any $L \in \mathfrak{M}_d^*(A)$ and $J \in \mathfrak{M}_b^*(A)$ with $J \subset L$, let $p = \mathrm{Emdim}[A,C,L]$ and $n = \mathrm{Emdim}[A,C,J]$.

In view of the relations

$$(1')\quad \begin{cases} L = T_d([A,C],L,V) \supset T_{d+1}([A,C],L,V) \supset \ldots \supset T_p([A,C],L,V)\ , \\ T_s([A,C],L,V) \in \mathfrak{M}_s^*(A) \ \underline{\text{for}}\ d \le s \le p \ \underline{\text{and}} \\ L \supset J \in \mathfrak{M}_b^*(A)\ , \end{cases}$$

we clearly get a unique sequence

$$(2')\qquad d \le r(b) < r(b+1) < \ldots < r(\nu) = p$$

of integers such that

$$\begin{cases} \Delta(A,J,T_d([A,C],L,V)) = \Delta(A,J,T_s([A,C],L,V)) \ \underline{\text{for}}\ d \le s \le r(b)\ , \\ \underline{\text{and}} \\ \Delta(A,J,T_{(r-1)}([A,C],L,V)) \ne \Delta(A,J,T_s([A,C],L,V)) \\ = \Delta(A,J,T_{r(i)}([A,C],L,V)) \\ \qquad \text{for } b < i \le \nu \ \underline{\text{and}}\ r(i-1) < s \le r(i)\ . \end{cases}$$

Equivalently, in view of (1'), the sequence (2') can be characterized recursively thus:

$$r(b) = \max\{s \in \{d,d+1,\ldots,p\}:\ T_s([A,C],L,V) \supset J\}\ ,$$

$$\left.\begin{array}{l} r(i) = \max\{s \in \{r(i-1)+1, r(i-1)+2, \ldots, p\} \\ \quad T_s([A,C],L,V) \supset \Delta(A,J,T_{r(i-1)+1}([A,C],L,V))\} \end{array}\right\} \text{ for } b < i \le \nu ,$$

and

$$r(\nu) = p .$$

[Alternatively, we can instead define the complement of $\{r(b), r(b+1), \ldots, r(\nu)\}$ in $\{d, d+1, \ldots, p\}$ thus. In view of the relations (1') and the equation

$$J \cap T_p([A,C],L,V) \cap H_1(A) = T_n([A,C],J,V) \cap H_1(A) ,$$

it follows that: upon letting

$$G_i = \{s \in \{d, d+1, \ldots, p\}: J + T_s([A,C],L,V) \in \mathfrak{M}^*_{b+p-n-i}(A)\}$$

we have $G_i \neq \emptyset$, for $1 \le i \le b + p - n - d$, and then upon letting

$$g_i = \max\{s: s \in G_i\} \quad \text{for} \quad 1 \le i \le b + p - n - d$$

we have that

$$p > g_1 > g_2 > \ldots > g_{b+p-n-d} \ge d$$

and

$$\{d, d+1, \ldots, p\} \setminus \{r(b), r(b+1), \ldots, r(\nu)\} = \{g_1, g_2, \ldots, g_{b+p-n-d}\} .$$

Whence, in particular, we must have $\nu = n$. We also note that in case of $p - d > n - b$, upon letting

$$u = \begin{cases} \text{the unique integer with } d \le u < p \text{ such that} \\ T_{u+1}([A,C],L,V) \subset J + T_p([A,C],L,V) \\ \text{and} \\ T_u([A,C],L,V) \not\subset J + T_p([A,C],L,V) \end{cases}$$

and

$$u^* = \max\{s \in \{d,d+1,\dots,p-1\}: \Delta(A,J,T_s([A,C],L,V)) = \Delta(A,J,T_{s+1}([A,C],L,V))\} ,$$

we clearly have

$$u = u^* = g_1 .$$

It follows that, if $p - d = n - b + 1$, then: $b - 1 \leq u + n - p < n$,

$$r(i) = p - n + i - 1 \quad \text{for} \quad b \leq i \leq u + n - p ,$$

and

$$r(i) = p - n + i \quad \text{and} \quad T_{p-n+i}([A,C],L,V) \subset J + T_p([A,C],L,V) \quad \text{for} \quad u + n - p < i \leq n.]$$

<u>We claim that</u>, <u>with the sequence</u> (2') <u>as defined above</u>, <u>we have</u> $\nu = n$ <u>and</u>, <u>upon letting</u> $r(b-1) = d - 1$, <u>we have</u>

$$\left.\begin{array}{l} T_i([A,C],J,V) = \Delta(A,J,T_s([A,C],L,V)) \\ \text{for} \quad r(i-1) < s \leq r(i) \\ \tau_i([A,C],J,V) = \tau_{r(i)}([A,C],L,V) \end{array}\right\} \underline{\text{for}} \quad b \leq i \leq n .$$

<u>In particular</u>, <u>if</u> $p - d = n - b + 1$ <u>then</u>, <u>with</u> u <u>as defined in the above bracketed remark</u>, <u>we have</u>

$$\left.\begin{array}{l} T_i([A,C],J,V) = \Delta(A,J,T_{p-n+i-1}([A,C],L,V)) \\ \tau_i([A,C],J,V) = \tau_{p-n+i-1}([A,C],L,V) \end{array}\right\} \underline{\text{for}} \quad b \leq i \leq u + n - p$$

<u>and</u>

$$\left.\begin{array}{l} T_i([A,C],J,V) = \Delta(A,J,T_{p-n+i}([A,C],L,V)) \\ \tau_i([A,C],J,V) = \tau_{p-n+i}([A,C],L,V) \end{array}\right\} \underline{\text{for}} \quad u + n - p < i \leq n .$$

(We note that, in case $\Delta(A,C,L) = \{0\}$, we have $p = \text{Emdim } A = n$ and $T_i([A,C],J,V) = T_i([A,C],L,V)$, for $u < i \leq n$.)

Conversely, given any $L \in \mathfrak{M}_d^*(A)$ *and given any sequence* $d \leq r(b) < r(b+1) < \ldots < r(n) = p$ *of integers with* $\text{Emdim}[A,C,L] = p \leq n \leq \text{Emdim } A$, *there exists* $J \in \mathfrak{M}_b^*(A)$ *with* $J \subset L$ *and* $\text{Emdim}[A,C,J] = n$ *such that the given sequence is the sequence derived from the pair* J *and* L *as above.*

(21.4) LEMMA. *Given any* $\{0\} \neq N \in \mathfrak{M}_e^*(A)$ *and any* $J \in \mathfrak{M}_b^*(A)$ *with* $J \subset N$, *let* $q = \text{Emdim}[A,C,N]$ *and* $n = \text{Emdim}[A,C,J]$. *Let* $B = A^N$, $D = C^N$, *and* $K = J^N$.

(Note that now: $K \in \mathfrak{M}_{b-e-1}^*(B)$ and $\text{Emdim}[B,D,K] = n - e - 1$.)

Assume that $D \in \mathfrak{D}_1(B)$.

(Note that, if $H_0(A)$ is algebraically closed, then: $D \in \mathfrak{D}_1(B) \Leftrightarrow q - e \geq 2$.)

Let $f: A \to A/C$ *be the canonical epimorphism and let* $h: \mathfrak{R}([B,D]) \to \mathfrak{R}(f(B))$ *be the canonical isomorphism. Let* $V \in \mathfrak{Z}([A,C])$ *be given such that* $[V/M(V): H_0(A)] = 1$. *Let* $W = h^{-1}(V \cap \mathfrak{R}(f(B)))$.

(Note that now: $\text{ord}([A,C],N,V) \neq \infty$, $W \in \mathfrak{Z}([B,D])$, $h(W) = V \cap \mathfrak{R}(f(B))$, and $\text{ord}_V M(h(W))$ is a positive integer.)

Then we have

$$T_i([B,D],K,W) = T_{e+1+i}([A,C],J,V)^N \quad \textit{for} \quad b - e - 1 \leq i \leq n - e - 1,$$

and

$$\tau_i([B,D],K,W) = [\tau_{e+1+i}([A,C],J,V) - \text{ord}([A,C],N,V)]/\text{ord}_V M(h(W)) \quad \textit{for} \quad b - e - 1 \leq i < n - e - 1.$$

(In connection with (21.4), observe that given any $L \in \mathfrak{M}_d^*(A)$ with $J \subset L$, (21.3) gives recipes for obtaining the sequences

$T_i([A,C],J,V)_{b\le i\le n}$ and $\tau_i([A,C],J,V)_{b\le i\le n}$ from the sequences $T_i([A,C],L,V)_{d\le i\le p}$ and $\tau_i([A,C],L,V)_{d\le i\le p}$, where $p = \text{Emdim}[A,C,L]$. This is especially significant when $L = \Delta(A)$. We would also like to draw the reader's attention to the case of (21.3) when $L = \Delta(A)$ and the case of (21.4) when $J = N$.)

(21.5). REMARK. Several concrete corollaries can be drawn from Lemmas (21.3) and (21.4). As a sample, we write down the following information.

Assume that $H_0(A)$ is algebraically closed. Let $p = \text{Emdim}[A,C]$. Given $\{0\} \ne N \in \mathfrak{M}_e^*(A)$, let $q = \text{Emdim}[A,C,N]$, $B = A^N$, and $D = C^N$. Note that now $\text{Emdim}[B,D] = q - e - 1$. Let

$$\Omega = \{I \in \mathfrak{M}_{e+1}^*(A) : I \subset N\}$$

and recall that

$$J \to J^N \text{ gives a bijection of } \Omega \text{ onto } \mathfrak{M}_0^*(B) .$$

Assume that $D \in \mathfrak{Q}_1(B)$, i.e., equivalently: $q - e \ge 2$. Let $f: A \to A/C$ be the canonical epimorphism and let $h: \mathfrak{R}([B,D]) \to \mathfrak{R}(f(B))$ be the canonical isomorphism. For every $V \in \mathfrak{Z}([A,C])$ let $V^N = h^{-1}(V \cap f(B))$.

We observe that for every $V \in \mathfrak{Z}([A,C])$, upon letting

$$m(V) = \{s \in \{0,1,\dots,p\}: T_s([A,C],V) \not\supset N\} ,$$

in view of (21.3) and (21.4) (with $J = N$ and $L = \Delta(A)$), we get:

$$T_{e+1}([A,C],N,V) = \Delta(A,N,T_{m(V)}([A,C],V)) \in \Omega$$

$$T_{e+1}([A,C],N,V)^N = \mathfrak{Z}^*([B,D],V^N)$$

and

$$m(V) = 0 \Leftrightarrow T_{m(V)}([A,C],V) = \mathfrak{Z}^*([A,C],V) \Leftrightarrow V \notin \mathfrak{Z}([A,C],N) .$$

We also observe that for any $J \in \Omega$, upon letting

$$\Gamma(J) = \{V \in \mathfrak{Z}([A,C]) : T_{e+1}([A,C],N,V) = J\} ,$$

we have

$$\Gamma(J) = \{V \in \mathfrak{Z}([A,C]) : \operatorname{ord}([A,C],J,V) \neq \operatorname{ord}([A,C],N,V\}$$

$$= \{V \in \mathfrak{Z}([A,C],J) : \operatorname{ord}([A,C],J,V) > \operatorname{ord}([A,C],N,V)\}$$

$$= \text{a finite set}$$

and, in view of (21.4), we get

$$\{V^N : V \in \Gamma(J)\} = \mathfrak{Z}([B,D] , J^N) ,$$

Finally we observe that, if $\Delta(A,C) = \{0\}$ and $e = 0$, then $p = \operatorname{Emdim} A = q = (\operatorname{Emdim} B) + 1$ and for any $V \in \mathfrak{Z}[A,C])$, upon letting

$$u(v) = \begin{cases} \text{the unique integer with } -1 \le u(v) < p \text{ such that} \\ T_{u(V)+1}([A,C],V) \subset N \text{ and } T_{u(V)}([A,C],V) \not\subset N , \end{cases}$$

in view of (21.3) and (21.4), with $J = N$ and $L = \Delta(A)$, we get:

$$u(V) = \max\{s \in \{-1,0,\dots,p-1\}: \Delta(A,N,T_s([A,C],V))$$
$$= \Delta(A,N,T_{s+1}([A,C],V))\} ,$$

$$u(V) = 1 \Leftrightarrow N = \mathfrak{Z}^*([A,C],V)$$

$$\Rightarrow \left.\begin{cases} T_i([B,D],V^N) = T_{i+1}([A,C],V)^N \\ \tau_i([B,D],V^N) = \tau_{i+1}([A,C],V) - \tau_0([A,C],V) \end{cases}\right\} \text{for } -1 \le i \le q-1 ,$$

and

$$u(V) \neq -1 \Leftrightarrow N \neq \mathfrak{Z}^*([A,C],V)$$

$$\Rightarrow \begin{cases} \left.\begin{array}{l} T_i([B,D],V^N) = \Delta(A,N,T_i([A,C],V))^N \\ \\ \tau_i([B,D],V^N) = \tau_i([A,C],V) \end{array}\right\} \text{ for } -1 \le i < u(V) \\ \text{and} \\ \left.\begin{array}{l} T_i([B,D],V^N) = T_{i+1}([A,C],V)^N \\ \\ \tau_i([B,D],V^N) = \tau_{i+1}([A,C],V) \end{array}\right\} \text{ for } u(V) \le i \le q-1 . \end{cases}$$

(21.6) REMARK. Assume that $H_0(A)$ is algebraically closed. Let $\pi \in H_1^*(A)$ and $N \in \mathfrak{M}_e^*(A)$ be given with $\pi \subset N$. Let $q = \text{Emdim}[A,C,N]$. We note that then clearly:

$$(21.6.1) \begin{cases} \text{the projection of } C \text{ from } N \text{ in } A \text{ is } \pi\text{-integral} \\ \Leftrightarrow q-e \ge 2 \text{ and } \pi \subset T_{e+1}([A,C],N,V) \text{ for all } V \in \mathfrak{Z}([A,C],\pi) \\ \Leftrightarrow q-e \ge 2 \text{ and } \pi \subset T_{e+1}([A,C],N,V) \text{ for all } V \in \mathfrak{Z}([A,C],N). \end{cases}$$

We also note that,

$$(21.6.2) \text{ if } e = 0, \text{then: } \begin{cases} \text{the projection of } C \text{ from } N \text{ in } A \\ \text{is } \pi\text{-integral} \\ \\ \Leftrightarrow q \ge 2 \text{ and } \pi \subset T_1([A,C],V) \text{ for all} \\ V \in \mathfrak{Z}([A,C],N). \end{cases}$$

§22. Osculating flats and integral projections in an abstract curve.

Let R be a homogeneous domain with $\text{Dim } R = 1$.

(22.1) LEMMA-DEFINITION. Let $p = \text{Emdim } R$. Let $V \in \mathfrak{Z}(R)$ be such that $[V/M(V) : H_0(R)] = 1$. For any $L \in \mathfrak{M}_d^*(R)$, by (19.3) and (20.8) we get the following:

<u>There exists a unique sequence</u> $L = L_d \supset L_{d+1} \supset \ldots \supset L_p$ <u>such that</u> $L_i \in \mathfrak{M}_i^*(A)$ <u>and</u> $\operatorname{ord}(R, L_{i-1}, V) < \operatorname{ord}(R, L_i, V)$ <u>for</u> $d < i \le p$. <u>Moreover we have</u>

$$L_i \cap H_1(R) = \{x \in L_{i-1} \cap H_1(A) : \operatorname{ord}(R,x,V) > \operatorname{ord}(R, L_{i-1}, V)\}$$

$$\text{for} \quad d < i \le p ,$$

and clearly $L_p = \{0\}$ and $\operatorname{ord}(R, L_p, V) = \infty$.

<u>More generally we can say that given any</u> α <u>with</u> $d \le \alpha \le p$, <u>there exists a unique sequence</u> $L = L_{\alpha,d} \supset L_{\alpha,d+1} \supset \ldots \supset L_{\alpha,\alpha}$ <u>such that</u> $L_{\alpha,i} \in \mathfrak{M}_i^*(R)$ <u>and</u> $\operatorname{ord}(R, L_{\alpha,i-1}, V) < \operatorname{ord}(R, L_{\alpha,i}, V)$ <u>for</u> $d < i \le \alpha$. <u>Moreover we have</u> $L_{\alpha,i} = L_{\beta,i}$ <u>whenever</u> $d \le i \le \alpha \le \beta \le p$.

<u>Alternatively, the said sequence</u> $L = L_d \supset L_{d+1} \supset \ldots \supset L_p$ <u>can be completely characterized by saying the following</u>: <u>There exists a unique sequence</u> $L = L'_d \supset L'_{d+1} \supset \ldots \supset L'_p$ <u>such that</u> $L'_i \in \mathfrak{M}_i^*(R)$ <u>and</u> $\operatorname{ord}(R, L'_{i-1}, V) < \operatorname{ord}(R, L'_i, V)$ <u>for</u> $d < i \le p'$ <u>and</u> $\operatorname{ord}(R, L'_{p'}, V) = \infty$. <u>Moreover we have</u> $p' = p$ <u>and</u> $L'_i = L_i$ <u>for</u> $d \le i \le p$.

We <u>define</u>

$$\left.\begin{aligned} T_i(R,L,V) &= L_i \\ \tau_i(R,L,V) &= \operatorname{ord}(R, L_i, V) \end{aligned}\right\} \quad \text{for} \quad d \le i \le p$$

We note that, then $\tau_p(R,L,V) = \infty$, $\tau_i(R,L,V)$ is a nonnegative integer for $d \le i < p$, and in view of (19.3) we have

$$\tau_d(R,L,V) > 0 \Leftrightarrow L \subset \mathfrak{Z}^*(R,V) .$$

We also <u>define</u>

$$\left.\begin{aligned} T_i(R,V) &= T_i(R, \Delta(R), V) \\ \tau_i(R,V) &= \tau_i(R, \Delta(R), V) \end{aligned}\right\} \quad \text{for} \quad -1 \le i \le p$$

and we note that, in view of (19.3), we have,

$$\tau_{-1}(R,V) = 0 \quad \text{and} \quad T_0(R,V) = \mathfrak{Z}^*(R,V) .$$

$T_i(R,L,V)$ may be called the <u>osculating</u> i-<u>flat</u> in R <u>at</u> V <u>relative to</u> L. $T_i(R,V)$ may be called the <u>osculating</u> i-<u>flat</u> in R <u>at</u> V. From these phrases we may drop " in R " when the reference to R is clear from the context.

(22.2) to (22.6).

In view of (19.3), (20.1) to (20.11), and (20.13), we get assertions (22.2) to (22.6) which are respectively obtained from (21.2) to (21.6) by letting $S = R^N$ and replacing: A as well as [A,C] by R; C by $\{0\}$; B as well as [B,D] by S; D by $\{0\}$; f by the identity map $R \to R$; and h by the identity map $\mathfrak{R}(S) \to \mathfrak{R}(S)$.

(22.7) REMARK. Given any $C \in \mathfrak{D}_1(A)$, let $r = \text{Emdim}[A,C]$ and let $f: A \to A/C$ be the canonical epimorphism. Let $V \in \mathfrak{Z}([A,C])$ be such that $[V/M(V) : H_0(A)] = 1$.

We note that, then clearly: $\text{Emdim } f(A) = r$ and

$$\left.\begin{aligned} f(T_i([A,C],V)) &= T_i(F(A),V) , \\ T_i([A,C],V) &= f^{-1}(T_i(f(A),V)) , \\ \tau_i([A,C],V) &= \tau_i(f(A),V), \end{aligned}\right\} \quad \text{for} \quad -1 \le i \le r .$$

More generally, we note that, for any $L \in \mathfrak{M}_d^*(A)$, upon letting $p = \text{Emdim}[A,C,L]$, we clearly have: $f(L) \in \mathfrak{M}^*_{d+r-p}(f(A))$ and

$$\left.\begin{aligned} f(T_i([A,C],L,V)) &= T_{i+r-p}(f(A),f(L),V) , \\ T_i([A,C],L,V) &= L \cap f^{-1}(T_{i+r-p}(f(A),f(L),V)), \\ \tau_i([A,C],L,V) &= \tau_{i+r-p}(f(A),f(L),V) , \end{aligned}\right\} \quad \text{for} \quad d \le i \le p .$$

§23. Intersection multiplicity with an embedded curve.

Given $C \in \mathfrak{D}_1(A)$, let $f: A \to A/C$ be the canonical epimorphism. For any $I \in H(A)$ or $I \subset A$ and any $Q \in H(A)$ or $Q \subset A$ we <u>define</u>:

$$\mu([A,C],I,Q) = \sum_{V \in \mathfrak{Z}([A,C]),Q)} \operatorname{ord}([A,C],I,V)[V/M(V): \mathfrak{R}([A,\mathfrak{Z}^*([A,C],V)])] ,$$

$$\mu([A,C],I,\backslash Q) = \sum_{V \in \mathfrak{Z}([A,C],\backslash Q)} \operatorname{ord}([A,C],I,V)[V/M(V): \mathfrak{R}([A,\mathfrak{Z}^*([A,C],V)])]$$

$$\mu^*([A,C],I,Q) = \sum_{V \in \mathfrak{Z}([A,C],Q)} \operatorname{ord}([A,C],I,V)[V/M(V): H_0(A)] ,$$

and

$$\mu^*([A,C],I,\backslash Q) = \sum_{V \in \mathfrak{Z}([A,C],\backslash Q)} \operatorname{ord}([A,C],I,V)[V/M(V): H_0(A)] .$$

For any $I \in H(A)$ or $I \subset A$ we <u>define</u>:

$$\mu([A,C],I) = \mu([A,C],I,0) \quad \text{and} \quad \mu^*([A,C],I) = \mu^*([A,C],I,0) .$$

For any $Q \in H(A)$ or $Q \subset A$ we <u>define</u>:

$$\mu_{\mathfrak{C}}([A,C],Q) = \sum_{P \in \mathfrak{D}_0([A,C],Q)} \lambda_{\mathfrak{C}}(\mathfrak{R}([A,C],P)) ,$$

$$\mu_{\mathfrak{C}}([A,C],\backslash Q) = \sum_{P \in \mathfrak{D}_0([A,C],\backslash Q)} \lambda_{\mathfrak{C}}(\mathfrak{R}([A,C],P)) ,$$

$$\mu^*_{\mathfrak{C}}([A,C],Q) = \sum_{P \in \mathfrak{D}_0([A,C],Q)} \lambda_{\mathfrak{C}}^{f(k)}(\mathfrak{R}([A,C],P) \quad \text{where} \quad k = H_0(A) ,$$

$$\mu^*_{\mathfrak{C}}([A,C],\backslash Q) = \sum_{P \in \mathfrak{D}_0([A,C],\backslash Q)} \lambda_{\mathfrak{C}}^{f(k)}(\mathfrak{R}([A,C],P) \quad \text{where} \quad k = H_0(A) ,$$

$$\operatorname{Adj}([A,C],Q) = \{\Phi \in H^*(A): \mathfrak{R}([A,C],\Phi,P) \in \operatorname{adj}(\mathfrak{R}([A,C],P)) \text{ for all } P \in \mathfrak{D}_0([A,C],Q)\} ,$$

$$\operatorname{Adj}([A,C],\backslash Q) = \{\Phi \in H^*(A): \mathfrak{R}([A,C],\Phi,P) \in \operatorname{adj}(\mathfrak{R}([A,C],P)) \text{ for all } P \in \mathfrak{D}_0([A,C],\backslash Q)\} ,$$

$$\mathrm{Tradj}([A,C],Q) = \{\Phi \in H^*(A): \mathfrak{R}([A,C],\Phi,P) \in \mathrm{Tradj}(\mathfrak{R}([A,C],P)) \text{ for all } P \in \mathfrak{D}_0([A,C],Q)\}, \text{ and}$$

$$\mathrm{Tradj}([A,C],\backslash Q) = \{\Phi \in H^*(A): \mathfrak{R}([A,C],P) \in \mathrm{tradj}(\mathfrak{R}([A,C],P) \text{ for all } P \in \mathfrak{D}_0([A,C],\backslash Q)\} ,$$

By an <u>adjoint</u> of C in A <u>at</u> Q, we mean a member of Adj([A,C],Q). By an <u>adjoint</u> of C in A <u>outside</u> Q, we mean a member of Adj([A,C],\Q). By a <u>true adjoint</u> of C in A <u>at</u> Q, we mean a member of Tradj([A,C],Q). By a <u>true adjoint</u> of C in A <u>outside</u> Q, we mean a member of Tradj([A,C],\Q). From these phrases we may drop " in A ", when the reference to A is clear from the context.

Finally we <u>define</u>

$$\mu_{\mathfrak{C}}([A,C]) = \mu_{\mathfrak{C}}([A,C],0) \quad \text{and} \quad \mu^*_{\mathfrak{C}}([A,C]) = \mu^*_{\mathfrak{C}}([A,C],0) \quad \text{and}$$

$$\mathrm{Adj}([A,C]) = \mathrm{Adj}([A,C],0) \quad \text{and} \quad \mathrm{Tradj}([A,C]) = \mathrm{Tradj}([A,C],0).$$

By an <u>adjoint of</u> C in A, we mean a member of Adj([A,C]), and by a <u>true adjoint of</u> C in A, we mean a member of Tradj([A,C]). Again from these phrases we may drop " in A ", when the reference to A is clear from the context.

In view of (5.1), (5.6), (5,8), (5.10), (5.11), (17.4), (18.1), (18.2), (18.3), (18.4), (18.5), and (18.10), we clearly get (23.1) to (23.7):

(23.1) Let $k = H_0(A)$. Then for any $I \in H(A)$ or $I \subset A$ and any $P \in \mathfrak{D}_0([A,C])$, upon letting $I' = IA$ in case $I \in H(A)$ and $I' = (I \cap H(A))A$ in case $I \subset A$, we have:

$$\begin{aligned}\mu([A,C],I,P) &= \lambda(\mathfrak{R}([A,C],P), \mathfrak{R}([A,C],I,P)) \\ &= \lambda(\mathfrak{R}(A,P),\mathfrak{R}(A,C,P)],\mathfrak{R}(A,I,P)) \\ &= \begin{cases} 0 & , \text{ if } I' \not\subset P , \\ \text{a positive integer,} & \text{if } I' \subset P \text{ and } I' \not\subset C , \\ \infty & , \text{ if } I' \subset C; \text{ and} \end{cases}\end{aligned}$$

$$\begin{aligned}\mu^*([A,C],I,P) &= \mu([A,C],I,P)\mathrm{Deg}[A,P] \\ &= \lambda^{f(k)}(\mathfrak{R}([A,C],P),\mathfrak{R}([A,C],I,P)) \\ &= \lambda^k([\mathfrak{R}(A,P),\mathfrak{R}(A,C,P)],\mathfrak{R}(A,I,P))\end{aligned}$$

$$= \begin{cases} 0 & \text{, if } I' \not\subset P\,, \\ \text{a positive integer,} & \text{if } I' \subset P \text{ and } I' \not\subset C\,, \\ & \text{, if } I' \subset C\,. \end{cases}$$

For any $P \in \mathfrak{D}_0([A,C])$, we also have:

$$\begin{aligned} 1 \le \operatorname{card} \mathfrak{Z}([A,C],P) \le \mu([A,C],P) &= \lambda(\mathfrak{R}([A,C],P)) \\ &= \lambda([\mathfrak{R}(A,P),\mathfrak{R}(A,C,P)],\mathfrak{R}(A,P,P)) \\ &= \text{a positive integer}\,, \end{aligned}$$

$$\begin{aligned} \mu^*([A,C],P) &= \mu([A,C],P)\operatorname{Deg}[A,P] = \lambda^{f(k)}(\mathfrak{R}([A,C],P)) \\ &= \lambda^{k}([R(A,P),\mathfrak{R}(A,C,P)],\mathfrak{R}(A,P,P)) = \text{a positive integer}, \end{aligned}$$

$$\begin{aligned} \mu_{\mathfrak{C}}([A,C],P) = \lambda_{\mathfrak{C}}(\mathfrak{R}([A,C],P)) &= \lambda_{\mathfrak{C}}([\mathfrak{R}(A,P),\mathfrak{R}(A,C,P)]) \\ &= \text{a nonnegative integer}, \end{aligned}$$

$$\begin{aligned} \mu^*_{\mathfrak{C}}([A,C],P) = \mu_{\mathfrak{C}}([A,C],P)\operatorname{Deg}[A,P] &= \lambda^{f(k)}_{\mathfrak{C}}(\mathfrak{R}([A,C],P)) \\ &= \lambda^{k}_{\mathfrak{C}}([\mathfrak{R}(A,P),\mathfrak{R}(A,C,P)]) \\ &= \text{a nonnegative integer}. \end{aligned}$$

and

$$\begin{aligned} & \mu_{\mathfrak{C}}([A,C],P) = 0 \\ \Leftrightarrow\ & \mu^*_{\mathfrak{C}}([A,C],P) = 0 \\ \Leftrightarrow\ & \mathfrak{R}([A,C],P) \text{ is normal} \\ \Leftrightarrow\ & \mathfrak{R}([A,C],P) \text{ is regular} \\ \Leftrightarrow\ & \mu([A,C],I,P) = \operatorname{ord}_{\mathfrak{R}([A,C],P)} \mathfrak{R}([A,C],I,P) \\ & \qquad \text{for every } I \in H(A) \text{ or } I \subset A\,. \end{aligned}$$

(23.2) For any $I \in H(A)$ or $I \subset A$ and any $Q \in H(A)$ or $Q \subset A$, upon letting

$$I' = \begin{cases} IA + C & \text{, in case } I \in H(A) \\ (I \cap H(A))A + C, & \text{in case } I \subset A\,, \end{cases}$$

and $Q' = \begin{cases} QA + C, & \text{in case } Q \in H(A), \\ (Q \cap H(A))A + C, & \text{in case } Q \subset A, \end{cases}$

we have:

$$\mu([A,C],I,Q) = \mu([A,C],I',(\mathrm{rad}_A(I'+Q')) \cap (H_1(A)A))$$
$$= \sum_{P \in \mathfrak{D}_0([A,C],Q)} \mu([A,C],I,P)$$
$$= \text{a nonnegative integer or } \infty .$$

$$\mu^*([A,C],I,Q) = \mu([A,C],I',(\mathrm{rad}_A(I'+Q') \cap (H_1(A)A))$$
$$= \sum_{P \in \mathfrak{D}_0([A,C],Q)} \mu^*([A,C],I,P)$$
$$= \text{a nonnegative integer or } \infty ,$$

$$\mu([A,C],I,Q) = \infty \Leftrightarrow \mu^*([A,C],I,Q) = \infty \Leftrightarrow \mathfrak{D}([A,C],I) = \mathfrak{D}([A,C])$$
$$\text{and } \mathfrak{D}([A,C],Q) \neq \emptyset \Leftrightarrow I' = C \text{ and } H_1(A) \not\subset \mathrm{rad}_A Q'$$

$$\mu([A,C],I,Q) = 0 \Leftrightarrow \mu^*([A,C],I,Q) = 0 \Leftrightarrow \mathfrak{D}([A,C],I) \cap \mathfrak{D}([A,C],Q) =$$
$$\Leftrightarrow H_1(A) \subset \mathrm{rad}_A(I'+Q') ,$$

$$\mu([A,C],I) = \mu([A,C],I,Q) \Leftrightarrow \mu^*([A,C],I) = \mu^*([A,C],I,Q)$$
$$\Leftrightarrow \mathfrak{D}([A,C],I) \subset \mathfrak{D}([A,C],Q)$$
$$\Leftrightarrow Q' \cap (H_1(A)A) \subset \mathrm{rad}_A I' ,$$

$$\mu([A,C],I,\backslash Q) = \mu([A,C],I',\backslash(\mathrm{rad}_A Q') \cap (H_1(A)A))$$
$$= \sum_{P \in \mathfrak{D}_0([A,C],\backslash Q)} \mu([A,C],I,P)$$
$$= \text{a nonnegative integer or } \infty ,$$

$$\mu^*([A,C],I,\backslash Q) = \mu^*([A,C],I',\backslash(\mathrm{rad}_A Q') \cap (H_1(A)A))$$
$$= \sum_{P \in \mathfrak{D}_0([A,C],\backslash Q)} \mu^*([A,C],I,P)$$
$$= \text{a nonnegative integer or } \infty ,$$

$\mu([A,C],I,\backslash Q)$

$$= \infty \Leftrightarrow \mu^*([A,C],\backslash Q) = \infty \Leftrightarrow \mathfrak{D}([A,C],I) = \mathfrak{D}([A,C]) \text{ and } \mathfrak{D}([A,C],\backslash Q) \neq \emptyset$$
$$\Leftrightarrow I' = C \neq Q' ,$$

and

$$\mu([A,C],I\backslash Q) = 0 \Leftrightarrow \mu^*([A,C],I,\backslash Q) = 0 \Leftrightarrow \mathfrak{D}([A,C],I) \cap \mathfrak{D}([A,C],\backslash Q) = \emptyset$$
$$\Leftrightarrow \mathfrak{D}' \cap (H_1(A)A) \subset \mathrm{rad}_A I' .$$

Moreover, for any $J \in H(A)$ with $J \in I'$ or $J \subset A$ with $J \cap H(A) \subset I'$ we have:

$$\mu([A,C],J,Q) \geq \mu([A,C],I,Q) \text{ and }$$
$$\mu^*([A,C],J,Q) \geq \mu^*([A,C],I,Q) ,$$

$$\mu([A,C],J,Q) = \mu([A,C],I,Q) \Leftrightarrow \mu^*([A,C],J,Q) = \mu^*([A,C],I,Q)$$
$$\Leftrightarrow \begin{cases} \mathfrak{R}([A,C],J,P)\mathfrak{R}([A,C],P)^* \\ \quad = \mathfrak{R}([A,C],I,P)\mathfrak{R}([A,C],P)^* \\ \\ \text{for all } P \in \mathfrak{D}_0([A,C],Q) , \end{cases}$$

$$\mu([A,C],J,\backslash Q) \geq \mu([A,C],I,\backslash Q) \text{ and }$$
$$\mu^*([A,C],J,\backslash Q) \geq \mu^*([A,C],I,\backslash Q) ,$$

and

$$\mu([A,C],J,\backslash Q) = \mu([A,C],I,\backslash Q) \Leftrightarrow \mu^*([A,C],J,\backslash Q) = \mu^*([A,C],I,\backslash Q)$$
$$\Leftrightarrow \begin{cases} \mathfrak{R}([A,C],J,P)\mathfrak{R}([A,C],P)^* \\ = \mathfrak{R}([A,C],I,P)\mathfrak{R}([A,C],P)^* \\ \\ \text{for all } P \in \mathfrak{D}_0([A,C],P) , \end{cases}$$

where $\mathfrak{R}([A,C],P)^*$ is the integral closure of $\mathfrak{R}([A,C],P)$ in $\mathfrak{K}([A,C])$.

For any $Q \in H(A)$ or $Q \subset A$ (in view of (15.4) and (15.5) we also have

$$\mu_{\mathfrak{C}}([A,C],Q) = \sum_{P\in\mathfrak{D}_0([A,C],Q)} \mu_{\mathfrak{C}}([A,C],P) = \text{a nonnegative integer} ,$$

$$\mu^*_{\mathfrak{C}}([A,C],Q) = \sum_{P\in\mathfrak{D}_0([A,C],Q)} \mu^*_{\mathfrak{C}}([A,C],P) = \text{a nonnegative integer} ,$$

$$\mu_{\mathfrak{C}}([A,C],Q) = 0 \Leftrightarrow \mu^*_{\mathfrak{C}}([A,C],Q) = 0 \Leftrightarrow \mathfrak{R}([A,C],P) \text{ is regular for all } P \in \mathfrak{D}_0([A,C],Q) ,$$

$$\mu([A,C],\Phi,Q) \geq \mu_{\mathfrak{C}}([A,C],Q) \text{ and } \mu^*([A,C],\Phi,Q) \geq \mu^*_{\mathfrak{C}}([A,C],Q) \text{ for all } \Phi \in \operatorname{Adj}([A,C],Q) ,$$

$$\operatorname{Adj}([A,C],Q) = \bigcap_{P\in\mathfrak{D}_0([A,C],Q)} \operatorname{Adj}([A,C],P) ,$$

$$\begin{aligned}\operatorname{Tradj}([A,C],Q) &= \bigcap_{P\in\mathfrak{D}_0([A,C],Q)} \operatorname{Tradj}([A,C],P) \\ &= \{\Phi \in \operatorname{Adj}([A,C],Q) : \mu([A,C],\Phi,Q)\} \\ &= \{\Phi \in \operatorname{Adj}([A,C],Q) : \mu^*([A,C],\Phi,Q) = \mu^*_{\mathfrak{C}}([A,C],Q)\} ,\end{aligned}$$

$$\mu_{\mathfrak{C}}([A,C],\backslash Q) = \sum_{P\in\mathfrak{D}_0([A,C],\backslash Q)} \mu_{\mathfrak{C}}([A,C],P) = \text{a nonnegative integer} ,$$

$$\mu^*_{\mathfrak{C}}([A,C],\backslash Q) = \sum_{P\in\mathfrak{D}_0([A,C],\backslash Q)} \mu^*_{\mathfrak{C}}([A,C],P) = \text{a nonnegative integer} ,$$

$$\mu^*_{\mathfrak{C}}([A,C],\backslash Q) = 0 \Leftrightarrow \mu^*_{\mathfrak{C}}([A,C],\backslash Q) = 0 \Leftrightarrow \mathfrak{R}([A,C],P) \text{ is regular for all } P \in \mathfrak{D}_0([A,C],\backslash Q) ,$$

$$\mu([A,C],\Phi,\backslash Q) \geq \mu_{\mathfrak{C}}([A,C],\backslash Q) \text{ and } \mu^*([A,C],\Phi,\backslash Q) \geq \mu^*_{\mathfrak{C}}([A,C],\backslash Q) \text{ for all } \Phi \in \operatorname{Adj}([A,C],\backslash Q) ,$$

$$\operatorname{Adj}([A,C],\backslash Q) = \bigcap_{P\in\mathfrak{D}_0([A,C],\backslash Q)} \operatorname{Adj}([A,C],P) ,$$

and

$$\operatorname{Tradj}([A,C],\backslash Q) = \bigcap_{P\in\mathfrak{D}_0([A,C],\backslash Q)} \operatorname{Tradj}([A,C],P)$$

$$= \{\Phi \in \mathrm{Adj}([A,C],\backslash Q) : \mu([A,C],\Phi,\backslash Q) = \mu_{\mathfrak{C}}([A,C],\backslash Q)\}$$

$$= \{\Phi \in \mathrm{Adj}([A,C],\backslash Q) : \mu^*([A,C],\Phi,\backslash Q) = \mu^*_{\mathfrak{C}}([A,C],\backslash Q)\} .$$

(23.3) For any $I \in H(A)$ or $I \subset A$, upon letting

$$I' = \begin{cases} IA + C & \text{, in case } I \in H(A) , \\ (I \cap H(A))A + C, & \text{in case } I \subset A , \end{cases}$$

we have:

$$\mu([A,C],I)$$

$$= \mu([A,C],I') = \sum_{V \in \mathfrak{Z}([A,C])} \mathrm{ord}([A,C],I,V)[V/M(V) : \mathfrak{R}([A,\mathfrak{Z}^*([A,C],V)])]$$

$$= \sum_{P \in \mathfrak{D}_0([A,C],I,P)} \mu([A,C],I,P)$$

$$= \text{a nonnegative integer or } \infty .$$

$$\mu^*([A,C],I) = \mu^*([A,C],I') = \sum_{V \in \mathfrak{Z}([A,C])} \mathrm{ord}([A,C],I,V)[V/M(V) : H_0(A)]$$

$$= \sum_{P \in \mathfrak{D}_0([A,C])} \mu^*([A,C],I,P)$$

$$= \text{a nonnegative integer or } \infty ,$$

$$\mu([A,C],I) = \infty \Leftrightarrow \mu^*([A,C],I) = \infty \Leftrightarrow \mathfrak{D}([A,C],I) = \mathfrak{D}([A,C]) \Leftrightarrow I' = C ,$$

$$\mu([A,C],I) = 0 \Leftrightarrow \mu^*([A,C],I) = 0 \Leftrightarrow \mathfrak{D}([A,C],I) = \emptyset \Leftrightarrow H_1(A) \subset \mathrm{rad}_A I' ,$$

and for any $J \in H(A)$, with $J \in I'$ or $J \subset A$ with $J \cap H(A) \subset I'$, we have

$$\mu([A,C],J) \geq \mu([A,C],I) \quad \text{and} \quad \mu^*([A,C],J) \geq \mu^*([A,C],I)$$

and

$$\mu([A,C],J) = \mu([A,C],I) \Leftrightarrow \mu^*([A,C],J) = \mu^*([A,C],I)$$

$$\Leftrightarrow \begin{cases} \mathfrak{R}([A,C],J,P)\mathfrak{R}([A,C],P)^* = \\ \mathfrak{R}([A,C],I,P)\mathfrak{R}([A,C],P)^* \\ \text{for all } P \in \mathfrak{D}_0([A,C]) \end{cases}$$

where $\mathfrak{R}([A,C],P)^*$ is the integral closure of $\mathfrak{R}([A,C],P)$ in $\mathfrak{K}([A,C])$.

Finally (in view of (15.4) and (15.5) we also have:

$$\mu_{\mathfrak{C}}([A,C]) = \sum_{P\in\mathfrak{D}_0([A,C])} \mu_{\mathfrak{C}}([A,C],P) = \text{a nonnegative integer },$$

$$\mu^*_{\mathfrak{C}}([A,C]) = \sum_{P\in\mathfrak{D}_0([A,C])} \mu^*_{\mathfrak{C}}([A,C],P) = \text{a nonnegative integer },$$

$$\mu_{\mathfrak{C}}([A,C]) = 0 \Leftrightarrow \mu^*_{\mathfrak{C}}([A,C]) = 0 \Leftrightarrow \mathfrak{R}([A,C],P) \text{ is regular for all } P \in \mathfrak{D}_0([A,C]) ,$$

$$\mu([A,C],\Phi) \geq \mu_{\mathfrak{C}}([A,C]) \text{ and } \mu^*([A,C],\Phi) \geq \mu^*_{\mathfrak{C}}([A,C]) \text{ for all } \Phi \in \mathrm{Adj}([A,C]) ,$$

$$\mathrm{Adj}([A,C]) = \bigcap_{P\in\mathfrak{D}_0([A,C])} \mathrm{Adj}([A,C],P) ,$$

and

$$\begin{aligned} \mathrm{Tradj}([A,C]) &= \bigcap_{P\in\mathfrak{D}_0([A,C])} \mathrm{Tradj}([A,C],P) \\ &= \{\Phi \in \mathrm{Adj}([A,C]) : \mu([A,C],\Phi) = \mu_{\mathfrak{C}}([A,C])\} \\ &= \{\Phi \in \mathrm{Adj}([A,C]) : \mu^*([A,C],\Phi) = \mu^*_{\mathfrak{C}}([A,C])\} . \end{aligned}$$

(23.4). If $H_0(A)$ is algebraically closed, then:

$$\left.\begin{aligned} \mu^*([A,C],I,Q) &= \mu([A,C],I,Q) \\ \mu^*([A,C],I,\backslash Q) &= \mu([A,C],I,\backslash Q) \end{aligned}\right\} \begin{aligned} &\text{for any } I \in H(A) \text{ or } I \subset A \\ &\text{and any } Q \in H(A) \text{ or } Q \subset A ; \end{aligned}$$

$$\mu^*([A,C],I) = \mu_{\mathfrak{C}}([A,C],I) \quad \text{for any } I \in H(A) \text{ or } I \subset A ,$$

$$\left.\begin{aligned} \mu^*_{\mathfrak{C}}([A,C],Q) &= \mu_{\mathfrak{C}}([A,C],Q) , \\ \mu^*_{\mathfrak{C}}([A,C],\backslash Q) &= \mu_{\mathfrak{C}}([A,C],\backslash Q) , \end{aligned}\right\} \text{ for any } Q \in H(A) \text{ or } Q \subset A ,$$

and

$$\mu^*_{\mathfrak{C}}([A,C]) = \mu_{\mathfrak{C}}([A,C]) .$$

(23.5). We have

$$\left.\begin{aligned} \mu([A,C],xy,Q) &= \mu([A,C],x,Q) + \mu([A,C],y,Q) , \\ \mu^*_{\mathfrak{C}}([A,C],xy,Q) &= \mu^*([A,C],x,Q) + \mu^*([A,C],y,Q) , \\ \mu([A,C],xy,\backslash Q) &= \mu([A,C],x,\backslash Q) + \mu([A,C],y,\backslash Q) , \\ \mu^*([A,C],xy,\backslash Q) &= \mu^*([A,C],x,\backslash Q) + \mu^*([A,C],y,\backslash Q), \end{aligned}\right\} \begin{array}{l} \text{for any } x \text{ and } y \\ \text{in } H(A)\backslash C \\ \text{and any} \\ Q \in H(A) \text{ or } Q \subset A \end{array}$$

and

$$\left.\begin{aligned} \mu([A,C],xy) &= \mu([A,C],x) + \mu([A,C],y) , \\ \mu^*([A,C],xy) &= \mu^*([A,C],x) + \mu^*([A,C],y) , \end{aligned}\right\} \begin{array}{l} \text{for any } x \text{ and } y \\ \text{in } H(A)\backslash C . \end{array}$$

(23.6) Let $P \in \mathfrak{D}_0([A,C])$, and let $I \in H(A)$ or $I \subset A$. Either assume that $\mathfrak{R}([A,C],I,P)\mathfrak{R}([A,C],P)$ is principal (note that this is certainly so if $I \in H(A)$ or $I \in H^*(A)$); or assume that: $\mu([A,C],P) = 1$. Then,

$$\begin{aligned} \mu([A,C],I,P) &= [\mathfrak{R}([A,C],P)/\mathfrak{R}([A,C],I,P)\mathfrak{R}([A,C],P) : \mathfrak{R}([A,C],P)] \\ &= [\mathfrak{R}(A,P)/(\mathfrak{R}(A,I,P)\mathfrak{R}(A,P) + \mathfrak{R}(A,C,P)) : \mathfrak{R}(A,P)] \end{aligned}$$

and

$$\begin{aligned} \mu^*([A,C],I,P) &= [\mathfrak{R}([A,C],P)/\mathfrak{R}([A,C],I,P)\mathfrak{R}([A,C],P) : H_0(A)] \\ &= [\mathfrak{R}(A,P)/(\mathfrak{R}(A,I,P)\mathfrak{R}(A,P) + \mathfrak{R}(A,C,P)) : H_0(A)] \end{aligned}$$

(23.7). If $\operatorname{Emdim}[A,C] = 1$, then $\mu([A,C],P) = 1$ for all $P \in \mathfrak{D}_0([A,C]$.

Next we claim that:

(23.8). LEMMA. <u>For any</u> $x \in H_m(A)\backslash C$ <u>and</u> $y \in H_n(A)\backslash C$ <u>we have</u>

$$n[\mu^*([A,C],x)] = m[\mu^*([A,C],y)] .$$

PROOF. Namely

$$n[\mu^*([A,C],x)] - m[\mu^*([A,C],y)]$$

$$= \sum_{V\in\mathfrak{Z}([A,C])} [n(\operatorname{ord}([A,C],x,V)) - m(\operatorname{ord}([A,C],y,V))][V/M(V) : H_0(A)]$$

$$= \sum_{V\in\mathfrak{Z}([A,C])} [\operatorname{ord}_V f(x^n)/f(y^m)][V/M(V) : H_0(A)] \qquad \text{by (18.10)}$$

$$= 0 \qquad \text{by (4.1), since } 0 \neq f(x^n)/f(y^m) \in \mathfrak{R}([A,C]) .$$

(23.9) DEFINITION-BEZOUT'S LITTLE THEOREM. In view of (23.8), there exists a unique positive integer, <u>to be denoted by</u> Deg[A,C], such that

$$\mu^*([A,C],\Phi) = (\operatorname{Deg}[A,C])(\operatorname{Deg}_A\Phi) \quad \underline{\text{for all}} \quad \Phi \in H^*(A) \quad \underline{\text{with}} \quad \Phi \not\subset C .$$

(23.10) REMARK. In view of the above formula, we can clearly characterize Deg[A,C] also in the following way. Let any $Q \in H(A)\backslash C$ or $Q \subset A$ with $Q \cap H(A) \not\subset C$ be given. Then

$$\begin{aligned}\operatorname{Deg}[A,C] &= \max\{\mu^*([A,C],\Phi,\backslash Q)/\deg_A\Phi : \Phi \in H^*(A) \text{ with } A \neq \Phi \not\subset C\}\\ &= \max\{\mu^*([A,C],\Phi,\backslash Q)/n : \Phi \in H_n^*(A) \text{ with } \Phi \not\subset C\}\\ &\qquad \text{for all large enough } n.\end{aligned}$$

Moreover, if $H_0(A)$ is algebraically closed, then

$$\operatorname{Deg}[A,C] = \max\{\mu^*([A,C],\Phi,\backslash Q)/n : \Phi \in H_n^*(A) \text{ with } \Phi \not\subset C\}$$

$$\text{for all } n > 0 .$$

(23.11) LEMMA. <u>Assume that</u>

$$\mathrm{card}\{P \in \mathfrak{D}_0([A,C]) : \mathrm{Deg}[A,P] = 1\} \geq 1 + \mathrm{Deg}[A,C] \ .$$

(Note that this assumption is automatically satisfied if $H_0(A)$ is algebraically closed; namely, $\mathfrak{D}_0([A,C])$ is always an infinite set, and, if $H_0(A)$ is algebraically closed, then $\mathrm{Deg}[A,P] = 1$ for all $P \in \mathfrak{D}_0([A,C])$.)

<u>Then</u> $\mathrm{Emdim}[A,C] \leq \mathrm{Deg}[A,C]$.

PROOF. Let $d = \mathrm{Deg}[A,C]$, and let

(1) $$s = [H_1(A) : H_0(A)] \ .$$

By assumption there exist pairwise distinct members $L_0, L_1, \ldots, L_d$ of $\mathfrak{M}(A)$ such that

(2) $$[L_i : H_0(A)] = s - 1 \ , \text{ for } 0 \leq i \leq d \ ,$$

and

(3) $$L_i A \in \mathfrak{D}_0([A,C]), \text{ for } 0 \leq i \leq d \ .$$

Suppose, if possible, that $L_0 \cap L_1 \cap \ldots \cap L_d \not\subset C$; then there exists $\Phi \in H_1^*(A)$ such that $\Phi \not\subset C$ and $\Phi \subset L_i A$ for $0 \leq i \leq d$. Now $L_0A, L_1A, \ldots, L_dA$ are pairwise distinct members of $\mathfrak{D}_0([A,C]) \cap \mathfrak{D}_0([A,\Phi])$ and hence, in view of (23.1) and (23.3), we see that $\mu^*([A,C],\Phi) \geq d + 1$. However, since $\Phi \in H_1^*(A)$ and $\Phi \not\subset C$, by (23.9) we get $\mu^*([A,C],\Phi) = d$, which is a contradiction. Therefore $L_0 \cap L_1 \cap \ldots \cap L_d \subset C$, and hence in view of (3) we have

(4) $$L_0 \cap L_1 \cap \ldots \cap L_d = C \cap H_1(A) \ .$$

By (1), (2) and (4) we see that

(5) $$[C \cap H_1(A) : H_0(A)] \geq s - d - 1 \ .$$

By (1) and (5) we get $\operatorname{Emdim}[A,C] \le d = \operatorname{Deg}[A,C]$.

(23.12) LEMMA. $\operatorname{Deg}[A,C] = 1 \Leftrightarrow \operatorname{Emdim}[A,C] = 1$.

PROOF. First suppose that $\operatorname{Emdim}[A,C] = 1$. Then there exist $P \in \mathfrak{D}_0([A,C])$ with $\operatorname{Deg}[A,P] = 1$, and there exists $\Phi \in H_1^*(A)$ with $\Phi \subset P$ and $\Phi \not\subset C$. Clearly $\Phi + C = P$ and hence, in view of (23.9), (23.3), (23.1) and (23.6) we get

$$\begin{aligned}\operatorname{Deg}[A,C] = \mu^*([A,C],\Phi) = \mu([A,C],\Phi,P) &= [\mathfrak{R}(A,P)/\mathfrak{R}(A,\Phi+C,P) : \mathfrak{R}(A,P)] \\ &= [\mathfrak{R}(A,P)/M(\mathfrak{R}(A,P)) : \mathfrak{R}(A,P)] \\ &= 1 .\end{aligned}$$

Conversely suppose that $\operatorname{Deg}[A,C] = 1$. Let $e = 1 + \operatorname{Emdim}[A,C]$. Then $e \ge 2$ and clearly there exist $\Phi_1,\Phi_2,\ldots,\Phi_e$ in $H_1^*(A)$ such that $\Phi_i \not\subset C$ for $1 \le i \le e$ and

(1) $$\Delta(A,C) + \Phi_1 + \Phi_2 + \ldots + \Phi_e = \Delta(A) .$$

Since $\operatorname{Deg}[A,C] = 1$, in view of (23.1), (23.3) and (23.9), we see tha for $1 \le i \le e$, we have $\operatorname{card} \mathfrak{D}_0([A,C],\Phi_i) = 1$ and, upon letting $\mathfrak{D}_0([A,C],\Phi_i) = \{P_i\}$, we have $\operatorname{Deg}[A,P_i] = 1$. By (1) we get

$$\mathfrak{D}_0([A,C],\Phi_1) \cap \ldots \cap \mathfrak{D}_0([A,C],\Phi_e) = \emptyset$$

and hence

$$\operatorname{card}\{P \in \mathfrak{D}_0([A,C]) : \operatorname{Deg}[A,P] = 1\} \ge \operatorname{card}\{P_1,P_2,\ldots,P_e\} \ge 2 .$$

Therefore by (23.11) we conclude that $\operatorname{Emdim}[A,C] = 1$.

(23.13) REMARK. As a consequence of (23.12), we see that, if $\operatorname{Deg}[A,C] = 1$, then $\mu([A,C],P) = 1$ for all $P \in \mathfrak{D}_0([A,C])$.

As a reformulation of (23.12) we have:

$$\mathfrak{D}_1^1(A) = \{E \in \mathfrak{D}_1(A) : \operatorname{Deg}[A,E] = 1\} ;$$

this motivates the notation $\mathfrak{D}_1^1(A)$.

As a consequence of (23.12) we also see that:

if Emdim A = Dim A, then $\mathfrak{M}_1^*(A) = \{E \in \mathfrak{D}_1(A): \mathrm{Deg}[A,E] = 1\}$.

The above observations may henceforth be used tacitly.

(23.14) PROJECTION FORMULA. <u>Let any</u> $N \in \mathfrak{M}^*(A)$ <u>and</u> $J \in \mathfrak{M}^*(A)$ <u>be given such that</u> $J \subset N$ <u>and</u> $J \subset C$. <u>Assume that</u> $C^N \in \mathfrak{D}_1(A^N)$.

(Note that, if $H_0(A)$ is algebraically closed, then:

$C^N \in \mathfrak{D}_1(A^N) \Leftrightarrow \mathrm{Emdim}[A,C,N] - \mathrm{Emdim}[A,N] \geq 2$.)

<u>Then we have</u>

$$\mu^*([A,C],J) - \mu^*([A,C],N) = [\mathfrak{K}([A,C]): \mathfrak{K}([A^N,C^N])]\ \mu^*([A^N,C^N],J^N) .$$

PROOF. Let $B = A^N$, $D = C^N$, and $K = J^N$. Let $h: \mathfrak{K}([B,D]) \to \mathfrak{K}(f(B))$ be the canonical isomorphism. For every $W \in \mathfrak{Z}([B,D])$ let

$$G(W) = \{V \in \mathfrak{Z}([A,C]): V \cap \mathfrak{K}(f(B)) = h(W)\} .$$

For every $W \in \mathfrak{Z}([B,D])$ and $V \in G(W)$, via the canonical isomorphism $h: \mathfrak{K}([B,D]) \to \mathfrak{K}(f(B))$ and the canonical epimorphisms $W \to W/M(W)$ and $V \to V/M(V)$, $V/M(V)$ becomes a $(W/M(W))$-vector-space and, upon letting

$$g(V) = [V/M(V): W/M(W)] \quad \text{and} \quad p(W) = [W/M(W): H_0(B)] ,$$

we have

(1) $$[V/M(V): H_0(A)] = g(V)p(W) .$$

Now

$$\mu^*([A,C],J) - \mu^*([A,C],N)$$

$$= \sum_{V \in \mathfrak{Z}([A,C])} [\mathrm{ord}([A,C],J,V) - \mathrm{ord}([A,C],N,V)][V/M(V): H_0(A)]$$

$$= \sum_{W\in \mathfrak{Z}([B,D])} \sum_{V\in G(W)} [\mathrm{ord}([A,C],J,V) - \mathrm{ord}([A,C],N,V)]g(v)p(W) \quad \text{by (1}$$

$$= \sum_{W\in \mathfrak{Z}([B,D])} \sum_{V\in G(W)} [\mathrm{ord}_V M(h(W))][\mathrm{ord}([B,D],K,W]g(V)p(W) \quad \text{by (18.12)}$$

$$= [\mathfrak{R}([A,C]): \mathfrak{R}([B,D])] \sum_{W\in \mathfrak{Z}([B,D])} [\mathrm{ord}([B,D],K,W)]p(W) \quad \text{by (4.3)}$$

$$= [\mathfrak{R}([A,C]): \mathfrak{R}([B,D])]\, \mu^*([B,D],K) .$$

$$= [\mathfrak{R}([A,C]) : \mathfrak{R}([A^N,C^N])]\mu^*([A^N,C^N],J^N).$$

(23.15) SPECIAL PROJECTION FORMULA. Let any $\{0\} \neq N \in \mathfrak{M}^*(A)$ be given. Assume that $C^N \in \mathfrak{D}_1(A^N)$.

(Note that, if $H_0(A)$ is algebraically closed, then:

$$C^N \in \mathfrak{D}_1(A^N) \Leftrightarrow \mathrm{Emdim}[A,C,N] - \mathrm{Emdim}[A,N] \geq 2.)$$

Then we have

(23.15.1) $\mathrm{Deg}[A,C] - \mu^*([A,C],N)$
$$= [\mathfrak{R}([A,C]) : \mathfrak{R}([A^N,C^N])]\mathrm{Deg}[A^N,C^N]$$

and

$$(23.15.2) \quad \begin{cases} \mathrm{Deg}[A,C] - \mu^*([A,C],N) = [\mathfrak{R}([A,C]) : \mathfrak{R}([A^N,C^N])] \\ \Leftrightarrow \mathrm{Deg}[A^N,C^N] = 1 \\ \Leftrightarrow \mathrm{Emdim}[A^N,C^N] = 1 \\ \Leftrightarrow \mathrm{Emdim}[A,C,N] - \mathrm{Emdim}[A,N] = 2 . \end{cases}$$

PROOF. In view of (23.9), (23.15.1) follows from (23.14) by taking $J = xA$ with $x \in (N \cap H_1(A))\backslash C$. Now (23.15.2) follows from (23.15.1), (23.12) and (14.3).

(23.16) REMARK. Assume that $H_0(A)$ is algebraically closed. Let

$$N^* = \{N \in \mathfrak{M}^*(A): \mathrm{Emdim}[A,C,N] - \mathrm{Emdim}[A,N] = 2\}$$

i.e.,

$$N^* = \{0 \neq N \in \mathfrak{M}^*(A) : \operatorname{Emdim}[A^N, C^N] = 1\} .$$

Let

$$N' = \{N \in N^* : \mathfrak{D}_0([A,C]) \cap \mathfrak{D}_0([A,N]) = \phi\} .$$

Then clearly $N' \neq \phi$ and by (23.15) we see that

$$\begin{aligned} \operatorname{Deg}[A,C] &= \max\{[\mathfrak{K}([A,C]) : \mathfrak{K}([A^N,C^N])] : N \in N^*\} \\ &= [\mathfrak{K}([A,C]) : \mathfrak{K}([A^N,C^N])] \quad \text{for all} \quad N \in N' . \end{aligned}$$

(23.17) LEMMA. _Let_ $\{0\} \neq N \in \mathfrak{M}_1^*(A)$. _Assume that_ $C^N \in \mathfrak{D}_1(A)$. (Note that, if $H_0(A)$ is algebraically closed, then:

$$C^N \in \mathfrak{D}_1(A^N) \Leftrightarrow \operatorname{Emdim}[A,C,N] \geq 3.)$$

Let

$$\Omega = \{P \in \mathfrak{M}_0^*(A) : N \subset P \quad \text{and} \quad [\mathfrak{K}([A,C]) : \mathfrak{K}([A^P,C^P])] \neq 1\}$$

and

$$\Omega' = \begin{cases} \text{the set of all subfields of } \mathfrak{K}([A,C]) \text{ which contain} \\ \mathfrak{K}(f(A^N)) \text{ and which are different from } \mathfrak{K}([A,C]) \end{cases}$$

Then we have the following:

(23.17.1) $\operatorname{card} \Omega \leq \operatorname{card} \Omega'$.

(23.17.2) _If_ $\operatorname{Deg}[A,C] - \mu^*([A,C],N) \leq 3$, _then_ $\operatorname{card} \Omega \leq 1$.

(Although we shall not use this remark in this book, we note that by a well-known fact from algebra (namely, if K^* is an algebraic overfield of a field K', then: K^* has a primitive element over $K' \Leftrightarrow$ there are only a finite number of subfields of K^* containing K') we know that Ω' is finite $\Leftrightarrow \mathfrak{K}([A,C])$ has a primitive element over $\mathfrak{K}(f(A^N))$. Hence by (23.17.1) we see that if $\mathfrak{K}([A,C])$ has a primitive element over $\mathfrak{K}(f(A^N))$ then Ω is finite. So, in particular, if $H_0(A)$ has zero characteristic, or, more generally if $\mathfrak{K}([A,C])$ is separable over $\mathfrak{K}(f(A^N))$, then Ω is finite.)

PROOF. Clearly

$$(1') \qquad \mathfrak{K}(f(A^N)) \subset \mathfrak{K}(f(A^P)) \subset \mathfrak{K}([A,C]) \text{ for all } P \in \mathfrak{M}_0^*(A) \text{ with } N \subset P$$

and hence to prove (23.17.1) it suffices to show that

$$(*) \qquad \text{card}\{P \in \mathfrak{M}_0^*(A) : \mathfrak{K}(f(A^P)) \subset K\} \leq 1 \text{ for every } K \in \Omega' .$$

In view of (1'), (*) is equivalent to:

$$(**) \quad \begin{cases} \mathfrak{K}(f(A^N))(\mathfrak{K}(f(A^P)), \mathfrak{K}(f(A^Q))) = \mathfrak{K}([A,C]) , \\ \text{for every } P \neq Q \text{ in } \mathfrak{M}_0^*(A) \text{ with } N \subset P \text{ and } N \subset Q . \end{cases}$$

To prove (**) let any $P \neq Q$ in $\mathfrak{M}_0^*(A)$ with $N \subset P$ and $N \subset Q$ be given. Then we can take elements X, Y, Z in $H_1(A)$ such that

$$N + XA = P, \quad N + YA = Q, \quad \text{and} \quad Z \in N \setminus C .$$

Now clearly

$$\mathfrak{K}(f(A^P)) = \mathfrak{K}(f(A^N))(f(X)/f(Z)) \text{ and } \mathfrak{K}(f(A^Q)) = \mathfrak{K}(f(A^N))(f(Y)/f(Z)) ;$$

also clearly

$$(N \cap H_1(A)) + XH_0(A) + YH_0(A) = H_1(A) ;$$

and hence

$$\mathfrak{K}([A,C]) = \mathfrak{K}(f(A^N))(f(X)/f(Z), f(Y)/f(Z)) ;$$

consequently

$$\mathfrak{K}(f(A^N))(\mathfrak{K}(f(A^P)), \mathfrak{K}(f(A^Q))) = \mathfrak{K}([A,C])$$

and this proves (**).

By (23.15) we have,

$$(\mathfrak{K}([A,C]) : \mathfrak{K}(f(A^N))] \le \mathrm{Deg}[A,C] - \mu^*([A,C],N)$$

and hence

$$\mathrm{Deg}[A,C] - \mu^*([A,C],N) \le 3$$
$$\Rightarrow [\mathfrak{K}([A,C]) : \mathfrak{K}(f(A^N))] = 1 \text{ or } 2 \text{ or } 3$$
$$\Rightarrow \mathrm{card}\ \Omega' \le 1 ;$$

therefore (23.17.2) follows from (23.17.1).

(23.18) DEFINITION. Let $P \in \mathfrak{Q}_0([A,C])$. Assume that

(*) $$[V/M(V) : H_0(A)] = 1 \quad \text{for all} \quad V \in \mathfrak{Z}([A,C],P) .$$

(Note that condition (*) is automatically satisfied if, either: $\mu^*([A,C],P) = 1$, or: $H_0(A)$ is algebraically closed. Also note that: (*) $\Rightarrow P \in \mathfrak{M}_0^*(A)$, $\mathrm{Deg}[A,P] = 1$, and $\mu^*([A,C],P) = \mu([A,C],P)$.)

We <u>define</u>:

$$T_1^*([A,C],P) = \{L \in \mathfrak{M}_1^*(A) : L \subset P \text{ and } \mu^*([A,C],L,P) \neq \mu^*([A,C],P)\}.$$

We note that, then clearly,

(23.18.1) $$\begin{cases} T_1^*([A,C],P) = \{L \in \mathfrak{M}_1^*(A) : \mu^*([A,C],L,P) > \mu^*([A,C],P)\} \\ \qquad = \{L \in \mathfrak{M}_1^*(A) : L = T_1([A,C],V) \text{ for some } V \in \mathfrak{Z}([A,C],P)\} \end{cases}$$

and hence

(23.18.2) $$1 \le \mathrm{card}\ T_1^*([A,C],P) \le \mathrm{card}\ \mathfrak{Z}([A,C],P) < \infty$$

and so in particular

(23.18.3) $\mu^*([A,C],P) = 1 \Rightarrow \text{card}\, T_1^*([A,C],P) = 1$.

By a tangent 1-flat of C in A at P we mean a member of $T_1^*([A,C],P)$: From this phrase we may drop " in A ", when the reference to A is clear from the context.

Finally we observe that by (23.2) and (23.6) we get (23.19) and (23.20):

(23.19) COMMUTING LEMMA. Let $D \in \mathfrak{D}_1(A)$ and $Q \in H(A)$ or $Q \subset A$ be such that $\mu([A,C],P) = 1 = \mu([A,D],P)$ for all $P \in \mathfrak{D}_0([A,C]) \cap \mathfrak{D}_0([A,D]) \cap \mathfrak{D}_0([A,Q])$. Then

$$\mu([A,C],D,Q) = \mu([A,C],C,Q) \text{ and } \mu^*([A,C],D,Q) = \mu^*([A,D],C,Q)$$

(23.20) COMMUTING LEMMA. If $\text{Emdim}[A,C] = 2$, then for any $D \in \mathfrak{D}_1(A)$ and any $Q \in H(A)$ or $Q \subset A$ we have

$$\mu([A,C],D,Q) = \mu([A,D],C,Q) \text{ and } \mu^*([A,C],D,Q) = \mu^*([A,D],C,Q)$$

in view of the fact that, then, $\mathfrak{R}([A,C],D,P)$ is principal for all $P \in \mathfrak{D}_0([A,C])$.

§24. Intersection multiplicity with an abstract curve.

(24.1) to (24.17).

Let R be a homogeneous domain with $\text{Dim}\, R = 1$.

For any $I \in H(R)$ or $I \subset R$ and any $Q \in H(R)$ or $Q \subset R$, we define

$$\mu(R,I,Q) = \sum_{V \in \mathfrak{Z}(R,Q)} \text{ord}(R,I,V)[V/M(V) : \mathfrak{K}([R,\mathfrak{Z}(R,V)])] ,$$

$$\mu(R,I,\backslash Q) = \sum_{V \in \mathfrak{Z}(R,\backslash Q)} \text{ord}(R,I,V)[V/M(V) : \mathfrak{K}([R,\mathfrak{Z}^*(R,V)])] ,$$

$$\mu^*(R,I,Q) = \sum_{V \in \mathfrak{Z}(R,Q)} \text{ord}(R,I,V)[V/M(V) : H_0(R)] ,$$

and

$$\mu^*(R,I,\backslash Q) = \sum_{V\in\mathfrak{Z}(R,\backslash Q)} \mathrm{ord}(R,I,V)[V/M(V) : H_0(R)] .$$

For any $I \in H(R)$ or $I \subset R$, we <u>define</u>:

$$\mu(R,I) = \mu(R,I,0) \quad \text{and} \quad \mu^*(R,I) = \mu^*(R,I,0) .$$

For any $Q \in H(R)$ or $Q \subset R$, we <u>define</u>:

$$\mu_{\mathfrak{C}}(R,\backslash Q) = \sum_{P\in\mathfrak{Q}_0(R,Q)} \lambda_{\mathfrak{C}}(\mathfrak{R}(R,P)) ,$$

$$\mu_{\mathfrak{C}}(R,\backslash Q) = \sum_{P\in\mathfrak{Q}_0(R,\backslash Q)} \lambda_{\mathfrak{C}}(\mathfrak{R}(R,P)) ,$$

$$\mu^*_{\mathfrak{C}}(R,Q) = \sum_{P\in\mathfrak{Q}_0(R,Q)} \lambda^k_{\mathfrak{C}}(\mathfrak{R}(R,P)), \text{ where } k = H_0(A) ,$$

$$\mu^*_{\mathfrak{C}}(R,\backslash Q) = \sum_{P\in\mathfrak{Q}_0(R,\backslash Q)} \lambda^k_{\mathfrak{C}}(\mathfrak{R}(R,P)), \text{ where } k = H_0(A) ,$$

$$\mathrm{Adj}(R,Q) = \{\Phi \in H^*(R) : \mathfrak{R}(R,\Phi,P) \in \mathrm{adj}(\mathfrak{R}(R,P)) \text{ for all } P \in \mathfrak{Q}_0(R,Q)\} ,$$

$$\mathrm{Adj}(R,\backslash Q) = \{\Phi \in H^*(R) : \mathfrak{R}(R,\Phi,P) \in \mathrm{adj}(\mathfrak{R}(R,P)) \text{ for all } P \in \mathfrak{Q}_0(R,\backslash Q)\} ,$$

$$\mathrm{Tradj}(R,Q) = \{\Phi \in H^*(R) : \mathfrak{R}(R,\Phi,P) \in \mathrm{tradj}(\mathfrak{R}(R,P)) \text{ for all } P \in \mathfrak{Q}_0(R,Q)\} ,$$

and

$$\mathrm{Tradj}(R,\backslash Q) = \{\Phi \in H^*(R) : \mathfrak{R}(R,\Phi,P) \in \mathrm{tradj}(\mathfrak{R}(R,P)) \text{ for all } P \in \mathfrak{Q}_0(R,\backslash Q)\}$$

By an <u>adjoint</u> in R at Q, we mean a member of $\mathrm{Adj}(R,Q)$. By an <u>adjoint</u> in R <u>outside</u> Q, we mean a member of $\mathrm{Adj}(R,\backslash Q)$. By a <u>true adjoint</u> in R at Q, we mean a member of $\mathrm{Tradj}(R,Q)$. By a <u>true adjoint</u> in R <u>outside</u> Q, we mean a member of $\mathrm{Tradj}(R,\backslash Q)$. From these phrases we may drop " in R ", when the reference to R is clear from the context.

Finally we <u>define</u>:

$$\mu_{\mathfrak{C}}(R) = \mu_{\mathfrak{C}}(R,0) \quad , \quad \mu_{\mathfrak{C}}^*(R) = \mu_{\mathfrak{C}}^*(R,0) \ ;$$

and

$$\mathrm{Adj}(R) = \mathrm{Adj}(R,0) \quad , \quad \mathrm{Tradj}(R) = \mathrm{Tradj}(R,0) \ .$$

By an _adjoint in_ R, we mean a member of Adj(R), and by _true adjoint in_ R, we mean a member of Tradj(R).

We clearly get assertions (24.1) to (24.13) where, for $1 \le i \le 13$, (24.i) is obtained from (23.i) by replacing: A as well as [A,C] by R; C by {0}; and f by the identity map $R \to R$. Note that (24.9) now gives the definition of Deg R. We also get assertions (24.14) to (24.17) from (23.14) to (23.17) respectively, where, in addition to the above replacements, we also replace $[A^N,C^N]$ by R^N, and $[A^P,C^P]$ by R^P.

(24.18) DEFINITION. Let $P \in \mathfrak{D}_0(R)$. Assume that

(*) $$[V/M(V) : H_0(R)] = 1 \quad \text{for all} \quad V \in \mathfrak{Z}(R,P) \ .$$

(Note that condition (*) is automatically satisfied if, either: $\mu^*(R,P) = 1$, or: $H_0(R)$ is algebraically closed. Also note that: (*) $\Rightarrow P \in \mathfrak{M}_0^*(R)$, $\mathrm{Deg}[R,P] = 1$, and $\mu^*(R,P) = \mu(R,P)$.)

We _define_

$$T_1^*(R,P) = \{L \in \mathfrak{M}_1^*(R) : L \subset P \ \text{and} \ \mu^*(R,L,P) \neq \mu^*(R,P)\} \ .$$

We note that then clearly

(24.18.1) $$\begin{cases} T_1^*(R,P) = \{L \in \mathfrak{M}_1^*(R) : \mu^*(R,L,P) > \mu^*(R,P)\} \\ \qquad = \{L \in \mathfrak{M}_1^*(R) : L = T_1(R,V) \ \text{for some} \ V \in \mathfrak{Z}(R,P)\} \end{cases}$$

and hence

(24.18.2) $$1 \le \mathrm{card}\, T_1^*(R,P) \le \mathrm{card}\, \mathfrak{Z}(R,P) < \infty$$

and so in particular

(24.18.3) $\mu^*(R,P) = 1 \Rightarrow \text{card } T_1^*(R,P) = 1$.

By a <u>tangent</u> 1-<u>flat</u> <u>in</u> R <u>at</u> P we mean a member of $T_1^*(R,P)$. From this phrase we may drop " in R " when the reference to R is clear from the context.

(24.19) REMARK. Let any $C \in \mathfrak{D}_1(A)$ be given.

We note that then, upon letting $f: A \to A/C$ to be the canonical epimorphism, $g(x) = f(x)$ for any $x \in H(A)$, and $g(J) = f((J \cap H(A))A)$ for any $J \subset A$, we clearly have:

$$\left.\begin{aligned}
&\mu([A,C],I,Q) = \mu(f(A),g(I),g(Q)) ,\\
&\mu([A,C],I,\backslash Q) = \mu(f(A),g(I),\backslash g(Q)) ,\\
&\mu^*([A,C],I,Q) = \mu^*(f(A),g(I),g(Q)) ,\\
&\mu^*([A,C],I,\backslash Q) = \mu^*(f(A),g(I),\backslash g(Q)),
\end{aligned}\right\} \begin{array}{l}\text{for any } I \in H(A) \text{ or } I \subset A\\ \text{and any } Q \in H(A)\\ \text{or } Q \subset A\end{array}$$

$$\left.\begin{aligned}
&\mu([A,C],I) = \mu(f(A),g(I)) ,\\
&\mu^*([A,C],I) = \mu^*(f(A),g(I)) ,
\end{aligned}\right\} \text{ for any } I \in H(A) \text{ or } I \subset A ;$$

$$\left.\begin{aligned}
&\mu_{\mathfrak{C}}([A,C],Q) = \mu_{\mathfrak{C}}(f(A),g(Q)) ,\\
&\mu_{\mathfrak{C}}([A,C],\backslash Q) = \mu_{\mathfrak{C}}(f(A),\backslash g(Q)) ,\\
&\mu^*_{\mathfrak{C}}([A,C],Q) = \mu^*_{\mathfrak{C}}(f(A),g(Q)) ,\\
&\mu^*_{\mathfrak{C}}([A,C],\backslash Q) = \mu^*_{\mathfrak{C}}(f(A),\backslash g(Q)) ,\\
&\{f(\Phi): \Phi \in \text{Adj}([A,C],Q)\} = \text{Adj}(f(A),g(Q)) ,\\
&\{f(\Phi): \Phi \in \text{Adj}([A,C],\backslash Q)\} = \text{Adj}(f(A),\backslash g(Q)) ,\\
&\{f(\Phi): \Phi \in \text{Tradj}([A,C],Q)\} = \text{Tradj}(f(A),g(Q)) ,\\
&\{f(\Phi): \Phi \in \text{Tradj}([A,C],\backslash Q)\} = \text{Tradj}(f(A),\backslash g(Q)),
\end{aligned}\right\} \begin{array}{l}\text{for any}\\ Q \in H(A) \text{ or}\\ Q \subset A :\end{array}$$

$$\left.\begin{aligned}
&\mu_{\mathfrak{C}}([A,C]) = \mu_{\mathfrak{C}}(f(A)),\\
&\mu^*_{\mathfrak{C}}([A,C]) = \mu^*_{\mathfrak{C}}(f(A)) ,\\
&\{f(\Phi): \Phi \in \text{Adj}([A,C])\} = \text{Adj}(f(A)) ,\\
&\{f(\Phi): \Phi \in \text{Tradj}([A,C])\} = \text{Tradj}(f(A)) ;
\end{aligned}\right\}$$

and

$$\left.\begin{array}{l}\{f(L): L \in T_1^*([A,C],P)\} \\ \quad = T_1^*(f(A),f(P)) \ ,\end{array}\right\} \begin{array}{l}\text{for any } P \in \mathfrak{D}_0([A,C]) \text{ such that} \\ [V/M(V): H_0(A)] = 1 \\ \text{for all } V \in \mathfrak{Z}([A,C],P).\end{array}$$

We also note that, if $C = \{0\}$, then clearly:

$$\left.\begin{array}{l}\mu([A,C],I,Q) = \mu(A,I,Q) \ , \\ \mu([A,C],I,\backslash Q) = \mu(A,I,\backslash Q) \ , \\ \mu^*([A,C],I,Q) = \mu^*(A,I,Q) \ , \\ \mu^*([A,C],I,\backslash Q) = \mu^*(A,I,\backslash Q)\end{array}\right\} \begin{array}{l}\text{for any } I \in H(A) \text{ or } I \subset A \\ \text{and any } Q \in H(A) \text{ or } Q \subset A \ ;\end{array}$$

$$\left.\begin{array}{l}\mu([A,C],I) = \mu(A,I) \ , \\ \mu^*([A,C],I) = \mu^*(A,I) \ ,\end{array}\right. \quad \text{for any } I \in H(A) \text{ or } I \subset A \ ;$$

$$\left.\begin{array}{l}\mu_{\mathfrak{C}}([A,C],Q) = \mu_{\mathfrak{C}}(A,Q) \ , \\ \mu_{\mathfrak{C}}([A,C],\backslash Q) = \mu_{\mathfrak{C}}(A,\backslash Q) \ , \\ \mu^*_{\mathfrak{C}}([A,C],Q) = \mu^*_{\mathfrak{C}}(A,Q), \\ \mu^*_{\mathfrak{C}}([A,C],\backslash Q) = \mu^*_{\mathfrak{C}}(A,\backslash Q), \\ \mathrm{Adj}([A,C],Q) = \mathrm{Adj}(A,Q) \ , \\ \mathrm{Adj}([A,C],\backslash Q) = \mathrm{Adj}(A,\backslash Q \\ \mathrm{Tradj}([A,C],Q) = \mathrm{Tradj}(A,Q) \ , \\ \mathrm{Tradj}([A,C],\backslash Q) = \mathrm{Tradj}(A,\backslash Q) \ ,\end{array}\right\} \text{for any } Q \in H(A) \text{ or } Q \subset A \ ;$$

$$\left.\begin{array}{l}\mu_{\mathfrak{C}}([A,C]) = \mu_{\mathfrak{C}}(A) \ , \\ \mu^*_{\mathfrak{C}}([A,C]) = \mu^*_{\mathfrak{C}}(A) \ , \\ \mathrm{Adj}([A,C]) = \mathrm{Adj}(A) \ , \\ \mathrm{Tradj}([A,C]) = \mathrm{Tradj}(A) \ ,\end{array}\right\}$$

and

$$T_1^*([A,C],P) = T_1^*(A,P) \left\{ \begin{array}{l} \text{for any } P \in \mathfrak{O}_0(A) \text{ such that} \\ [V/M(V): H_0(A)] = 1 \\ \text{for all } V \in \mathfrak{Z}(A,P) . \end{array} \right.$$

(24.20) LEMMA ON OVERADJOINTS. Let $V \in \mathfrak{Z}(R)$ be residually rational over $H_0(R)$. Given any nonnegative integers m,e,e' with $e \geq e'$ let

$$E(m,e) = \{x \in H_m(R): \operatorname{ord}(R,x,V) \geq e + \operatorname{ord}_V \mathfrak{C}(\mathfrak{R}(R,\mathfrak{Z}^*(R,V))), \text{ and } \operatorname{ord}(R,x,W) \geq \operatorname{ord}_W \mathfrak{C}(\mathfrak{R}(R,\mathfrak{Z}^*(W))) , \text{ whenever } V \neq W \in \mathfrak{Z}(R)\}$$

and similarly

$$E(m,e') = \{x \in H_m(R): \operatorname{ord}(R,x,V) \geq e' + \operatorname{ord}_V \mathfrak{C}(\mathfrak{R}(R,\mathfrak{Z}^*(R,V))), \text{ and } \operatorname{ord}(R,x,W) \geq \operatorname{ord}_W \mathfrak{C}(\mathfrak{R}(R,\mathfrak{Z}^*(R,W))) \text{ whenever } V \neq W \in \mathfrak{Z}(R)\} .$$

Then:

(1) $E(M,e) \subset E(m,e')$ are $H_0(R)$-vector subspaces of $H_m(R)$,

(2)
$$\left\{ \begin{array}{l} [E(m,e): H_0(R)] + e + \sum_{P \in \mathfrak{O}_0(R)} [\mathfrak{R}(R,P)/\mathfrak{C}(\mathfrak{R}(R,P)): H_0(R)] \\ \geq [E(m,e'): H_0(R)] + e' + \sum_{P \in \mathfrak{O}_0(R)} [\mathfrak{R}(R,P)/\mathfrak{C}(\mathfrak{R}(R,P)): H_0(R)] \\ \geq [H_m(R): H_0(R)] , \end{array} \right.$$

(3) $\mu^*(R,x) \geq e + \mu_{\mathfrak{C}}^*(R)$, for all $x \in E(m,e)$,

and

(4) $\mu^*(R,x) \geq e' + \mu_{\mathfrak{C}}^*(R)$, for all $x \in E(m,e')$.

PROOF. Let

$$P_0 = \mathfrak{Z}^*(R,V)$$

and

$$\mathfrak{R}(R,P_0)^* = \text{the integral closure of } \mathfrak{R}(R,P_0) \text{ in } \mathfrak{K}(R) .$$

Then in view of (5.12), the first two assertions follow from (15.10) by taking, for all $P \in \mathfrak{O}_0(R)$,

$$I(P) = \begin{cases} \mathfrak{C}(\mathfrak{R}(R,P_0))(\mathfrak{R}(R,P_0)^* \cap M(V))^e & \text{if } P = P_0 \\ \mathfrak{C}(\mathfrak{R}(R,P)) & \text{if } P \neq P_0 \end{cases}$$

and

$$I'(P) = \begin{cases} \mathfrak{C}(\mathfrak{R}(R(R,P_0))(\mathfrak{R}(R,P_0)^* \cap M(V))^{e'} & \text{if } P = P_0 \\ \mathfrak{C}(\mathfrak{R}(R,P_0)) & \text{if } P \neq P_0 . \end{cases}$$

The last two assertions follow from the definition of μ^* .

(24.21) LEMMA ON UNDERADJOINTS. <u>Assume that</u>

(*) $\mu(R,P) \le 2$ <u>for all</u> $P \in \mathfrak{O}_0(R)$.

<u>Let</u> $n = \text{Deg } R$, <u>and let</u> π <u>be a homogeneous ideal in</u> R. <u>Let</u> m <u>and</u> s(P), <u>for every</u> $P \in \mathfrak{O}_0(R,\pi)$, <u>be nonnegative integers such that</u>

(**) $s(P) \le (1/2)\mu_{\mathfrak{C}}(R,P)$ for all $P \in \mathfrak{O}_0(R,\pi)$

(***) $[H_m(R): H_0(R)] > (1/2)\mu^*_{\mathfrak{C}}(R) - \sum_{P\in\mathfrak{O}_0(R,\pi)} s(P)\text{Deg}[R,P]$

<u>and</u>

(****) $mn \le \mu^*_{\mathfrak{C}}(R) - \sum_{P\in\mathfrak{O}_0(R,\pi)} 2s(P)\text{Deg}[R,P]$.

<u>Then</u>

$$\text{Tradj}(R,\backslash\pi) \cap H^*_m(R) \neq \emptyset .$$

<u>Moreover</u>, <u>if</u>

$$m \le 2 \text{ and } n \not\equiv 0(2) .$$

<u>then there exists</u>

$$\Theta \in \mathrm{Tradj}(R,\backslash\pi) \cap H_m^*(R)$$

<u>such that</u> Θ <u>is irreducible in the sense that</u>:

$$\Theta = \Theta_1\Theta_2 \ \underline{\text{with}}\ \Theta_1 \ \underline{\text{and}}\ \Theta_2 \ \underline{\text{in}}\ H^*(R) \Rightarrow \Theta_1 = R \ \underline{\text{or}}\ \Theta_2 = R.$$

PROOF. Let

$$k = H_0(R) .$$

In view of (*), by (10.1.10) we have that

(1) $[\mathfrak{R}(R,P)/\mathfrak{C}(\mathfrak{R}(R,P)) : k] = (1/2)\mu_{\mathfrak{C}}^*(R,P)$ for all $P \in \mathfrak{Q}_0(R)$.

and by (10.1.13) we have that,

$$(2) \begin{cases} \Phi \in H^*(R) \ \text{ with } \ \mu^*(R,\Phi,P) \le \mu_{\mathfrak{C}}^*(R,P) \ \text{ for all } \ P \in \mathfrak{Q}_0(R) \\ \Rightarrow \mu^*(R,\Phi) \equiv 0(2) . \end{cases}$$

Let

(3) $$\Omega = \{P \in \mathfrak{Q}_0(R,\pi) : \mu(R,P) \neq 2\} .$$

Then by (*) we have $\mu_{\mathfrak{C}}(R,P) = 0$ for all $P \in \mathfrak{Q}_0(R,\pi)\backslash\Omega$, and hence by (**) we get

(4) $$s(P) = 0 \quad \text{for all} \quad P \in \mathfrak{Q}_0(R,\pi)\backslash\Omega .$$

For each $P \in \Omega$, in view of (**), by (10.1.12) there exists an ideal $J(P)$ in $\mathfrak{R}(R,P)$ with $\mathfrak{C}(\mathfrak{R}(R,P)) \subset J(P)$ such that

(5) $[\mathfrak{R}(R,P)/J(P) : k] = [\mathfrak{R}(R,P)/\mathfrak{C}(\mathfrak{R}(R,P)) : k] - s(P)\mathrm{Deg}[R,P]$

and

(6) $$\lambda^k(\mathfrak{R}(R,P),J(P)) = \lambda_{\mathfrak{C}}^k(\mathfrak{R}(R,P)) - 2s(P)\mathrm{Deg}[R,P] .$$

For each $P \in \mathfrak{Q}_0(R)$ we get an ideal $I(P)$ in $\mathfrak{R}(R,P)$ by setting

$$(7)\qquad I(P)\begin{cases} J(P) & \text{if } P \in \Omega \\ \mathfrak{C}(\mathfrak{R}(R,P)) & \text{if } P \notin \Omega\ . \end{cases}$$

By (1), (3), (4), (5) and (7) we then have

$$(8)\qquad \sum_{P\in\mathfrak{D}_0(R)} [\mathfrak{R}(R,P)/I(P) : k] = (1/2)\mu^*_{\mathfrak{C}}(R) - \sum_{P\in\mathfrak{D}_0(R,\pi)} s(P)\mathrm{Deg}[R,P]\ .$$

Upon letting

$$(9)\qquad E = \{x \in H_m(R) : \mathfrak{R}(R,x,P) \in I(P) \quad \text{for all} \quad P \in \mathfrak{D}_0(R)\}$$

by (15.10) we see that E is a k-vector-subspace of $H_m(R)$ with

$$(10)\qquad [E:k] + \sum_{P\in\mathfrak{D}_0(R)} [\mathfrak{R}(R,P)/I(P) : k] \geq [H_m(R) : k]\ .$$

By (***), (8) and (10) we get

$$(11)\qquad [E:k] > 0$$

Upon letting

$$(12)\qquad \Lambda = \{\Theta \in H^*_m(R) : \Theta = xR \quad \text{for some} \quad x \in E\}$$

by (9) we obviously get

$$\mu^*(R,\Phi,P) \geq \lambda^k(\mathfrak{R}(R,P),I(P)) \quad \text{for all} \quad \Theta \in \Lambda \quad \text{and} \quad P \in \mathfrak{D}_0(R)$$

and hence in view of (3), (4), (6) and (7) we see that

$$(13)\qquad \begin{cases} \text{for every } \Theta \in \Lambda \text{ we have:} \\ \mu^*(R,\Theta,P) \geq \mu^*_{\mathfrak{C}}(R,P) - 2s(P)\mathrm{Deg}[R,P] & \text{for all } P \in \mathfrak{D}_0(R,\pi) \\ \text{and} \\ \mu^*(R,\Theta,P) \geq \mu^*_{\mathfrak{C}}(R,P) & \text{for all } P \notin \mathfrak{D}_0(R,\pi) \end{cases}$$

By (13) we get

(14) $\mu^*(R,\Theta) \geq \mu^*_{\mathfrak{C}}(R) - \sum_{P\in\mathfrak{D}_0(R,\pi)} 2s(P)\mathrm{Deg}[R,P]$ for all $\Theta \in \Lambda$.

By (3), (7), (9) and (12) we also get

(15) $$\Lambda \subset \mathrm{Adj}(R,\backslash\pi) \cap H^*_m(R) \ .$$

By (9), (11) and (12) we have

(16) $$\Lambda \neq \emptyset \ .$$

In view of (12), by Bezout's Little Theorem (24.9) we also have

(17) $$\mu^*(R,\Theta) = mn \quad \text{for all} \quad \Theta \in \Lambda \ .$$

By (****), (13), (14) and (17) it follows that

(18) $$\begin{cases} \text{for every } \Theta \in \Lambda \text{ we have:} \\ \mu^*(R,\Theta,P) = \mu^*_{\mathfrak{C}}(R,P) - 2s(P)\mathrm{Deg}[R,P] \quad \text{for all} \quad P \in \mathfrak{D}_0(R,\pi) \\ \text{and} \\ \mu^*(R,\Theta,P) = \mu^*_{\mathfrak{C}}(R,P) \quad \text{for all} \quad P \in \mathfrak{D}_0(R,\backslash\pi). \end{cases}$$

By (15) and (18) we see that

(19) $$\Lambda \subset \mathrm{Tradj}(R,\backslash\pi) \cap H^*_m(R) \ .$$

Now let $\Theta \in \Lambda$ such that $\Theta = \Theta_1\Theta_2$ with $\Theta_i \in H^*(R)$ for i = 1, 2. In view of (2) and (18) we see that,

$$\mu^*(R,\Theta_i) \equiv 0(2) \ ,$$

and hence by Bezout's Little Theorem (24.9) we get that

(20) $$\begin{cases} \Theta = \Theta_1\Theta_2 \ \text{with} \ \Theta \in \Lambda \ \text{and} \ \Theta_i \in H^*(R) \ \text{for} \ i = 1,2 \\ \Rightarrow n(\mathrm{Deg}_R\Theta_i) \equiv 0(2) \ \text{for} \ i = 1,2 \ . \end{cases}$$

Now, if $m \leq 2$ and $n \not\equiv 0(2)$, then by (20) it follows that

(21) $$\mathrm{Deg}_R\Theta_i \equiv 0(2), \text{ and } \mathrm{Deg}_R\Theta_1 + \mathrm{Deg}_R\Theta_2 \leq 2 \ .$$

It follows that $\text{Deg}_R \Theta_i = 0$ for some i and hence $\Theta_1 = R$ or $\Theta_2 = R$. This together with (16) and (19) finishes the proof.

§25. Tangent cones and quasihyperplanes.

Assume that $\text{Emdim } A = \text{Dim } A = r$.

(25.1) DEFINITION. For any $\Phi \in H^*(A)$ and any $P \in \mathfrak{Q}[A,\Phi])$ we <u>define</u>

$$\mu'([A,\Phi],P) = \lambda'([\mathfrak{R}(A,P),\mathfrak{R}(A,\Phi,P)],M(\mathfrak{R}(A,P)))$$

We note that then:

$$\mu'([A,\Phi],P) = \text{a nonnegative integer} ,$$

$$\mu'([A,\Phi],P) = 0 \Leftrightarrow \Phi \not\subset P ,$$

$$\mu'([A,\Phi,P]) = 1 \Leftrightarrow \mathfrak{R}(A,P)/\mathfrak{R}(A,\Phi,P) \text{ is a regular local domain} ,$$

and so, in particular;

if $\Phi \in \mathfrak{Q}(A)$, then: $\mu'([A,\Phi],P) = 1 \Leftrightarrow \Phi \subset P$ and $\mathfrak{R}([A,\Phi],P)$ is regular.

(25.2) LEMMA-DEFINITION. Let $\Phi \in H_n^*(A)$ and $P \in \mathfrak{M}_0^*(A)$ be given. We can take $\varphi \in H(A)$ with $\varphi A = \Phi$, and $X_r \in H_1(A)$ with $P + X_r A = H_1(A)A$. Now we can uniquely write

$$\varphi = \sum_{i=0}^{n} \varphi_i X_r^{n-i} \quad \text{with} \quad \varphi_i \in H_i(A^P) .$$

Let d be the unique integer with $0 \le d \le n$ such that

$$\varphi_i = 0 \quad \text{for} \quad 0 \le i < d, \text{ and } \varphi_d \ne 0 .$$

<u>We claim that</u> $\varphi_d A$ (<u>whence in particular</u> d) <u>is independent of the choice of</u> φ <u>and</u> X_r.

Namely, let any $\varphi' \in H(A)$ with $\varphi' A = \Phi$ and any $X_r' \in H_1(A)$

with $P + X_r'A = H_1(A)A$ be given, and let

$$\varphi' = \sum_{i=0}^{n} \varphi_i' X_r'^{\,n-i} \quad \text{with} \quad \varphi_i' \in H_i(A^P)$$

be the corresponding expression. Then

$$\varphi' = a\varphi \quad \text{with} \quad 0 \neq a \in H_0(A)$$

and

$$X_r' = bX_r + X^* \quad \text{with} \quad 0 \neq b \in H_0(A) \quad \text{and} \quad X^* \in H_1(A^P) .$$

Substituting in the above expression of φ' we get

$$\varphi' = a\varphi = \sum_{i=0}^{n} a\varphi_i (bX_r + X^*)^{n-i}$$

and hence it follows that

$$\varphi_i' = 0 \quad \text{for} \quad 0 \leq i < d, \; \varphi_d' = ab^{n-d}\varphi_d \neq 0, \text{ and } \varphi_d'A = \varphi_d A ,$$

which proves our claim.

We <u>define</u>

$$T([A,\Phi],P) = \varphi_d A$$

and we note that then $T([A,\Phi],P) \in H_d^*(A,P)$.

(25.3) LEMMA. <u>Let</u> $\Phi \in H_n^*(A)$, $P \in \mathfrak{M}_0^*(A)$, <u>and</u> $d = \mathrm{Deg}_A T([A,\Phi],P)$. <u>Then we have the following</u>

(25.3.1) $\mu'([A,\Phi],P) = d$.

(25.3.2) <u>For any</u> $L \in \mathfrak{M}_{r-1}^*(A)$ <u>with</u> $L \subset P$ <u>we have</u>:

$$T([A,\Phi],P) \subset L \Leftrightarrow \mathfrak{R}(A,\Phi,P) \subset \mathfrak{R}(A,L,P) + M(\mathfrak{R}(A,P))^{d+1} .$$

(25.3.3) $\Phi \in H^*(A,P) \Leftrightarrow d = n \Leftrightarrow \Phi = T([A,\Phi],P)$
$\Leftrightarrow \Phi^P \in H_n^*(A^P) \Leftrightarrow \Phi^P \neq \{0\}$.

(25.3.4) If $\Phi \notin H^*(A,P)$ and $\Phi \in \mathfrak{D}(A)$, then
$[\mathfrak{R}([A,\Phi]) : \mathfrak{R}([A^P,\Phi^P])] = n - d$.

PROOF. Let $k = H_0(A)$ and let $Y_0, Y_1, \ldots, Y_{r-1}$ be indeterminat over k. We can take a homogeneous coordinate system $(X_0, X_1, \ldots, X_r)$ in A such that $P = (X_0, X_1, \ldots, X_{r-1})A$ and then we have

$$\Phi = \varphi A \quad \text{with} \quad \varphi = \sum_{i=d}^{n} \varphi_i(X_0, X_1, \ldots, X_{r-1}) X_r^{n-i}$$

where $\varphi_i(Y_0, Y_1, \ldots, Y_{r-1}) \in k[Y_0, Y_1, \ldots, Y_{r-1}]$ is either zero or a nonzero homogeneous polynomial of degree i for $d \le i \le n$; we alsc have

$$\varphi_d(Y_0, Y_1, \ldots, Y_{r-1}) \ne 0 \quad \text{and} \quad T([A, \Phi], P) = \varphi_d(X_0, X_1, \ldots, X_{r-1})A \ .$$

(25.3.3) is now obvious. To prove (25.3.4), assume that $\Phi \notin H^*(A,P)$ and $\Phi \in \mathfrak{Q}(A)$. Then $r > 0$, $A^P = k[X_0, X_1, \ldots, X_{r-1}]$, anc $\Phi^P = \{0\}$; so in particular $X_0 \notin \Phi$. Let $f\colon A \to A/\Phi$ be the canonical epimorphism. Now

$$\mathfrak{K}(f(A^P)) = f(k)(f(X_1)/f(X_0), f(X_2)/f(X_0), \ldots, f(X_{r-1})/f(X_0)) \ ,$$

$$\mathfrak{K}([A, \Phi]) = \mathfrak{K}(f(A^P))(f(X_r)/f(X_0)) \ ,$$

and $f(X_r)/f(X_0)$ satisfies the irreducible equation

$$\sum_{i=d}^{n} [f(\varphi_i(X_0, X_1, \ldots, X_{r-1}))/f(X_0)^i][f(X_r)/f(X_0)]^{n-i} = 0 \ .$$

of degree $n - d$ over $\mathfrak{K}(f(A^P))$. Therefore

$$[\mathfrak{K}([A, \Phi]) : \mathfrak{K}([A^P, \Phi^P])] = n - d \ ,$$

which proves (25.3.4)

To prove (25.3.1) and (25.3.2), we drop the assumption (made for proving (25.3.4)) that $\Phi \notin H^*(A,P)$ and $\Phi \in \mathfrak{Q}(A)$. Let $A' = k[X_0/X_r, X_1/X_r, \ldots, X_{r-1}/X_r]$ and $P' = (X_0/X_r, X_1/X_r, \ldots, X_{r-1}/X_r)A$ Now the elements $X_0/X_r, X_1/X_r, \ldots, X_{r-1}/X_r$ are algebraically indepenc ent over k, $A'_{P'}$ is an r-dimensional regular local ring, k is a coefficient set for $A'_{P'}$, $M(A'_{P'}) = (X_0/X_r, X_1/X_r, \ldots, X_{r-1}/X_r)A'_{P'}$, and (in view of (15.1), (15.2) and (15.4) we have $\mathfrak{R}(A,P) = A'_{P'}$ and $\mathfrak{R}(A,\Phi,P) = \varphi^*\mathfrak{R}(A,P)$ where $\varphi^* = \varphi/X_r^n = \sum_{i=d}^{n} \varphi_i(X_0/X_r, X_1/X_r, \ldots, X_{r-1}/X_r)$.

Therefore $\mu'([A,\Phi],P) = d$ which proves (25.3.1). To prove (25.3.2), let any $L \in \mathfrak{M}^*_{r-1}(A)$ with $L \subset P$ be given. Now $L = (a_0X_0+a_1X_1+\ldots+a_{r-1}X_{r-1})A$ where $a_0, a_1, \ldots, a_{r-1}$ are elements in k at least one of which is not zero. In view of (15.1) and (15.2) we have

$$\mathfrak{R}(A,L,P) = (a_0(X_0/X_r) + a_1(X_1/X_r) + \ldots + a_{r-1}(X_{r-1}/X_r))\mathfrak{R}(A,P)$$

and hence it follows that

$$T([A,\Phi),P) \subset L \Leftrightarrow \mathfrak{R}(A,\Phi,P) \subset \mathfrak{R}(A,L,P) + M(\mathfrak{R}(A,P))^{d+1}$$

which proves (25.3.2).

(25.4) DEFINITION. Let $\pi \in H_1^*(A)$ and $\Phi \in H_n^*(A)$. We call Φ a π-<u>quasihyperplane</u> in A, if, for some $P \in \mathfrak{M}_0^*(A)$ with $\pi \subset P$, we have

$$\mu'([A,\Phi],P) = n-1 \quad \text{and} \quad T([A,\Phi],P) = \pi^{n-1} .$$

We note that then, in view of (25.3.1), we have:

Φ is a π-quasihyperplane in A

$$\Leftrightarrow \begin{cases} \text{for some } P \in \mathfrak{Q}_0(A,\Phi) \text{ with } \mathrm{Deg}[A,P] = 1 \text{ we have } \mu'([A,\Phi],P) \\ = n-1 \\ \text{and } \mathfrak{Q}(A,T([A,\Phi],P)) = \mathfrak{Q}(A,\pi) \end{cases}$$

From the phrase "π-quasihyperplane in A " we may drop "hyper" in case $r = 3$.

(25.5) LEMMA. _Let_ $\pi \in H_1^*(A)$ _and_ $\Phi \in H_n^*(A)$. _Then_ Φ _is a_ π-_quasihyperplane in_ $A \Leftrightarrow n > 1$ _and there exists a homogeneous coordinate system_ $X_0, X_1, \ldots, X_r$ _in_ A _such that_ $\pi = X_0A$ _and_

$\Phi = (X_0^{n-1}X_r + \varphi_n)A$ with $\varphi_n \in H_n(A^P)$ <u>where</u> $P = (X_0, X_1, \ldots, X_{r-1})A$

PROOF. Obvious in view of (25.3.1).

(25.6) LEMMA. <u>If</u> $\pi \in H_1^*(A)$ <u>and</u> $\psi \in H_1^*(A)$ <u>are such that</u> $\pi \neq \psi$, <u>then</u> $\pi\psi \in H_2^*(A)$ <u>and</u> $\pi\psi$ <u>is a</u> π-<u>quasihyperplane in</u> A.

PROOF. Clearly $\pi\psi \in H_2^*(A)$, and we can take a homogeneous coordinate system $X_0, X_1, \ldots, X_r$ in A such that $\pi = X_0A$ and $\psi = X_rA$. Now $\pi\psi = (X_0X_r)A$ and hence by (25.5) we see that $\pi\psi$ is a π-quasihyperplane in A.

(25.7) LEMMA. <u>Suppose</u> $r = 3$ <u>and let</u> $\pi \in H_1^*(A)$ <u>and</u> $\Phi \in H_2^*($ <u>be such that</u> $L \in \mathfrak{Q}(A,\pi) \cap \mathfrak{Q}(A,\Phi)$ <u>for some</u> $L \in \mathfrak{M}_1^*(A)$. <u>Then either</u> Φ <u>is a</u> π-<u>quasiplane in</u> A, <u>or</u> $\Phi \in H^*(A,P)$ <u>for some</u> $P \in \mathfrak{M}_0^*(A)$ <u>with</u> $L \subset P$.

PROOF. Since $L \in \mathfrak{Q}(A,\pi)$, we can take elements X_0 and X_1 in $H_1(A)$ such that $\pi = X_0A$ and $L = (X_0, X_1)A$. Since $L \in \mathfrak{Q}(A,\Phi)$, we get

$$\Phi = (X_0X_3 + X_1X_2)A \text{ with } X_2 \text{ and } X_3 \text{ in } H_1(A) .$$

If

$$[(X_0,X_1,X_2,X_3)H_0(A) : H_0(A)] > 3$$

then, upon taking $P = (X_0,X_1,X_2)A$, by (25.5) we see that Φ is a π-quasiplane in A. If

$$[(X_0,X_1,X_2,X_3)H_0(A) : H_0(A)] \leq 3$$

then we can take $P \in \mathfrak{M}_0^*(A)$ such that $X_i \in P$ for $0 \leq i \leq 3$, and hence $L \subset P$ and $\Phi \in H^*(A,P)$.

(25.8) LEMMA. <u>Suppose</u> $r = 2$. <u>Then for any</u> $C \in \mathfrak{Q}_1(A)$ <u>and an</u> $P \in \mathfrak{Q}_0(A)$ <u>we have</u>: $\mu'([A,C],P) = \mu([A,C],P)$.

PROOF. Follows from (9.1), (23.1) and (23.2).

(25.9) LEMMA. _Suppose_ $r = 2$. _Then for any_ $C \in \mathfrak{D}_1(A)$ _we have_ $\mathrm{Deg}[A,C] = \mathrm{Deg}_A C$.

PROOF. Take any $P \in \mathfrak{M}_0^*(A)$. If $C \in H^*(A,P)$ then clearly $\mathrm{Deg}_A C = 1$ and $\mathrm{Emdim}[A,C] = 1$, and hence also $\mathrm{Deg}[A,C] = 1$ by (23.12). If $C \notin H^*(A,P)$ then by (25.3) and (25.8) we have

$$[\mathfrak{K}([A,C]) : \mathfrak{K}([A^P,C^P])] = (\deg_A C) - \mu^*([A,C],P)$$

and hence by (23.15) we get $\mathrm{Deg}[A,C] = \deg_A C$.

(25.10) LEMMA. _Suppose_ $r = 2$ _and let any_ $C \in \mathfrak{D}_1(A)$ _and any_ $P \in \mathfrak{M}_0^*(A)$ _be given_. _Then for any_ $L \in \mathfrak{M}_1^*(A)$ _with_ $L \subset P$ _we have_

$$T([A,C],P) \subset L \Leftrightarrow \mu^*([A,C],L,P) > \mu([A,C],P) \ .$$

Moreover, if $P \in \mathfrak{D}_0([A,C])$ _and_

(*) $\qquad [V/M(V) : H_0(A)] = 1$ for all $V \in \mathfrak{Z}([A,C],P)$,

then

$$\mathfrak{D}_1(A,T([A,C],P)) = \mathfrak{D}_1^1(A,T([A,C],P)) = T_1^*([A,C],P) \ .$$

(Note that condition (*) is automatically satisfied, if either $\mu^*([A,C],P) = 1$, or $H_0(A)$ is algebraically closed.)

PROOF. In view of (23.18), the second assertion follows from the first assertion. To prove the first assertion, let any $L \in \mathfrak{M}_1^*(A)$ with $L \subset P$ be given. Now

$$\begin{aligned} \mu^*([A,C],L,P) &= \mu^*([A,L],C,P) && \text{by (23.20)} \\ &= \mathrm{ord}_{\mathfrak{R}([A,L],P)} \mathfrak{R}([A,L],C,P) && \text{by (23.1) and (23.7)} \end{aligned}$$

and by (25.3) and (25.8) we have

$$T([A,C],P) \subset L \Leftrightarrow \mathrm{ord}_{\mathfrak{R}([A,L],P)}\mathfrak{R}([A,L],C,P) > \mu([A,C],P) .$$

Therefore

$$T([A,C],P) \subset L \Leftrightarrow \mu^*([A,C],L,P) > \mu([A,C],P) .$$

(25.11) DEFINITION. For any Φ and ψ in $H^*(A)$ and any $P \in \mathfrak{D}_{r-2}(A)$ we <u>define</u>

$$\mu([A,\Phi,\psi],P) = \lambda([\mathfrak{R}(A,P),\mathfrak{R}(A,\Phi,P),\mathfrak{R}(A,\psi,P)], M(\mathfrak{R}(A,P)))$$

and we note that then:

$$\mu([A,\Phi,\psi],P) = \mu([A,\psi,\Phi],P) = \text{a nonnegative integer or } \infty ,$$

$$\mu([A,\Phi,\psi],P) = 0 \Leftrightarrow P \not\subset \mathfrak{D}(A,\Phi) \cap \mathfrak{D}(A,\psi) ,$$

and

$$\mu([A,\Phi,\psi],P) = \infty \Leftrightarrow P \in \mathfrak{D}(A,P') \quad \text{for some}$$

$$P' \in \mathfrak{D}_{r-1}(A,\Phi) \cap \mathfrak{D}_{r-1}(A,\psi) .$$

(25.12) LEMMA. <u>For any</u> Φ, Φ' <u>and</u> ψ <u>in</u> $H^*(A)$ <u>and any</u> $P \in \mathfrak{D}_{r-2}(A)$ <u>we have</u>

$$\mu([A,\Phi\Phi',\psi],P) = \mu([A,\Phi,\psi],P) + \mu([A,\Phi',\psi],P)$$

PROOF. Follows from (8.3).

(25.13) LEMMA. <u>Suppose</u> $r = 2$. <u>Then for any</u> $C \in \mathfrak{D}_1(A)$, <u>any</u> $D \in H^*(A)$, <u>and any</u> $P \in \mathfrak{D}_0(A)$, <u>we have</u> $\mu([A,C,D],P) = \mu([A,C],D,P)$.

PROOF. Follows from (23.6).

(25.14) BEZOUT'S THEOREM FOR PLANE CURVES. <u>Suppose</u> $r = 2$. <u>Then for any</u> C <u>and</u> D <u>in</u> $H^*(A)$ <u>with</u> $\mathfrak{D}_1(A,C) \cap \mathfrak{D}_1(A,D) = \emptyset$ <u>we have</u>

$$\sum_{P\in\mathfrak{D}_0(A)} \mu([A,C,D],P)\mathrm{Deg}[A,P] = (\deg_A C)(\deg_A D) .$$

PROOF. Follows from (23.9), (25.9), (25.12), and (25.13).

§26. 2-equimultiple plane projections of projective space quintics.

Assume that Emdim A = Dim $A = 3$ and $H_0(A)$ is algebraically closed. Let $\pi \in H_1^*(A)$. Let $C \in \mathfrak{D}_1(A,\backslash\pi)$ and let $n = \mathrm{Deg}[A,C]$.

We want to prove Theorem (26.11) and its detailed version Proposition (26.10). These will be preceded by Lemmas (26.1) to (26.9). To avoid repetition we first introduce some suggestive terms. These definitions are meant to be local, i.e., they are to be in force only in this section; we shall not even use them in the statement of Theorem (26.11).

By a d-<u>chord</u> (of C in A lying on π) we mean a member L of $\mathfrak{D}_1^1(A,\pi)$ such that $\mu([A,C],L) \geq d$.

By a d-<u>secant</u> (of C in A lying on π) we mean a member L of $\mathfrak{D}_1^1(A,\pi)$ such that $\mu([A,C],L) = d$.

By a <u>good projecting center</u> (for C in A relative to π) we mean a member N of $\mathfrak{D}_0([A,\pi],C)$ such that: $\mu([A,C],N) = 1$, the projection of C from N is birational, the projection of C from N is π-integral, $\mu([A^N,C^N],Q) \leq 2$ for all $Q \in \mathfrak{D}_0(A^N)$, $\mu([A^N,C^N],Q_0) = 2$ for some $Q_0 \in \mathfrak{D}_0([A^N,C^N],\pi^N)$, and $\mathrm{Deg}[A^N,C^N] = n - 1$.

By a <u>better projecting center</u> (for C in A relative to π) we mean a member N of $\mathfrak{D}_0([A,\pi],\backslash C)$ such that: the projection of C from N is birational, the projection of C from N is π-integral, $\mu([A^N,C^N],Q) \leq 2$ for all $Q \in \mathfrak{D}_0(A^N)$, $\mu([A^N,C^N],Q_0) = 2$ for some $Q_0 \in \mathfrak{D}_0([A^N,C^N],\pi^N)$, and $\mathrm{Deg}[A^N,C^N] = n$.

By a <u>best projecting center</u> (for C in A relative to π) we mean a member N of $\mathfrak{D}_0([A,\pi],\backslash C)$ such that: the projection of C from N is birational, the projection of C from N is π-integral, $\mu([A^N,C^N],Q) = 1$ for all $Q \in \mathfrak{D}_0([A^N,C^N],\backslash\pi^N)$, and $\mathrm{Deg}[A^N,C^N] = n$.

(26.1) LEMMA. Given any $P_0 \in \mathfrak{D}_0(A,\pi)$, let

$$d = \mu([A,C],P_0) ,$$

$$\Lambda = \{L \in \mathfrak{D}_1^1(A,\pi) : L \subset P_0\} , \text{ and}$$

$$\Gamma = \{L \in \Lambda : L \text{ is a d-secant}\} .$$

Then

$$\text{card}(\Lambda \setminus \Gamma) < \infty = \text{card}\,\Gamma .$$

Moreover, if b is any integer such that $b \leq d + 1$ and $b \leq n$, then there exists a b-chord $L \in \Lambda$.

PROOF. By Bezout's Little Theorem (23.9) we know that $\text{card}\,\mathfrak{D}_0([A,C],\pi) < \infty$. Now let

$$\Gamma^* = \{L : L = \Delta(A,P_0,P) \text{ for some } P \in \mathfrak{D}_0([A,C],\pi)\setminus\{P_0\}\} .$$

Now $\text{card}\,\mathfrak{D}_0([A,C],\pi) < \infty$ implies that

$$\text{card}\,\Gamma^* < \infty .$$

Now obviously, (or say in view of (23.18)) we have

$$\Lambda\setminus\Gamma = \begin{cases} \Gamma^* & \text{if } d = 0 \\ \Gamma^* \cup (T_1^*([A,C],P_0) \cap \mathfrak{D}_1^1(A,\pi)) , & \text{if } d \neq 0 \end{cases}$$

Hence $\text{card}(\Lambda \setminus \Gamma) < \infty$ and since $\text{card}\,\Lambda = \infty$ we must have $\text{card}\,\Gamma = \infty$.

Now let any integer b be given such that $b \leq d + 1$ and $b \leq n$. If $b \neq d + 1$ then we would have $b \leq d$ and so it would suffice to take L to be any member of Γ. If $b = d + 1$ and

$\mathfrak{D}_0([A,C],\pi) \not\subset \{P_0\}$ then there exists some $P \in \mathfrak{D}_0([A,C],\pi)\setminus\{P_0\}$ and it suffices to take $L = \Delta(A,P_0,P)$. Finally, if $b = d + 1$ and $\mathfrak{D}_0([A,C],\pi) \subset \{P_0\}$ then: in view of Bezout's Little Theorem (23.9) we see that $\mathfrak{D}_0([A,C],\pi) = \{P_0\}$ and

$$\mu([A,C],\pi,P_0) = \mu([A,C],\pi) = n > d = \mu([A,C],P_0) .$$

Consequently there exists $V \in \mathfrak{Z}([A,C],P_0)$ such that

$$\mathrm{ord}([A,C],\pi,V) > \mathrm{ord}([A,C],P_0,V)$$

and hence upon letting

$$L = T_1([A,C],P_0,V) ,$$

in view of (21.1), we have

$$L \in \mathfrak{D}_1^1(A,\pi) \quad \text{and} \quad P_0 \in \mathfrak{D}_0(A,L) ;$$

now obviously (or say in view of (23.18))

$$\mu([A,C],L) > \mu([A,C],P_0) = d = b - 1 ,$$

and hence L is a b-chord.

(26.2) LEMMA. _If_ $n \geq 2$, _then there exists a_ 2-_chord_ L.

PROOF. Clearly (or say by Bezout's Little Theorem (23.9)) there exists $P_0 \in \mathfrak{D}_0(A,\pi)$ with $\mu([A,C],P_0) \geq 1$, and hence our assertion follows by (26.1).

(26.3) _Assume that_ $n \geq 4$ _and_

$$(*) \qquad \sum_{P\in\mathfrak{D}_0([A,C],\pi)} [\mu([A,C],P) - 1] \geq 2 .$$

Then there exists a 4-_chord_ L.

PROOF. If $\mu([A,C],P_0) \geq 3$ for some $P_0 \in \mathfrak{D}_0([A,C],\pi)$, then our assertion follows from (26.1). If $\mu([A,C],P) < 3$ for all

$P \in \mathfrak{D}_0([A,C],\pi)$ then, in view of (*), we can find two distinct members P_1 and P_2 of $\mathfrak{D}_0([A,C],\pi)$ such that

$$\mu([A,C],P_i) \geq 2 \text{ , for } i = 1,2,$$

and now it clearly suffices to take $L = \Delta(A,P_1,P_2)$.

(26.4). <u>Assume that</u> $n = 5$ <u>and</u>

(*) $$\mu([A,C],\pi,P) \leq 1 \text{ for all } P \in \mathfrak{D}_0(A,\pi) \text{ .}$$

<u>Then there exists a</u> 2-<u>chord</u> L <u>which is not a</u> 3-<u>secant</u>.

PROOF. By (26.2) there exists a 2-chord L'. If L' is not a 3-secant, then it suffices to take $L = L'$. So now assume that L' is a 3-secant. Then, in view of (*), we get $\operatorname{card} \mathfrak{D}_0([A,C],L') = 3$. Let P_1, P_2, P_3 be the distinct members of $\mathfrak{D}_0([A,C],L')$. In view of (*), by Bezout's Little Theorem (23.9) we get $\operatorname{card} \mathfrak{D}_0([A,C],\pi)\setminus \mathfrak{D}_0(A,L') = 2$. Let N_2 and N_3 be the distinct members of $\mathfrak{D}_0([A,C],\pi)\setminus\mathfrak{D}_0(A,L')$. Let $L^* = \Delta(A,P_1,N_2)$. Then clearly L^* is a 2-chord. If L^* is not a 3-secant then it suffices to take $L = L^*$. So now also assume that L^* is a 3-secant. Then in view of (*) we must have $N_3 \in \mathfrak{D}_0(A,L^*)$. For $i = 1,2$ let $L_i \in \mathfrak{D}_1^1(A,\pi)$ be as indicated by the following suggestive figure

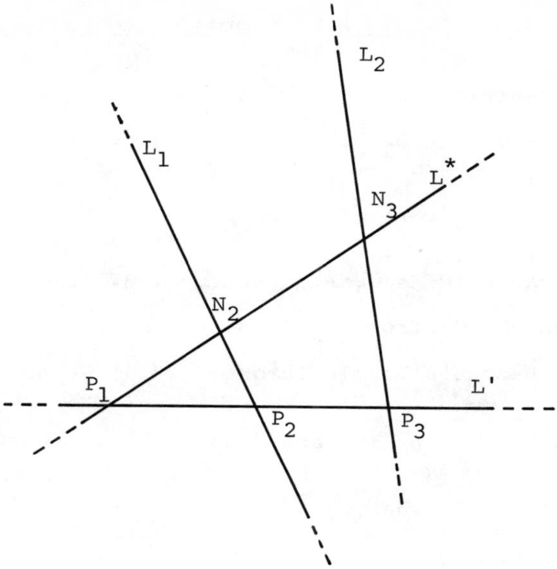

i.e., let $L_i = \Delta(A, N_i, P_i)$. Then in view of (*) we see that L_i is 2-secant for $i = 1,2$. So it suffices to take $L = L_1$ or $L = L_2$.

REMARK. Note that this lemma is just an elementary combinatorial fact:

Namely, $Q_0(A,\pi)$ is a set of five distinct points in the plane π, and we are looking for a line which joins two of them and does not contain exactly three points. There are $\binom{5}{2} = 10$, not necessarily distinct, lines to choose from. If a line contains exactly three of the five points then it appear exactly three times in the ten lines; and since ten is not divisible by three, there must exist a line in the ten lines which does not contain exactly three points.

(26.5) LEMMA. _Assume that_ Emdim[A,C] = 3, _and let_ L _be any_ (n-1)-_chord_. _Then_ L _is an_ (n-1)-_secant_,

$$\text{card}\,\mathfrak{Q}_0([A,L],C) < \infty = \text{card}\,\mathfrak{Q}_0([A,L],\backslash C) \ .$$

and every $N \in \mathfrak{Q}_0([A,L],\backslash C)$ is a best projecting center.

PROOF. The assertions about card follow from our assumption that $L \in \mathfrak{Q}_1^1(A,\pi)$ and $C \in \mathfrak{Q}_1(A,\backslash\pi)$. Now let any

(1) $$N \in \mathfrak{Q}_0([A,L],\backslash C)$$

be given. Then clearly $C^N \in \mathfrak{Q}_1(A^N)$, and by (18.13.1) we also know that the projection of C from N is π-integral. Since $\text{Emdim}[A,C] = 3$, by Bezout's Little Theorem (23.9) we get

(2) $$\mu([A,C],\psi) \le n \quad \text{for any } \psi \in H_1^*(A) \ .$$

Since $N \notin \mathfrak{Q}_0(A,C)$, we see that,

(3) $$\begin{cases} \text{for any } J \in \mathfrak{Q}_1^1(A) \text{ with } N \in \mathfrak{Q}_0(A,J) \text{ and } J \neq L, \text{ we have} \\ \Delta(A,L,J) \in H_1^*(A) \text{ and} \\ \mu([A,C],L) + \mu([A,C],J) \le \mu([A,C],\Delta(A,L,J)) \end{cases}$$

because clearly

$$L \in \mathfrak{Q}(A,\Delta(A,L,J)), J \in \mathfrak{Q}(A,\Delta(A,L,J)), \text{ and } \mathfrak{Q}_0(A,L) \cap \mathfrak{Q}_0(A,J) = \{N\}.$$

By (2) and (3) we get that

(4) $$\begin{cases} \text{for every } J \in \mathfrak{Q}_1^1(A) \text{ with } N \in \mathfrak{Q}_0(A,J) \text{ and } J \neq L, \text{ we have} \\ \mu([A,C],L) + \mu([A,C],J) \le n \ . \end{cases}$$

By assumption

(5) $$\mu([A,C],L) \ge n - 1 \ .$$

Clearly there exists $P \in \mathfrak{Q}_0([A,C],\backslash L)$ and then upon letting $J' = \Delta(A,P,N)$ we get $J' \in \mathfrak{Q}_1^1(A)$ with $N \in \mathfrak{Q}_0(A,J')$, $J' \neq L$, and $\mu([A,C],J') \ge 1$. Consequently by (4) and (5) we see that

(6) $$\mu([A,C],L) = n - 1$$

i.e., L is an (n-1)-secant. In view of (4) and (6), by the Projection Formulas (23.14) and (23.15) we deduce that the projection of C from N is birational,

$$\mathrm{Deg}[A^N,C^N] = \mathrm{Deg}[A,C] = n ,$$

and

$$\mu([A^N,C^N],Q) = 1 \quad \text{for all} \quad Q \in \mathfrak{Q}_0([A^N,C^N],\backslash L^N) .$$

Also clearly

$$\mathfrak{Q}_0([A^N,C^N],\backslash\pi^N) \subset \mathfrak{Q}_0([A^N,C^N],\backslash L^N) .$$

Therefore N is a best projecting center.

(26.6) LEMMA. Assume that $n = 4$. Let L be a 2-secant and let

(*) $$N \in \mathfrak{Q}_0([A,L],\backslash C)$$

be such that

(**) the projection of C from N is birational.

Then N is a better projecting center.

PROOF. Now $N \in \mathfrak{Q}_0([A,\pi],\backslash C)$ and hence by (18.13.1) we know that the projection of C from N is π-integral. By assumption

(1) $\mathrm{Deg}[A,C] = 4$, and $L \in \mathfrak{Q}_1^1(A,\pi)$ with $\mu([A,C],L) = 2$.

In view of (1), (*) and (**), by the Projection Formula (23.14), we see that

(2) $$\mathrm{Deg}[A^N,C^N] = 4$$

and

(3) $\mu([A^N,C^N],L^N) = 2$ and $L^N \in \mathfrak{O}_0([A^N,C^N],\pi^N)$

Given any $Q \in \mathfrak{O}_0(A^N,\backslash L^N)$, upon letting

$$D = \Delta(A^N,L^N,Q)$$

we clearly have

(4) $$D \in H_1^*(A^N)$$

and

(5) $$\mu([A^N,C^N],Q) + \mu([A^N,C^N],L^N) \leq \mu([A^N,C^N],D) .$$

In view of (2) and (4), by Bezout's Little Theorem (23.9) we get

(6) $$\mu([A^N,C^N],D) = 4 ;$$

now by (3), (5) and (6), we get

(7) $$\mu([A^N,C^N],Q) \leq 2 .$$

Therefore N is a better projecting center.

(26.7) LEMMA. _Assume that_ $n = 5$. _Let_ L _be a_ 3-_secant, and let_ $N \in \mathfrak{O}_0(A,L)$ _be such that_

(*) $$\mu([A,C],N) \geq 1$$

(**) $$\mu([A,C],\pi,N) > 1$$

and

(***) _the projection of_ C _from_ N _is birational_.

Then N _is a good projecting center_.

PROOF. By (***) we have

(1) $$\text{Emdim}[A,C,N] \geq 2 .$$

Since

(2) $$N \in \mathfrak{D}_0(A,L)$$

and

(3) $$L \in \mathfrak{D}_1^1(A,\pi) \ ,$$

we get

(4) $$N \in \mathfrak{D}_0(A,\pi) \ .$$

By (*), (**) and (4) we get that

(5) $$\mu([A,C],N) = 1$$

and

(6) card $\mathfrak{Z}([A,C],N) = 1$ and $\pi \subset T_1([A,C],V)$

$$\text{where } \{V\} = \mathfrak{Z}([A,C],N) \ .$$

In view of (1), (4) and (6), by (21.6.2) we see that the projection of C from N is π-integral. Now by assumption

(7) $$\mu([A,C],L) = 3 \ .$$

In view of (***), (2), (3), (4) and (7), by the Projection Formulas (23.14) and (23.15) we see that

(8) $$\mathrm{Deg}[A^N,C^N] = 4$$

and

(9) $\mu([A^N,C^N],L^N) = 2$ and $L^N \in \mathfrak{D}_0([A^N,C^N],\pi^N) \ .$

Given any $Q \in \mathfrak{D}_0(A^N, \backslash L^N)$, upon letting

$$D = \Delta(A^N,L^N,Q)$$

we clearly have

(10) $$D \in H_1^*(A^N)$$

and

(11) $\mu([A^N, C^N], Q) + \mu([A^N, C^N], L^N) \le \mu([A^N, C^N], D)$.

In view of (7) and (9), by Bezout's Little Theorem (23.9) we get

(12) $\mu([A^N, C^N], D) = 4$;

now, by (8), (10) and (11), we get,

(13) $\mu([A^N, C^N], Q) \le 2$.

Therefore N is a good projecting center.

(26.8) CONE LEMMA. Let L be a 2-chord. Assume that $n \le 5$,

(1) $\operatorname{Emdim}[A, C] = 3$,

and

(*) $\mu([A, C], P) = 1$ for all $P \in \mathfrak{Q}_0([A, C])$.

Also assume that there exists $N' \in \mathfrak{Q}_0(A, L)$ such that

(**) the projection of C from N' is not birational.

Then every $N \in \mathfrak{Q}_0([A, L], \backslash C) \backslash \{N'\}$ is a best projecting center.

PROOF. The only composite positive integer ≤ 5 is 4, and $4 = 2$ times 2 is the only proper factorization of 4. Therefore, in view of (1) and (*), by the Special Projection Formula (23.15), we see that $C^{N'} \in \mathfrak{Q}_1(A^{N'})$ and

$$\operatorname{Deg}[A^{N'}, C^{N'}] = 2 \text{ and } \mu([A, C], N') \le 1 .$$

Upon letting

(2) $\Phi = C^{N'}A$

we clearly have

(3) $\Phi \in \mathfrak{Q}(A)$,

(4) $$\Phi \in H^*(A,N') ,$$

(5) $$C \in \mathfrak{D}_1(A,\Phi) ,$$

and, in view of (1), by (25.9) we also see that

(6) $$\Phi \in H_2^*(A) .$$

Now by assumption

$$N' \in \mathfrak{D}_0(A,L) \quad \text{and} \quad \mu([A,C],L) \geq 2 > 1 \geq \mu([A,C],N')$$

and hence by the Projection Formula (23.14) we see that

(7) $$L^{N'} \in \mathfrak{D}_0(A^{N'},C^{N'}) .$$

By (2) and (7) we get

(8) $$L \in \mathfrak{D}_1^1(A,\Phi)$$

By (3) and (6) we see that

(9) $$\mathfrak{D}(A,\Phi) \cap H_1^*(A) = \emptyset .$$

Now let any

(10) $$N \in \mathfrak{D}_0([A,L],\backslash C)\backslash\{N'\} .$$

be given. Let us keep in mind the following suggestive figure:

S,T : Points of C on L, not necessarily distinct and may contain N' .

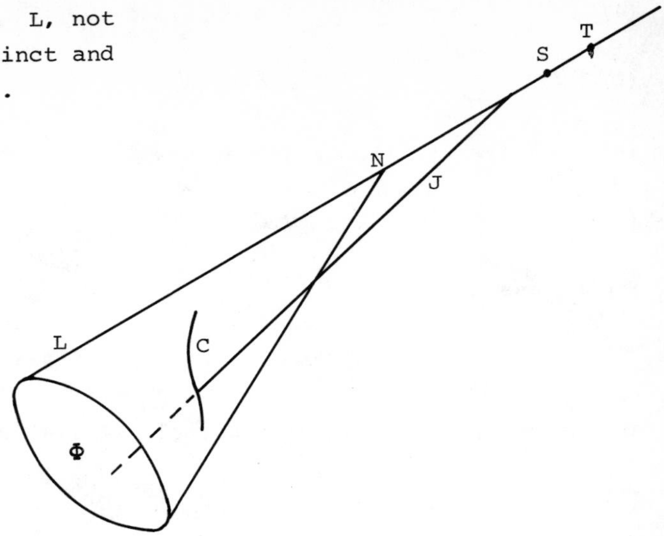

Now in view of (') and (18.3.1) we know that $C^N \in \mathfrak{D}_1(A^N)$ and the projection of C from N is π-integral. Since $N' \in \mathfrak{D}_0(A,L)$, by (4) and (10) we see that

(11) $\left\{ \begin{array}{l} \text{for any } J \in \mathfrak{D}_1^1(A) \text{ with } N \in \mathfrak{D}_0(A,J) \text{ and } J \neq L, \text{ we have} \\ \Delta(A,N',J) \in H_1^*(A) \text{ and if } J \in \mathfrak{D}(A,\Phi), \text{ then} \\ \Delta(A,N',J) \in \mathfrak{D}(A,\Phi) \end{array} \right.$

By (9) and (11) we get that

(12) $\left\{ \begin{array}{l} \text{for any } J \in \mathfrak{D}_1^1(A) \text{ with } N \in \mathfrak{D}_0(A,J) \text{ and } J \neq L, \text{ we have} \\ J \notin \mathfrak{D}(A,\Phi) . \end{array} \right.$

In view of (6) and (12), by Bezout's Little Theorem (23.9) we conclud that

$$(13)\quad \begin{cases} \text{for any } J \in \mathfrak{Q}_1^1(A) \text{ with } N \in \mathfrak{Q}_0(A,J) \text{ and } J \neq L, \text{ we have} \\ \mu([A,J],\Phi) \le 2\ . \end{cases}$$

By (8) and (10) we know that $N \in \mathfrak{Q}_0(A,\Phi)$ and hence by (13) we get that

$$(14)\quad \begin{cases} \text{for any } J \in \mathfrak{Q}_1^1(A) \text{ with } N \in \mathfrak{Q}_0(A,J) \text{ and } J \neq L, \text{ we have} \\ \mu([A,J],\Phi,\backslash N) \le 1\ . \end{cases}$$

By (10) we know that $N \notin \mathfrak{Q}_0(A,C)$, and hence by (5) and (14) we get that

$$(15)\quad \begin{cases} \text{for any } J \in \mathfrak{Q}_1^1(A) \text{ with } N \in \mathfrak{Q}_0(A,J) \text{ and } J \neq L, \text{ we have} \\ \mu([A,J],C) \le 1 \end{cases}$$

(because $\mu([A,J],C) = \mu([A,J],C,\backslash N)$ by (10)

$\le \mu([A,J],\Phi,\backslash N)$ by (5)

≤ 1 by (14).)

In view of (*), by the Commuting Lemma (23.19) we know that

$$\mu([A,J],C) = \mu([A,C],J) \quad \text{for any } J \in \mathfrak{Q}_1^1(A)$$

and hence by (15) we get that

$$(16)\quad \begin{cases} \text{for any } J \in \mathfrak{Q}_1^1(A) \text{ with } N \in \mathfrak{Q}_0(A,J) \text{ and } J \neq L, \text{ we have} \\ \mu([A,C],J) \le 1\ . \end{cases}$$

Now in view of (16), by the Projection Formula (23.14) we conclude that the projection of C from N is birational,

$$\mathrm{Deg}[A^N, C^N] = \mathrm{Deg}[A,C] = n$$

and

(17) $\mu([A^N,C^N],Q) = 1$ for all $Q \in \mathfrak{Q}_0([A^N,C^N],\backslash L^N)$.

Since $L \in \mathfrak{Q}_1^1(A,\pi)$, by (17) we get

$$\mu([A^N,C^N],Q) = 1 \quad \text{for all} \quad Q \in \mathfrak{Q}_0([A^N,C^N],\backslash \pi^N) .$$

Therefore N is a best projecting center.

(26.9) PLANE LEMMA. Assume that L, J_1, J_2 are distinct members of $\mathfrak{Q}_1^1(A)$ such that, for $\Delta(A,L,J_1,J_2) = \psi$, say, we have $\psi \in H_1^*(A)$. Further assume that, $\mathfrak{Q}_0(A,L) \cap \mathfrak{Q}_0(A,J_1) \cap \mathfrak{Q}_0(A,J_2) = \phi$.

Note that, then, there is a unique member, say N, common to $\mathfrak{Q}_0(A,J_1)$ and $\mathfrak{Q}_0(A,J_2)$; and then $N \notin \mathfrak{Q}_0(A,L)$.

Further assume that

(*) $$\mu([A,C],N) \leq 1$$

and

(**) $$t + \sum_{i=1}^{2} \mu([A,C],J_i) \geq n + 2, \text{ where}$$

$t = \Sigma\mu([A,C],L,P)$, the summation being extended over all $P \in \mathfrak{Q}_0(A,C)\backslash(\mathfrak{Q}_0(A,J_1) \cup \mathfrak{Q}_0(A,J_2))$.

Then we have

$$C \in \mathfrak{Q}_1(A,\psi) \quad \text{and hence} \quad \mathrm{Emdim}[A,C] \leq 2 .$$

PROOF. In view of (*), by (23.18.3) we see that

$$\mu([A,C],J_i,N) \leq 1 \quad \text{for} \quad i = 1 \text{ or } 2$$

and hence upon relabelling J_1 and J_2 suitably we may suppose that

(1) $$\mu([A,C],J_1,N) \leq 1.$$

By the definition of N we clearly have

(2) $$\mu([A,C],J_1,N) + \mu([A,C],J_1,\backslash J_2) = \mu([A,C],J_1) .$$

By (1) and (2) we get

(3) $$1 + \mu([A,C],J_1,\backslash J_2) \geq \mu([A,C],J_1) .$$

By the definition of ψ we have

(4) $$L \in \mathfrak{D}(A,\psi) \text{ and } J_i \in \mathfrak{D}(A,\psi) \text{ for } i = 1,2 .$$

Now clearly

$$\begin{aligned}\mu([A,C],\psi) &\geq t + \mu([A,C],J_1,\backslash J_2) + \mu([A,C],J_2) && \text{by (4)}\\ &\geq n + 1 && \text{by (**) and (3)}\end{aligned}$$

and hence by Bezout's Little Theorem (23.9) we get $C \in \mathfrak{D}_1(A,\psi)$.

(26.10) QUADRIC LEMMA. <u>Let</u> L <u>be a</u> 2-<u>secant</u>. <u>Assume that</u> $n = 5$,

(*) $$\mu([A,C],P) = 1 \quad \underline{\text{for all}} \quad P \in \mathfrak{D}_0([A,C],\backslash L) .$$

<u>and</u>

(**) $$\begin{cases} \underline{\text{there does not exist any}} \quad \Phi \in H_2^*(A) \quad \underline{\text{such that}} \\ \Phi \quad \underline{\text{is a}}\ \pi\text{-}\underline{\text{quasiplane and}} \quad C \in \mathfrak{D}(A,\Phi) . \end{cases}$$

Let

$$\Omega = \{N \in \mathfrak{D}_0([A,L],\backslash C): \ N \ \underline{\text{is a better projection center}}\}.$$

Then

$$\text{card } \mathfrak{D}_0([A,L],\backslash C)\backslash\Omega) \leq 2 \quad \text{and} \quad \text{card } \mathfrak{D}_0(A,L)\backslash\Omega < \infty = \text{card } \Omega.$$

PROOF. In view of (**), by (25.6) we see that

(1) $$\text{Emdim}[A,C] = 3.$$

Given any $N \in \mathfrak{Q}_0([A,L],\backslash C)$, by (18.13.1) we know that the projection of C from N is π-integral, and (because 5 is a prime number and $\operatorname{Emdim}[A,C] = 3$) in view of the Projection Formula (23.14) and (23.15) we see that the projection of C from N is birational,

$$\operatorname{Deg}[A^N,C^N] = 5 ,$$

$$\mu([A^N,C^N],J^N) = \mu([A,C],J) \quad \text{for every} \quad J \in \mathfrak{Q}_1^1(A) \quad \text{with} \quad N \in \mathfrak{Q}_0(A,J)$$

and hence in particular

$$\mu([A^N,C^N],L^N) = \mu([A,C],L) = 2 \quad \text{and} \quad L^N \in \mathfrak{Q}_0([A^N,C^N],\pi^N) .$$

Thus it only remains to prove that $\operatorname{card} \Omega^* \le 2$ where

$$\Omega^* = \{N \in \mathfrak{Q}_0([A,L],\backslash C) : \mu([A,C],J) \ge 3 \quad \text{for some} \quad J \in \mathfrak{Q}_1^1(A) \quad \text{with} \quad N \in \mathfrak{Q}_0(A,J)\} .$$

Suppose, if possible, that $\operatorname{card} \Omega^* \ge 3$. Then we can take

(2) three distinct members N_1,N_2,N_3 of $\mathfrak{Q}_0([A,L],\backslash C)$

for which we can find

(3) $J_i \in \mathfrak{Q}_1^1(A)$ with $N_i \in \mathfrak{Q}_0(A,J_i)$ for $i = 1,2,3$

such that

(4) $\mu([A,C],J_i) \ge 3$ for $i = 1,2,3$.

Since L is a 2-secant, in view of (2), (3) and (4) we get

(5) $\mathfrak{Q}_0(A,L) \cap \mathfrak{Q}_0(A,J_i) = \{N_i\}$ and $N_i \notin \mathfrak{Q}_0(A,C)$ for $i = 1,2,3$.

Now we can choose $P_{ij} \in \mathfrak{Q}_0(A)$ as indicated in the following suggestive figure

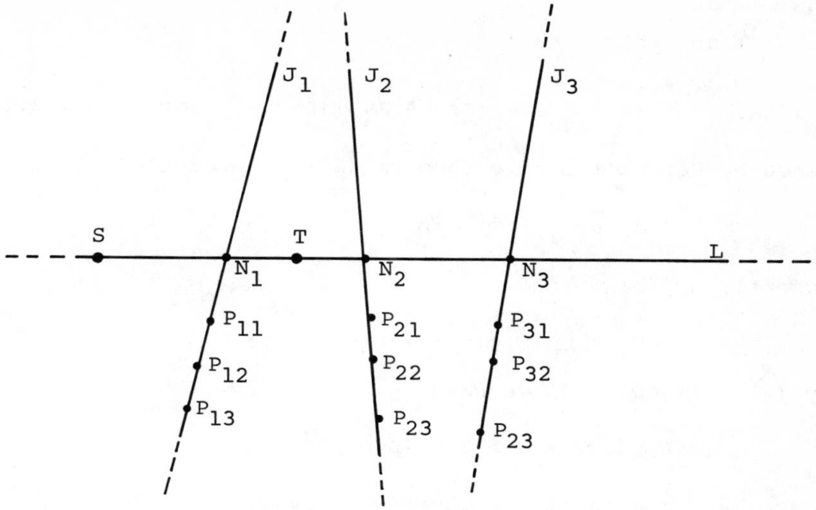

S,T : Points of C on L, not necessarily distinct.

$\mathfrak{D}_0([A,L],C) = \{S,T\}$.

i.e., we choose nine distinct members P_{ij}, $i,j = 1,2,3$, of $\mathfrak{D}_0(A)$ such that

(6) $$P_{ij} \in \mathfrak{D}_0(A,J_i) \quad \text{for} \quad i,j = 1,2,3 .$$

Now we have,

$[H_2(A) : H_0(A)]$

= number of distinct monomials of degree 2 in 4 indeterminates

= 10 ,

(i.e., geometrically speaking, there are ∞^9 quadric surfaces in projective 3-space) and hence we can find

(7) $$\Phi \in H_2^*(A)$$

such that

(8) $$P_{ij} \in \mathfrak{D}_0(A,\Phi) \quad \text{for} \quad i,j = 1,2,3.$$

By (6), (7) and (8) we have

$$\mu([A,J_i],\Phi) \geq 3 > 2 = (\mathrm{Deg}[A,J_i])(\mathrm{Deg}_A\Phi), \text{ for } i = 1,2,3,$$

and hence by Bezout's Little Theorem (23.9) we conclude that

(9) $$J_i \in \mathfrak{D}_1^1(A,\Phi) \quad \text{for} \quad i = 1,2,3.$$

Consequently by (3) we get

(10) $$N_i \in \mathfrak{D}_0(A,\Phi) \quad \text{for} \quad i = 1,2,3.$$

Now by (2), (7) and (10) we have

$$\mu([A,L],\Phi) \geq 3 > 2 = (\mathrm{Deg}[A,L])(\mathrm{Deg}_A\Phi)$$

and hence by Bezout's Little Theorem we conclude that

(11) $$L \in \mathfrak{D}_1^1(A,\Phi).$$

Since L is a 2-secant, we have

(12) $$\mu([A,C],L) = 2.$$

Now let

(13_1) $s = \sum \mu([A,C],\Phi,P)$, where the summation is extended over all $P \in \mathfrak{D}_0(A,C) \setminus \bigcup_{i=1}^{3} \mathfrak{D}_0(A,J_i)$.

In view of (5), (11), (12) and (13_1) we have

(13_2) $$s \geq \mu([A,C],L) = 2 .$$

Now suppose, if possible $\mathfrak{D}_0(A,J_i) \cap \mathfrak{D}_0(A,J_j) \neq \phi$ for some distinct $i,j \in \{1,2,3\}$. In view of (*),(2),(3),(4),(5),(12),(13_1) and (13_2), we see that (26.9) can be applied to L,J_i and J_j; and we get $\mathrm{Emdim}[A,C] \leq 1$, a contradiction to (1). Thus we have proved

(14) $$\mathfrak{D}_0(A,J_i) \cap \mathfrak{D}_0(A,J_j) = \phi, \text{ for distinct } i,j \in \{1,2,3\}.$$

Now (9) gives,

(15) $$\mu([A,C],\Phi,J_i) \geq \mu([A,C],J_i), \text{ for } i = 1,2,3.$$

Now

$\mu([A,C],\Phi)$

$\geq s + \sum_{i=1}^{3} \mu([A,C],\Phi,J_i)$ by (13_1) and (14)

$\geq s + \sum_{i=1}^{3} \mu([A,C],J_i)$ by (15)

≥ 11 by (4) and (13_2)

> 5 times 2

$= (\mathrm{Deg}[A,C](\mathrm{Deg}_A\Phi)$ by (7)

and hence by Bezout's Little Theorem (23.9) we must have

(16) $C \in \mathfrak{Q}_1(A,\Phi)$

In view of (**), (7), (11) and (16), by (25.7) we conclude that

(17) $\Phi \in H^*(A,P)$ for some $P \in \mathfrak{Q}_0(A,L)$

In view of (14), we can take a

(18) permutation $(e(1), e(2), e(3))$ of $(1,2,3)$

such that

(19) $P \notin \mathfrak{Q}_0(A,J_{e(i)})$ for $i = 1,2$.

Upon letting

(20) $\Psi_i = \Delta(A,P,J_{e(i)})$ for $i = 1,2$,

by (14),(18) and (19) we get

(21) $\Psi_i \in H_1^*(A)$ for $i = 1,2$ with $\Psi_1 \neq \Psi_2$

and

(22) $J_{e(3)} \notin \mathfrak{Q}(A,\Psi_1\Psi_2)$.

By (7), (9), (17), (18), (20) and (21) we see that

(23) $$\Phi = \psi_1 \psi_2 \ .$$

By (9) and (18) we also have

(24) $$J_{e(3)} \in \mathfrak{D}(A, \Phi) \ .$$

Now (22), (23) and (24) yield a contradiction; therefore we must have card $\Omega^* \le 2$.

(26.11) PROPOSITION.

(26.11.1) _If_ $\text{Emdim}[A,C] \neq 3$ _then_: _there exists_ $\psi \in H_1^*(A)$ _such that_ $C \in \mathfrak{D}(A,\psi)$, _and there exists_ $\Phi \in H_2^*(A)$ _such that_ Φ _is a_ π_-quasiplane and_ $C \in \mathfrak{D}(A,\Phi)$.

(26.11.2) _If_ $n \le 2$ _then_ $\text{Emdim}[A,C] \neq 3$.

(26.11.3) _If_ $n = 3 = \text{Emdim}[A,C]$, _then_: _there exists a_ 2-_chord_; _every_ 2-_chord is a_ 2-_secant_; _and for any_ 2-_secant_ L _we have that_

$$\text{card } \mathfrak{D}_0([A,L],C) < \infty = \text{card } \mathfrak{D}_0([A,L],\backslash C)$$

and every $N \in \mathfrak{D}_0([A,L],\backslash C)$ _is a best projecting center_.

(26.11.4) _If_ $n = 4$, $\text{Emdim}[A,C] = 3$, _and_

$$\mu([A,C],P_0) \neq 1 \text{ for some } P_0 \in \mathfrak{D}_0([A,C],\pi)$$

then: _there exists a_ 3-_chord_; _every_ 3-_chord is a_ 3-_secant_; _and for any_ 3-_secant_ L _we have that_

$$\text{card } \mathfrak{D}_0([A,L],C) < \infty = \text{card } \mathfrak{D}_0([A,L],\backslash C)$$

and every $N \in \mathfrak{D}_0([A,L],\backslash C)$ _is a best projecting center_.

(26.11.5) _Assume that_ $n = 4$, $\text{Emdim}[A,C] = 3$ _and_

$$\mu([A,C],P) = 1 \text{ \underline{for all} } P \in \mathfrak{D}_0([A,C],\pi) \ .$$

Then there exists a 2-*chord, and for any* 2-*chord* L *we have the following. If* L *is not a* 2-*secant then* L *is a* 3-*secant,*

$$\operatorname{card} \mathfrak{D}_0([A,L],C) < \infty = \operatorname{card} \mathfrak{D}_0([A,L],\backslash C) ,$$

and every $N \in \mathfrak{D}_0([A,L],\backslash C)$ *is a best projecting center. If* L *is a* 2-*secant and*

$$\mu([A,C],P) = 1 \quad \textit{for all} \quad P \in \mathfrak{D}_0([A,C],\backslash\pi) ,$$

then upon letting

$$\Omega = \{N \in \mathfrak{D}_0([A,L],\backslash C): \ N \text{ is a better projecting center}\}$$

we have

$$\operatorname{card} \mathfrak{D}_0([A,L],\backslash C)\backslash\Omega \le 1 \quad \textit{and} \quad \operatorname{card} \mathfrak{D}_0(A,L)\backslash\Omega < \infty = \operatorname{card} \Omega .$$

(26.11.6) *If* $n = 5$, $\operatorname{Emdim}[A,C] = 3$, *and*

$$\sum_{P\in\mathfrak{D}_0([A,C],\pi)} [\mu([A,C],P) - 1] \ge 2 ,$$

then: there exists a 4-*chord; every* 4-*chord is a* 4-*secant; and for any* 4-*secant* L *we have that*

$$\operatorname{card} \mathfrak{D}_0([A,L],C) < \infty = \operatorname{card} \mathfrak{D}_0([A,L],\backslash C)$$

and every $N \in \mathfrak{D}_0([A,L],\backslash C)$ *is a best projecting center.*

(26.11.7) *Assume that:* $n = 5$;

(*) $$2 > \sum_{P\in\mathfrak{D}_0([A,C],\pi)} [\mu([A,C],P) - 1] \ne 0 ;$$

(**) $$\mu([A,C],P) = 1 \quad \textit{for all} \quad P \in \mathfrak{D}_0([A,C],\backslash\pi) ;$$

and there does not exist any $\Phi \in H_2^*(A)$ *such that* Φ *is a* π-*quasiplane and* $C \in \mathfrak{D}_1(A,\Phi)$. [Note that (*) + (**) is equivalent to assuming that there exists $P_0 \in \mathfrak{D}_0([A,C],\pi)$ such that:

$\mu([A,C],P_0) = 2$, and $\mu([A,C],P) = 1$ for all $P \in \mathfrak{Q}_0([A,C],\backslash P_0)$.

Then there exists a 2-secant L such that

$$\mu([A,C],P) = 1 \text{ for all } P \in \mathfrak{Q}_0([A,C],\backslash L) ,$$

and for any such 2-secant L upon letting

$$\Omega = \{N \in \mathfrak{Q}_0([A,L],\backslash C): N \text{ is a better projecting center}\}$$

we have

$$\text{card } \mathfrak{Q}_0([A,L],\backslash C)\backslash\Omega \le 2 \text{ and } \text{card } \mathfrak{Q}_0(A,L)\backslash\Omega < \infty = \text{card } \Omega .$$

(26.11.8) Assume that: $n = 5$;

(*) $$\sum_{P\in \mathfrak{Q}_0([A,C],\pi)} [\mu([A,C],P) - 1] = 0 ;$$

(**) $$\mu([A,C],P) = 1 \text{ for all } P \in \mathfrak{Q}_0([A,C],\backslash\pi) ;$$

$$\mu([A,C],\pi,N') > 1 \text{ for some } N' \in \mathfrak{Q}_0(A,\pi) ;$$

and there does not exist any $\Phi \in H_2^*(A)$ such that Φ is a π-quasiplane and $C \in \mathfrak{Q}_1(A,\Phi)$. [Note that (*) + (**) is equivalent to the assumption that

$$\mu([A,C],P) = 1 \text{ for all } P \in \mathfrak{Q}_0([A,C]).]$$

Then there exists a 2-chord L such that $N' \in \mathfrak{Q}_0(A,L)$, and for any such 2-chord L we have the following. If L is a 2-secant then upon letting

$$\Omega = \{N \in \mathfrak{Q}_0([A,L],\backslash C): N \text{ is a better projecting center}\}$$

we have

$$\text{card } \mathfrak{Q}_0([A,L],\backslash C)\backslash\Omega \le 2 \text{ and } \text{card}\mathfrak{Q}_0(A,L)\backslash\Omega < \infty = \text{card } \Omega .$$

If L is a 3-secant and the projection of C from N' is birational *then* N' *is a good projecting center*. *If* L *is a* 3-*secant and the projection of* C *from* N' *is not birational then*

$$\text{card}\ \mathfrak{Q}_0([A,L],C) < \infty = \text{card}([A,L],\backslash C)$$

and every $N \in \mathfrak{Q}_0([A,L],\backslash C)$ *is a best projecting center*. *If* L *is neither a* 2-*secant nor a* 3-*secant*, *then*. L *is a* 4-*secant*,

$$\text{card}\ \mathfrak{Q}_0([A,L],C) < \infty = \text{card}\ \mathfrak{Q}_0([A,L],\backslash C) \ ,$$

and every $N \in \mathfrak{Q}_0([A,L],\backslash C)$ *is a best projecting center*.

(26.11.9) *Assume that*: $n = 5$;

(*) $$\sum_{P \in \mathfrak{Q}_0([A,C],\pi)} [\mu([A,C],P) - 1] = 0 \ ;$$

(**) $\mu([A,C],P) = 1$ *for all* $P \in \mathfrak{Q}_0([A,C],\backslash\pi)$;

(***) $\mu([A,C],\pi,P) \le 1$ *for all* $P \in \mathfrak{Q}_0(A,\pi)$;

and there does not exist any $\Phi \in H_2^*(A)$ *such that* Φ *is a* π-*quasiplane and* $C \in \mathfrak{Q}_1(A,\Phi)$. [Note that (***) $\Rightarrow$ (*) ; also note that (*) + (**) is equivalent to the assumption that

$$\mu[A,C],P) = 1 \quad \text{for all} \quad P \in \mathfrak{Q}_0([A,C]).]$$

Then there exists a 2-*chord* L *such that* L *is not a* 3-*secant*, *and for any such* L *we have the following*. *If* L *is a* 2-*secant then upon letting*

$$\Omega = \{N \in \mathfrak{Q}_0([A,L],\backslash C): \ N \text{ is a better projecting center}\}$$

we have

$$\text{card}\ \mathfrak{Q}_0([A,L],\backslash C)\backslash\Omega \le 2 \quad \textit{and} \quad \text{card}\ \mathfrak{Q}_0(A,L)\backslash\Omega < \infty = \text{card}\ \Omega \ .$$

If L *is not a* 2-*secant then* L *is a* 4-*secant*,

$$\text{card } \mathfrak{D}_0([A,L],C) < \infty = \text{card } \mathfrak{D}_0([A,L],\backslash C) ,$$

and every $N \in \mathfrak{D}_0([A,L],\backslash C)$ *is a best projecting center.*

PROOF. The first assertion in (26.11.1) is obvious and, in view of (25.6), the second assertion follows from the first assertion. (26.11.2) follows from (23.11). (26.11.3) follows from (26.2) and (26.5), (26.11.4) follows from (26.1) and (26.5). (26.11.5) follows from (26.2), (26.5), (26.6) and (26.8). (26.11.6) follows from (26.3) and (26.5). To prove (26.11.7), first note that by (*) there exists $P_0 \in \mathfrak{D}_0([A,C],\pi)$ such that $\mu([A,C],P_0) = 2$, then note that by (26.1) there exists a 2-secant L such that $P_0 \in \mathfrak{D}_0(A,L)$, and finally note that by (*) and (**) we must now have $\mu([A,C],P) = 1$ for all $P \in \mathfrak{D}_0([A,C],\backslash L)$; the rest of (26.11.7) follows from (26.10). (26.11.8) follows from (26.1), (26.10), (26.7), (26.8) and (26.5). Finally, (26.11.9) follows from (26.4), (26.10) and (26.5).

(26.12) THEOREM. *Assume that* $n \le 5$ *and*

$$\mu([A,C],P) = 1 \quad \text{for all} \quad P \in \mathfrak{D}_0([A,C],\backslash\pi) .$$

Then at least one of the following three situations prevails.

(1) *There exists* $\Phi \in H_2^*(A)$ *such that:* Φ *is a* π-*quasiplane and* $C \in \mathfrak{D}_1(A,\Phi)$.

(2) $3 \le n \le 5$ *and there exists* $N \in \mathfrak{D}_0(A,\pi)$ *such that: the projection of* C *from* N *is birational, the projection of* C *from* N *is* π-*integral,* $\mu([A^N,C^N],Q) = 1$ *for all* $Q \in \mathfrak{D}_0([A^N,C^N],\backslash\pi^N)$, *and* $\min(4,n) \le \text{Deg}[A^N,C^N] \le n$.

(3) $4 \le n \le 5$ *and there exists* $N \in \mathfrak{D}_0(A,\pi)$ *such that: the projection of* C *from* N *is birational, the projection of* C *from* N *is* π-*integral,* $\mu([A^N,C^N],Q) \le 2$ *for all* $Q \in \mathfrak{D}_0(A^N)$, $\mu([A^N,C^N],Q_0) = 2$ *for some* $Q_0 \in \mathfrak{D}_0([A^N,C^N],\backslash\pi^N)$, $\mu([A^N,C^N],Q_1) = 2$

for some $Q_1 \in \mathfrak{D}_0([A^N,C^N],\pi^N)$, and $\min(4,n) \le \mathrm{Deg}[A^N,C^N] \le n$.

PROOF. Follows from (26.11).

CHAPTER III: BIRATIONAL GEOMETRY OR GENUS

In this chapter X,Y,Z are indeterminates; for any polynomial $\xi(Y)$ in Y, with coefficients in some ring, $\xi'(Y)$ denotes the Y-derivative of $\xi(Y)$; for any polynomial in two or more of the indeterminates X,Y,Z, the partial derivatives are indicated by the corresponding subscript; thus for instance, for a polynomial $\eta(X,Y)$ in X and Y, with coefficients in some ring, $\eta_X(X,Y)$ denotes the partial derivative of $\eta(X,Y)$ with respect to X, and $\eta_Y(X,Y)$ denotes the partial derivative of $\eta(X,Y)$ with respect to Y.

§27. Different

We shall now review Dedekind's theory of the different.

(27.1) DEFINITION. Let S be a normal domain with quotient field L. Let R be the integral closure of S in a finite algebraic field extension of K of L. We define

$$\mathfrak{C}^*(S,K) = \underline{\text{the complementary module of}}\ S\ \underline{\text{in}}\ K$$
$$= \{\alpha \in K: \mathrm{Trace}_{K/L}(\alpha\beta) \in S \text{ for all } \beta \in R\}$$

and we note that then $\mathfrak{C}^*(S,K)$ is an R-submodule of K with $R \subset \mathfrak{C}^*(S,K)$. We define

$$\mathfrak{D}(S,K) = \underline{\text{the different of}}\ S\ \text{in}\ K$$
$$= \{a \in K: a\alpha \in R \text{ for all } \alpha \in \mathfrak{C}^*(S,K)\}$$

and we note that then $\mathfrak{D}(S,K) \subset R$ and $\mathfrak{D}(S,K)$ is an ideal in R. In (27.2) we shall see that, if K is separable over L, then $\mathfrak{D}(S,K) \neq \{0\}$; after proving (27.2) we shall use this observation tacitly.

(27.2) DEDEKINDS FORMULA FOR DIFFERENT AND CONDUCTOR. *Let* S *be a normal domain with quotient field* L. *Let* R *be the integral*

<u>closure of</u> S <u>in a finite algebraic seaprable field extension</u> K <u>of</u> L. <u>Let</u> $y \in R$ <u>be such that</u> $K = L(y)$, <u>and let</u>

$$\xi(Y) = Y^n + \xi_{n-1}Y^{n-1}+\ldots+\xi_0 \quad \underline{\text{with}} \quad \xi_0,\ldots,\xi_{n-1} \quad \underline{\text{in}} \quad L$$

<u>be the minimal monic polynomial of</u> y <u>over</u> L. (Note that since S is normal we have $\xi_i \in S$ for all i.) <u>Then we have</u>

$$\mathfrak{C}(S[y]) = \xi'(y)\mathfrak{C}^*(S,K) .$$

<u>Also we have</u>

$$\xi'(y) \in \mathfrak{C}(S[y]) \cap \mathfrak{D}(S,K) \quad \text{and} \quad \mathfrak{C}(S[y])\mathfrak{D}(S,K) \subset \xi'(y)R .$$

(Whence in particular $\mathfrak{D}(S,K) \neq \{0\}$, because $\xi'(y) \neq 0$.) <u>Moreover, if</u> S <u>is a Dedekind domain, then we have</u>

$$\mathfrak{C}(S[y])\mathfrak{D}(S,K) = \xi'(y)R .$$

PROOF. We can take elements $y_1,\ldots,y_n$ in an overfield of K such that, $y = y_1$ and

$$\xi(Y) = (Y - y_1)(Y - y_2)\ldots(Y-y_n).$$

For any $\zeta(Y) \in L[Y]$ of degree $< n$, by Lagrange interpolation, we have

$$(1) \qquad \zeta(Y) = \sum_{i=1}^{n} \frac{\zeta(y_i)}{\xi'(y_i)} \frac{\xi(Y)}{Y-y_i} .$$

For proof, note that both sides are polynomials in Y of degree $< n$ and their values coincide for the n distinct values $y_1,y_2,\ldots,y_n$ of Y. Since $\xi(Y) \in S[Y]$ is monic of degree n, by the division algorithm we have

$$(2) \qquad \xi(Y) = (Y-Z) \sum_{j=0}^{n-1} \eta_j(Z)Y^j + \xi(Z)$$

with

(3) $$\eta_j(Z) \in S[Z], \text{ for } 0 \le j \le n-1 ,$$

and

(4) $$\eta_{n-1}(Z) = 1.$$

By (2) we have

$$\frac{\xi(Y)}{Y-y_i} = \sum_{j=0}^{n-1} \eta_j(y_i)Y^j \quad , \quad \text{for } 1 \le i \le n ,$$

and substituting these in (1) and collecting the coefficients of Y^j we get

(5) $$\zeta(Y) = \sum_{j=0}^{n-1} \operatorname{Trace}_{K/L}\left(\frac{\zeta(y)\eta_j(y)}{\xi'(y)}\right)Y^j \text{ for every}$$

$$\zeta(Y) \in L[Y] \text{ of degree } < n .$$

Given any $z \in K$ we can write $z = \zeta(y)$ for some $\zeta(Y) \in L[Y]$ of degree $< n$, and then substituting y for Y in (5) we get that

(6) $$z = \sum_{j=0}^{n-1} \operatorname{Trace}\left(\frac{z\eta_j(y)}{\xi'(y)}\right) y^j \quad \text{for all } z \in K .$$

Given any $z \in S[y]$, we can write

$$z = \zeta(y), \text{ where } \zeta(Y) = \sum_{j=0}^{n-1} \zeta_j Y^j \text{ with } \zeta_j \in S ,$$

and then we get

$$\zeta_{n-1} = \operatorname{Trace}_{K/L}\left(\frac{z}{\xi'(y)}\right) \qquad \text{by (4) and (5).}$$

Thus we have

(7) $$\operatorname{Trace}_{K/L}\left(\frac{z}{\xi'(y)}\right) \in S, \text{ for all } z \in S[y] .$$

To prove that $\mathfrak{C}(S[y] \subset \xi'(y)\mathfrak{C}^*(S,K)$ it suffices to show that

$$\theta \in \mathfrak{C}(S[y]) \text{ and } \beta \in R \Rightarrow \operatorname{Trace}_{K/L}\left(\frac{\theta\beta}{\xi'(y)}\right) \in S$$

and this in turn follows by taking $z = \theta\beta$ in (7).

To prove the reverse inclusion $\mathfrak{C}(S[y]) \supset \xi'(y)\mathfrak{C}^*(S,K)$, it suffices to show that,

$$\alpha \in \mathfrak{C}^*(S,K) \text{ and any } \gamma \in R \Rightarrow \xi'(y)\alpha\gamma \in S[y].$$

So let any $\alpha \in \mathfrak{C}^*(S,K)$ and any $\gamma \in R$ be given. Since $\gamma \in R$, by (3) we get

(8) $$\gamma\eta_j(y) \in R \quad \text{for} \quad 0 \le j \le n-1.$$

Since $\alpha \in \mathfrak{C}^*(S,K)$, by (8) we get

(9) $$\text{Trace}_{K/L}(\alpha\gamma\eta_j(y)) \in S \quad \text{for} \quad 0 \le j \le n-1.$$

Upon letting $z = \xi'(y)\alpha\gamma$, by (6) and (9) we get $z \in S[y]$.

This completes the proof of the equation

(10) $$\mathfrak{C}(S[y]) = \xi'(y)\mathfrak{C}^*(S,K).$$

From (10) it immediately follows that

$$\xi'(y) \in \mathfrak{C}(S[y]) \cap \mathfrak{D}(S,K)$$

Upon multiplying both sides of (10) by $\mathfrak{D}(S,K)$, we get

$$\mathfrak{C}(S[y])\mathfrak{D}(S,K) \subset \xi'(y) \text{ , in the general case,}$$

and, in case S is a Dedekind domain, we further have,

$$\mathfrak{C}(S[y])\mathfrak{D}(S,K) = \xi'(y)R \text{ ;}$$

because then R is also a Dedekind domain and hence for any finite R-submodule I of K, upon letting $J = \{\alpha \in K\colon ab \in R$ for all $b \in I\}$, we have $IJ = R$; hence $\mathfrak{C}^*(S,K)\mathfrak{D}(S,K) = R$.

(27.3) LEMMA. <u>Let</u> S <u>be a domain with quotient field</u> L, <u>let</u> K <u>be a finite algebraic field extension of</u> L, <u>let</u> $y \in K$ <u>be such</u>

<u>that</u> $K = L(y)$, <u>and let</u> $\xi(Y)$ <u>be the minimal monic polynomial of</u> y <u>over</u> L. <u>Assume that</u> $\xi(Y) \in S[Y]$. <u>Let</u> Q <u>be a prime ideal in</u> $S[Y]$. <u>Then</u>: $\xi'(y) \notin Q \Leftrightarrow (Q \cap S)S[y]_Q = M(S[y]_Q)$ <u>and</u> $S[y]_Q$ <u>is residually separable over</u> S.

PROOF. Let $S^* = S_{S\cap Q}$; $R = S^*[y]$ and $P = QR$. Then S^* is a quasilocal domain and P is a maximal ideal in R with $P \cap S^* = M(S^*)$. Let $f\colon R \to R/P$ be the canonical epimorphism. Let $z = f(y)$ and $k = f(S^*)$. Then k is a field and $f(R) = k(z) =$ a finite algebraic field extension of k. Let $\eta(Y)$ be the minimal monic polynomial of z over k. Then our assertion is clearly equivalent to:

$$(*)\qquad \xi'(y) \notin P \Leftrightarrow M(S^*)R_P = M(R_P) \text{ and } \eta'(z) \neq 0 .$$

Let $u\colon S^*[Y] \to R$ be the unique S^*-epimorphism with $u(Y) = y$; then clearly $\operatorname{Ker} u = \xi(Y)S^*[Y]$. Let $v\colon S^*[Y] \to k[Y]$ be the unique epimorphism such that $v(Y) = Y$ and $v(a) = f(a)$ for all $a \in S^*$; then clearly $\operatorname{Ker} v = M(S^*)S^*[Y]$. Let $\zeta(Y) \in k[Y]$ be obtained by applying f to the coefficients of $\xi(Y)$, and let

$$(1)\qquad I = v(u^{-1}(M(S^*)R)) \text{ and } J = v(u^{-1}(P)) .$$

Then clearly

$$(2)\qquad \zeta(Y)k[Y] = I \subset J = \eta(Y)k[Y] = \text{a maximal ideal in } k[Y].$$

$$(3)\qquad \xi'(Y) \notin P \Leftrightarrow \zeta'(Y) \notin J$$

and

$$(4)\qquad \eta'(z) \neq 0 \Leftrightarrow \eta'(Y) \notin \eta(Y)k[Y] .$$

In view of (2) we have

$$\zeta(Y) = \eta(Y)^i\alpha(Y)$$

where i is a positive integer and $\alpha(Y) \in k[Y]$ is coprime with $\eta(Y)$; differentiating both sides we get ,

$$\zeta'(Y) = i\eta(Y)^{i-1}\eta'(Y)\alpha(Y) + \eta(Y)^{i}\alpha'(Y)$$

and hence

$$(5) \qquad \zeta'(Y) \notin \eta(Y)k[Y] \Leftrightarrow i = 1 \quad \text{and} \quad \eta'(Y) \notin \eta(Y)k[Y].$$

By (2) and (5) we see that

$$(6) \qquad \zeta'(Y) \notin J \Leftrightarrow Ik[Y]_J = M(k[Y]_J) \quad \text{and} \quad \eta'(Y) \notin \eta(Y)k[Y].$$

We observe that, if h is an epimorphism of a domain D onto a domain E and V and W are ideals in D such that $\operatorname{Ker} h \subset V \subset W$ and W is prime, then clearly: $VD_W = M(D_W) \Leftrightarrow h(V)E_{h(W)} = M(E_{h(W)})$. In view of (1), upon applying this observation twice we see that

$$(7) \qquad M(S^*)R_P = M(R_P) \Leftrightarrow Ik[Y]_J = M(k[Y]_J) .$$

Now (*) follows from (3), (4), (6) and (7) .

REMARK. Note that we have:

$$\begin{array}{ccccccc}
 & & M(S^*)S^*[Y] & \xrightarrow{u'} & P & & \\
 & & \downarrow & & \downarrow & & \\
\xi(Y)S^*[Y] & \longrightarrow & S^*[Y] & \xrightarrow{u} & S^*[y] = R & \longrightarrow & 0 \\
\downarrow{\scriptstyle v'} & & \downarrow{\scriptstyle v} & & \downarrow{\scriptstyle f} & & \\
\eta(Y)k[Y] & \hookrightarrow & k[Y] & \xrightarrow{\tilde{u}} & k[z] = R/P & \longrightarrow & 0 \\
 & & \downarrow & & \downarrow & & \\
 & & 0 & & 0 & &
\end{array}
\qquad
\begin{aligned} z &= f(y) \\ &= \tilde{u}(Y) . \end{aligned}$$

where the columns and last two rows are exact and u' and v' are restrictions of u and v respectively. We are proving (*) essentially by diagram chasing in the above diagram.

(27.4) LEMMA. <u>Let</u> S <u>be a domain with quotient field</u> L. <u>Let</u> R <u>be an overdomain of</u> S <u>such that</u> R <u>is a finite</u> S-<u>module</u>. <u>Let</u> P <u>be a prime ideal in</u> R <u>such that</u> $(S \cap P)R_P = M(R_P)$. <u>Let</u> $g: R_P \to R_P/M(R_P)$ <u>be the canonical epimorphism</u>. <u>Let</u> Ω <u>be the set of all prime ideals</u> P' <u>in</u> R <u>such that</u> $P' \cap S = P \cap S$ <u>and</u> $P' \neq P$. <u>Then we have the following</u>:

(1) <u>If</u> $y \in R$ <u>is such that</u> $y \notin P$, $g(R_P) = g(S_{S\cap P})(g(y))$, <u>and</u> $y \in P'$ <u>for all</u> $P' \in \Omega$, <u>then upon letting</u> $Q = S[y] \cap P$ <u>we have</u> $S[y]_Q = R_P$ <u>and hence in particular</u> $L(y) =$ <u>the quotient field of</u> R.

(2) <u>If</u> R_P <u>is residually separable over</u> S, <u>then there exists</u> $y \in R$ <u>such that</u> $y \notin P$, $g(R_P) = g(S_{S\cap P})(g(y))$, <u>and</u> $y \in P'$ <u>for all</u> $P' \in \Omega$.

PROOF. Let $E = S_{S\cap P}$, $D =$ the quotient ring of R with respect to the multiplicative set $S\backslash P$, $I = PD$, and $\Omega^* =$ the set of all maximal ideals in D different from I. Then D is a quasilocal ring, $E \subset D$, D is a finite E-module, I is a maximal ideal in D, $D_I = R_P$, and $P' \to P'D$ gives a bijection of Ω' onto Ω^*.

To prove (1), let any $y \in R\backslash P$ be given such that $g(R_P) = g(S_{S\cap P})(g(y))$ and $y \in P'$ for all $P' \in \Omega'$. Then, upon letting $J = E[y] \cap I$, we have that, J is a maximal ideal in $E[y]$, and I is the only maximal ideal in D which contains J; also D is a finite $E[y]$-module, $g(D_I) = g(E[y]_J)$, and $JD_I = M(D_I)$; therefore by Nakayama's lemma we get $E[y]_J = D_I$. Upon letting $Q = S[y] \cap P$ we clearly have $S[y]_Q = E[y]_J$; consequently $S[y]_Q = R_P$, and hence in particular $L(y) =$ the quotient field of R.

To prove (2), assume that R_P is residually separable over S. Then we can find $x \in D\backslash I$ such that $g(D_I) = g(E)(g(x))$. Now Ω^* is a finite set and hence by the Chinese remainder theorem we can find $z \in D$ such that $g(z) = g(x)$ and $z \in I^*$ for all $I^* \in \Omega^*$, and then we can find $t \in S\backslash P$ such that, upon letting $y = tz$, we have

$y \in R$. Clearly $y \notin P$, $g(R_P) = g(S_{S\cap P})(g(y))$, and $y \in P'$ for all $P' \in \Omega'$.

(27.5) LEMMA. Let S be a normal domain with quotient field L, let R be an overdomain of S such that R is a finite S-module, and let P be a prime ideal in R. Then the following two conditions are equivalent:

(*) $(S \cap P)R_P = M(R_P)$ and R_P is residually separable over S.

(**) There exists $y \in R$ such that, upon letting $Q = S[y] \cap P$ and $\xi(Y)$ = the minimal monic polynomial of y over L, we have $S[y]_Q = R_P$ and $\xi'(y) \notin P$; (note that, since S is normal we have $\xi(Y) \in S[Y]$).

PROOF. Follows from (27.3) and (27.4).

(27.6) Let S be a normal domain with quotient field L, and let R be the integral closure of S in a finite algebraic field extension K of L. Assume that R is a finite S-module. (Note that this condition is certainly satisfied, if S is a quotient ring, with respect to some multiplicative set, of an affine ring over a field.) Let P be a prime ideal in R such that $(S \cap P)R_P = M(R_P)$ and R_P is residually separable over S. Then: $\mathfrak{D}(S,K) \not\subset P$; K is separable over L ; and for every derivation $D: K \to K$ with $D(S_{S\cap P}) \subset S_{S\cap P}$, we have $D(R_P) \subset R_P$.

PROOF. By (27.5) there exists $y \in R$ such that, upon letting $Q = S[y] \cap P$ and $\xi(Y)$ = the minimal monic polynomial of y over L, we have $S[y]_Q = R_P$ and $\xi'(y) \notin P$; (note that since S is normal we have $\xi(Y) \in S[Y]$). By (27.2) we have $\xi'(y) \in \mathfrak{D}(S,K)$, and hence we get $\mathfrak{D}(S,K) \not\subset P$. Since $\xi'(y) \notin P$, we have $\xi'(y) \neq 0$; consequently K is separable over L. Given any derivation $D: K \to K$ such that $D(S_{S\cap P}) \subset S_{S\cap P}$, let $\eta(Y) \in S_{S\cap P}[Y]$ be obtained by applying D to the coefficients in $\xi(Y)$. Then

$$0 = D\xi(y) = \eta(y) + \xi'(y)Dy \;;$$

therefore, since $\xi'(y) \in R\backslash P$, we get $Dy \in R_P$. Now, since $R_P = S[y]_Q$, we conclude that $D(R_P) \subset R_P$.

(27.7) LEMMA. Let L be a function field of transcendence degree one over a field k, and let K be a finite algebraic separab field extension of L. Let Δ be the set of all $W \in \mathfrak{X}(K,k)$ such that either $(L \cap M(W))W \neq M(W)$ or W is not residually separable over $L \cap W$. Then Δ is a finite set, and for every $V \in \mathfrak{X}(K,k)\backslash\Delta$ we have $\operatorname{ord}_V \mathfrak{D}(L \cap V, K) = 0$.

PROOF. By (27.6) it follows that $\operatorname{ord}_V \mathfrak{D}(L \cap V,K) = 0$ for all $V \in \mathfrak{X}(K,k)\backslash\Delta$. To prove that Δ is a finite set, we can take $y \in K$ such that $K = L(y)$ and then, upon letting

$\xi(Y) =$ the minimal monic polynomial of y over L,

we have $\xi'(y) \neq 0$ and hence we can find a finite subset Δ^* of $\mathfrak{X}(K,k)$ such that

$$\xi(Y) \in (L \cap V)[Y] \text{ and } \xi'(y) \in V\backslash M(V) \text{ for all } V \in \mathfrak{X}(K,k)\backslash\Delta^* .$$

Now by (27.5) we get $\Delta \subset \Delta^*$, and hence Δ is a finite set.

(27.8) DEFINITION. Let K be a function field of transcendenc degree one over a field k. By a separating transcendental of K/k we mean an element x in K such that K is separable algebraic over $k(x)$. Note that then: K is separably generated over $k \Leftrightarrow$ there exists a separating transcendental of K/k. Also note that by F. K. Schmidt's theorem, if k is perfect then K is separably generated over k; this also follows from the following lemma:

(27.9) LEMMA. <u>For a function field</u> K <u>of transcendence degree one over a field</u> k, <u>the following three conditions are equivalent</u>:

(1) <u>For some</u> $V \in \mathfrak{X}(K,k)$ <u>we have that</u> V <u>is residually separable over</u> k.

(2) K <u>is separably generated over</u> k.

(3) <u>For infinitely many</u> $V \in \mathfrak{X}(K,k)$ <u>we have that</u> V <u>is residually separable over</u> k.

PROOF. If there exists $V \in \mathfrak{X}(K,k)$ such that V is residually separable over k then, upon taking $S = k[x]$ with $x \in M(V)\backslash M(V)^2$, by (27.6) we get that K is separable algebraic over k(x); thus (1) ⇒ (2). To show that (2) ⇒ (3), assume (2) and take a separating transcendental x of K/k; now, if k is infinite, then clearly $\{k[x]_{(x-a)k[x]} : a \in k\}$ is an infinite set of members of $\mathfrak{X}(k(x),k)$ each of which is residually rational over k, and, if k is finite, then clearly every member of $\mathfrak{X}(k(x),k)$ is residually separable over k; thus in either case there are infinitely many members of $\mathfrak{X}(k(x),k)$ which are residually separable over k; therefore, upon taking $L = k(x)$, by (27.7) we deduce that there are infinitely many members of $\mathfrak{X}(K,k)$ which are residually separable over k. The implication (3) ⇒ (1) is of course obvious.

§28. Differentials.

Let K be a <u>separably generated</u> function field of transcendence degree one over a field.

Note that if x is a separating transcendental of K/k, then there is a unique derivation K → K, <u>to be denoted by</u> $\frac{d}{dx}$, such that $\frac{d\eta(x)}{dx} = \eta'(x)$ for all $\eta(Y) \in k[Y]$.

We observe the <u>chain rule</u>: If x and y are any separating transcendentals of K/k then for any $z \in K$ we have $\frac{dz}{dx} = \frac{dz}{dy}\frac{dy}{dx}$.

We also note the <u>criterion for a separating transcendental</u>: If

x is any separating transcendental of K/k, then for any $z \in K$ we have that: z is a separating transcendental of $K/k \Leftrightarrow \frac{dz}{dx} \neq 0$.

As a consequence of the above criterion we observe that: if x and z are any elements in K such that x is a separating transcendenal of K/k and z is not a separating transcendental of K/k, then $x + z$ is a separating transcendental of K/k, and if also $z \neq 0$, then xz is a separating transcendental of K/k.

<u>The observations made in the above three paragraphs may be used tacitly.</u>

For the present situation of function fields, we can reformulate (27.2) thus:

(28.1) SPECIAL CASE OF DEDEKIND'S FORMULA FOR CONDUCTOR AND DIFFERENT. <u>Let</u> x <u>and</u> y <u>be any elements in</u> K <u>such that</u> x <u>is a separating transcendental of</u> K/k, <u>and</u> $k(x,y) = K$. <u>Let nonconstant irreducible</u> $\zeta(X,Y) \in k[X,Y]$ <u>be such that</u> $\zeta(x,y) = 0$. <u>Let</u> $V \in \mathfrak{X}(K,k)$ <u>be such that</u> $x \in V$ <u>and</u> $y \in V$. <u>Let</u> $Q = k[x,y] \cap M(V)$. <u>Assume that</u> y <u>is integral over</u> $k(x) \cap V$. <u>Then</u>

$$(*) \qquad \mathrm{ord}_V \mathfrak{C}(k[x,y]_Q) + \mathrm{ord}_V \mathfrak{D}(k(x) \cap V, K) = \mathrm{ord}_V \zeta_Y(x,y)$$

PROOF. Upon letting

$$(1) \qquad S = k(x) \cap V$$

and

$$\xi(Y) = \text{the minimal monic polynomial of } y \text{ over } k(x)$$

we clearly have

$$\xi(Y)/\zeta(x,Y) \in S \setminus M(V)$$

and hence

$$(2) \qquad \xi'(y)V = \zeta_Y(x,y)V .$$

Upon letting

$$R = \text{the integral closure of } S \text{ in } K,$$

by (27.2) we have

$$\mathfrak{C}(S[y])\mathfrak{D}(S,K) = \xi'(y)R .$$

and hence

(3) $$\mathrm{ord}_V\mathfrak{C}(S[y]) + \mathrm{ord}_V\mathfrak{D}(S,K) = \mathrm{ord}_V\xi'(y) .$$

Upon letting

$$P = S[y] \cap M(V) ,$$

in view of (3.1) we have

(4) $$\mathfrak{C}(S[y])V = \mathfrak{C}(S[y]_P)V ;$$

also clearly

(5) $$S[y]_P = k[x,y]_Q .$$

Now (*) follows from (1) to (5).

(28.2) SPECIAL CASE OF CONVERSION FORMULA FOR DIFFERENT. _Let any separating transcendentals_ x _and_ y _of_ K/k _and any_ $V \in \mathfrak{X}(K,k)$ _be given such that_ $x \in V$ _and_ $y \in V$. _Assume that_ $k(x,y) = K$. _Assume that_ y _is integral over_ $k(x) \cap V$, _and_ x _is integral over_ $k(y) \cap V$. _Then_

(**) $$\mathrm{ord}_V\mathfrak{D}(k(y) \cap V,K) = \mathrm{ord}_V \frac{dy}{dx} + \mathrm{ord}_V\mathfrak{D}(k(x) \cap V,K)$$

PROOF. We can take nonconstant irreducible $\zeta(X,Y) \in k[X,Y]$ such that $\zeta(x,y) = 0$. Let $Q = k[x,y] \cap M(V)$. Then by (28.1) we get

(1) $$\mathrm{ord}_V\mathfrak{C}(k[x,y]_Q) + \mathrm{ord}_V\mathfrak{D}(k(x) \cap V,K) = \mathrm{ord}_V\zeta_Y(x,y) .$$

By applying (28.1) with x and y interchanged, we also get

(2) $$\mathrm{ord}_V\mathfrak{C}(k[x,y]_Q) + \mathrm{ord}_V\mathfrak{D}(k(y) \cap V,K) = \mathrm{ord}_V\zeta_X(x,y)$$

By the chain rule we have

$$0 = \frac{d\zeta(x,y)}{dx} = \zeta\ (x,y) + \zeta_Y(x,y)\ \frac{dy}{dx}$$

and hence

(3) $$\mathrm{ord}_V \zeta_X(x,y) = \mathrm{ord}_V \zeta_Y(x,y) + \mathrm{ord}_V \frac{dy}{dx}\ .$$

Now (**) follows from (1), (2) and (3).

We want to show that (28.1) and (28.2) remain valid without the assumptions about integralness made in their last but one sentences. For this purpose we first prove two exchange lemmas

(28.3) FIRST EXCHANGE LEMMA. Let x and y be any elements in K such that x is a separating transcendental of K/k, and $k(x,y) = K$. Let nonconstant irreducible $\zeta(X,Y) \in k[X,Y]$ be such that $\zeta(x,y) = 0$. Let $V \in (K,k)$ be such that $x \in V$ and $y \in V$. Then there exists a separating transcendental z of K/k and a nonconstant irreducible $\eta(Z,Y) \in k[Z,Y]$ such that: $\eta(z,y) = 0$; y is integral over $k[z]$; $k[x,y] = k[z,y]$; $\mathrm{ord}_V \frac{dz}{dx} = 0$; and $\mathrm{ord}_V \eta_Y(z,y)$ $= \mathrm{ord}_V \zeta_Y(x,y)$.

PROOF. We can find $\gamma(Y) \in k[Y]\backslash k$ such that $\gamma(y) \in M(V)$, and then we can find a positive integer n_0 such that for every integer $n \geq n_0$ we have:

(1) $\zeta(Z + \gamma(Y)^n, Y)$ is a monic polynomial in Y over $k[Z]$,

(2) $$\mathrm{ord}_V[n\gamma(y)^{n-1}\gamma'(y)\ \frac{dy}{dx}] > 0$$

and

(3) $$\mathrm{ord}_V[n\gamma(y)^{n-1}\gamma'(y)\zeta_X(x,y)] > \mathrm{ord}_V\zeta_Y(x,y)$$

(note that (3) can be arranged because, since x is a separating transcendental of K/k, we have $\zeta_Y(x,y) \neq 0$).

Fix any integer $n \geq n_0$, and let

$$z = x - \gamma(y)^n$$

Now clearly

$$k[z,y] = k[x,y] .$$

Also

$$\frac{dz}{dx} = 1 - n\gamma(y)^{n-1}\gamma'(y)\,\frac{dy}{dx}$$

and hence by (2) we have

$$\mathrm{ord}_V \frac{dz}{dx} = 0 ;$$

consequently, in particular, $\frac{dz}{dx} \neq 0$, and hence z is a separating transcendental of K/k. Let

$$\eta(Z,Y) = \zeta(Z + \gamma(Y)^n, Y) \in k[Z,Y]$$

Then obviously $\eta(Z,Y)$ is a nonconstant irreducible polynomial. Clearly $\eta(z,y) = 0$, and hence in view of (1) we see that y is integral over $k[x]$. By the chain rule we get

$$\eta_Y(z,y) = \eta\gamma(y)^{n-1}\gamma'(y)\zeta_X(x,y) + \zeta_Y(x,y)$$

and hence by (3) we conclude that

$$\mathrm{ord}_V\eta_Y(z,y) = \mathrm{ord}_V\zeta_Y(x,y) .$$

(28.4) SECOND EXCHANGE LEMMA. _Let any separating transcendentals_ x _and_ y _of_ K/k _and any_ $V \in \mathfrak{X}(K,k)$ _be given such that_ $x \in V$ _and_ $y \in V$. _Then there exists a separating transcendental_ z _of_ K/k _with_ $z \in V$ _such that_: $k(x,z) = K$; z _is integral over_ $k(x) \cap V$; x _is integral over_ $k(z) \cap V$; $k(y,z) = K$; z _is integral over_ $k(y) \cap V$; _and_ y _is integral over_ $k(z) \cap V$.

PROOF. Let

$$\Delta(u) = \{W \in \mathfrak{X}(K,k) : u \notin W\} \quad \text{for every} \quad u \in K$$

and

$$\left.\begin{aligned} \Gamma(u,W) &= \{W' \in \mathfrak{X}(K,k) : W' \cap k(u) = W \cap k(u)\} \\ \Gamma^*(u,W) &= \Gamma(u,W) \setminus \{W\} \end{aligned}\right\} \quad \begin{aligned} &\text{for every } u \in K \\ &\text{and every} \\ &W \in \mathfrak{X}(K,k) \,. \end{aligned}$$

Let

(1) $$\Omega = \Delta(x) \cup \Gamma^*(x,V) \cup \Delta(y) \cup \Gamma^*(y,V) \,.$$

Then, since $x \in V$ and $y \in V$, we have

(2) $$V \notin \Omega \,.$$

Now $\Omega \cup \{V\}$ is a finite subset of $\mathfrak{X}(k,k)$ and hence, in view of (27.7), we can find I and J in $\mathfrak{X}(K,k)$ such that

(3) $$\Gamma(x,I) \cap \Gamma(y,J) = \emptyset = (\Gamma(x,I) \cup \Gamma(x,J)) \cap (\Omega \cup \{V\})$$

(4) $$(k(x) \cap M(I))I = M(I)$$

(5) $$(k(y) \cap M(J))J = M(J)$$

and such that, upon letting

$$f_I : I \to I/M(I) \quad \text{for} \quad f_J : J \to J/M(J)$$

to be the canonical epimorphisms, we have that

(6) $f_I(I)$ is a finite separable algebraic field extension of $f_I(k(x) \cap I)$

and

(7) $f_J(J)$ is a finite separable algebraic field extension of $f_J(k(y) \cap J)$.

Clearly

(8) $\Gamma(x,I) \cup \Gamma(y,J) \cup \{V\} \cup \Omega$ is a finite set

and hence we can find $\alpha \in k[x]\setminus k$ such that

$$\alpha \notin M(W) \quad \text{for all} \quad W \in \Gamma(X,I) \cup \Gamma(y,J) \cup \Omega \cup \{V\} ,$$

and then upon letting

$$D = \text{the integral closure of } k[1/\alpha] \text{ in } K$$

we have that D is a Dedekind domain with quotient field K such that

(9) $D \subset W$ for all $W \in \Gamma(x,I) \cup \Gamma(y,J) \cup \Omega \cup \{V\}$.

In view of (9) we have $D \subset I$ and hence in view of (6) we can find

(10) $a \in D\setminus M(I)$ such that $f_I(I) = f_I(k(x) \cap I)(f_I(a))$.

In view of (9) we have $D \subset J$ and hence in view of (7) we can find

(11) $b \in D\setminus M(J)$ such that $f_J(J) = f_J(k(x) \cap J)(f_J(b))$.

In view of (2), (3), (8) and (9), by the Chinese remainder theorem we can find

(12) $t \in D$

such that

(13) $t \in M(V)$ and $t \notin M(W)$ for all $W \in \Omega$,

(14) $t - a \in M(I)$ and $t \in M(W)$ for all $W \in \Gamma^*(x,I)$,

and

(15) $t - b \in M(J)$ and $t \in M(W)$ for all $W \in \Gamma^*(y,J)$.

Now by (1) and (9) to (15) we get that:

(16) $t \in M(V)$ and $t \in W\backslash M(W)$ for all $W \in \Gamma^*(x,V) \cup \Delta(\mathbf{x})$

(17) $t \in M(V)$ and $t \in W\backslash M(W)$ for all $W \in \Gamma^*(y,V) \cup \Delta(y)$

(18) $t \in M(W)$ for all $W \in \Gamma^*(x,I)$, $t \in I\backslash M(I)$,

and $f_I(I) = f_I(k(x) \cap I)(f_I(t))$

and

(19) $t \in M(W)$ for all $W \in \Gamma^*(y,J)$, $t \in J\backslash M(J)$,

and $f_J(J) = f_J(k(x) \cap J)(f_J(t))$.

In view of (8) we can find $\beta \in k(x)\backslash k$ such that

$$\beta \in M(W) \text{ for all } W \in \Gamma(x,I) \cup \Gamma(x,J) \cup \Omega \cup \{V\}$$

and then we can find a positive integer n such that

$$\beta^n x \in M(W) \text{ for all } W \in \Gamma(x,I) \cup \Gamma(x,J) \cup \Omega \cup \{V\} ;$$

since x is a separating transcendental of K/k, upon letting

$$\gamma = \begin{cases} \beta^n , & \text{if } K \text{ is separable over } k(\beta) \\ \beta^n x , & \text{if } K \text{ is not separable over } k(\beta) \end{cases}$$

we clearly get that

(20) $\gamma \in M(W)$ for all $W \in \Gamma(x,I) \cup \Gamma(x,J) \cup \Omega \cup \{V\}$

and

(21) γ is a separating transcendental of K/k.

Upon letting

$$z = \begin{cases} t & \text{, if } t \text{ is a separating transcendental of } K/k \\ t+\gamma & \text{, if } t \text{ is not a separating transcendental of } K/k \end{cases}$$

by (1) and (16) to (21) we get that:

(22) z is a separating transcendental of K/k

(23) $z \in M(V)$ and $z \in W\backslash M(W)$ for all $W \in \Gamma^*(x,V) \cup \Delta(x)$

(24) $z \in M(V)$ and $z \in W\backslash M(W)$ for all $W \in \Gamma^*(y,V) \cup \Delta(y)$

(25) $z \in M(W)$ for all $W \in \Gamma^*(x,I)$, $z \in I\backslash M(I)$,
and $f_I(I) = f_I(k(x) \cap I)(f_I(z))$

and

(26) $z \in M(W)$ for all $W \in \Gamma^*(y,J)$, $z \in J\backslash M(J)$
and $f_J(J) = f_J(k(y) \cap J)(f_J(z))$.

In view of (27.4), by (4) and (25) we get that $k(x,z) = K$. Since $x \in V$, by (23) we see that z is integral over $k(x) \cap V$, and x is integral over $k(z) \cap V$.

In view of (27.4), by (5) and (26) we get that $k(y,z) = K$. Since $y \in V$, by (24) we see that z is integral over $k(y) \cap V$, and y is integral over $k(z) \cap V$.

(28.5) CONVERSION FORMULA FOR DIFFERENT. <u>Let any separating transcendentals</u> x <u>and</u> y <u>of</u> K/k <u>and any</u> $V \in \mathfrak{X}(K,k)$ <u>be given such that</u> $x \in V$ <u>and</u> $y \in V$. <u>Then</u>

(**) $$\operatorname{ord}_V \mathfrak{D}(k(y) \cap V, K) = \operatorname{ord}_V \frac{dy}{dx} + \operatorname{ord}_V \mathfrak{D}(k(x) \cap V, K) .$$

PROOF. By (28.4) there exists a separating transcendental z of K/k with $z \in V$ such that: $k(x,z) = K$; z is integral over $k(x) \cap V$; x is integral over $k(z) \cap V$; $k(y,z) = K$; z is integral over $k(y) \cap V$; and y is integral over $k(z) \cap V$. By applying

(28.2) to the pair (x,z) we get

(1) $$\operatorname{ord}_V \mathfrak{D}(k(z) \cap V, K) = \operatorname{ord}_V \frac{dz}{dx} + \operatorname{ord}_V \mathfrak{D}(k(x) \cap V, K) .$$

By applying (28.2) to the pair (z,y) we get

(2) $$\operatorname{ord}_V \mathfrak{D}(k(y) \cap V, K) = \operatorname{ord}_V \frac{dy}{dz} + \operatorname{ord}_V \mathfrak{D}(k(z) \cap V, K) .$$

By the chain rule we have

$$\frac{dy}{dx} = \frac{dy}{dz}\frac{dz}{dx}$$

and hence

(3) $$\operatorname{ord}_V \frac{dy}{dx} = \operatorname{ord}_V \frac{dy}{dz} + \operatorname{ord}_V \frac{dz}{dx} .$$

Now (**) follows from (1), (2) and (3).

(28.6) DEDEKIND'S FORMULA FOR CONDUCTOR AND DIFFERENT. Let x and y be any elements in K such that x is a separating transcendental of K/k, and $k(x,y) = K$. Let nonconstant irreducible $\zeta(X,Y) \in k[X,Y]$ be such that $\zeta(x,y) = 0$. Let $V \in \mathfrak{X}(K,k)$ be such that $x,y \in V$. Let $Q = k[x,y] \cap M(V)$. Then

(*) $$\operatorname{ord}_V \mathfrak{C}(k[x,y]_Q) + \operatorname{ord}_V \mathfrak{D}(k(x) \cap V, K) = \operatorname{ord}_V \zeta_Y(x,y) .$$

PROOF. By (28.3) there exists a separating transcendental z o[f] K/k and a nonconstant irreducible $\eta(Z,Y) \in k[X,Y]$ such that $\eta(z,y) = 0$, y is integral over $k[z]$,

(1) $$k[x,y] = k[z,y] ,$$

and

(2) $$\operatorname{ord}_V \frac{dz}{dx} = 0 \quad \text{and} \quad \operatorname{ord}_V \eta_Y(z,y) = \operatorname{ord}_V \zeta_Y(x,y) .$$

Now in particular $k(z,y) = K$, $z \in V$, and y is integral over $k(z) \cap V$; consequently by applying (28.1) to the pair (z,y), in vie[w] of (1) we get

(3) $\quad \operatorname{ord}_V \mathfrak{C}(k[x,y]_Q) + \operatorname{ord}_V \mathfrak{D}(k(z) \cap V, K) = \operatorname{ord}_V \eta_Y(z,y)$.

By applying (28.5) to the pair (x,z) we also get

(4) $\quad \operatorname{ord}_V \mathfrak{D}(k(z) \cap V, K) = \operatorname{ord}_V \frac{dz}{dx} + \operatorname{ord}_V \mathfrak{D}(k(x) \cap V, K)$.

Now (*) follows from (2), (3) and (4).

(28.7) DEFINITION. For any $V \in \mathfrak{X}(K,k)$ and any (ordered) pair (α,x) of elements α and x in K, we <u>define</u>

$$\operatorname{ord}_V(\alpha,x) = \begin{cases} \operatorname{ord}_V \alpha + \operatorname{ord}_V \mathfrak{D}(k(x) \cap V, K) & \text{if } \alpha \neq 0,\ x \in V, \\ \quad \text{and } x \text{ is a separating transcendental of } K/k \\ \operatorname{ord}_V \alpha + \operatorname{ord}_V \mathfrak{D}(k(\frac{1}{x}) \cap V, K) + \operatorname{ord}_V x^2, & \text{if } \alpha \neq 0,\ x \notin V, \\ \quad \text{and } x \text{ is a separating transcendental of } K/k \\ \infty, & \text{if either } \alpha = 0 \\ \quad \text{or } x \text{ is not a separating transcendental of } K/k \end{cases}$$

and we note that then:

$$\operatorname{ord}_V(\alpha,x) = \text{an integer or } \infty ,$$

$\operatorname{ord}_V(\alpha,x) \neq \infty \Leftrightarrow \alpha \neq 0$ and x is a separating transcendental of K/k ,

and

$$\operatorname{ord}_V a + \operatorname{ord}_V(\alpha,x) = \operatorname{ord}_V(a\alpha,x) \quad \text{for all } a \in K .$$

(When we want to make an explicit reference to k we may write $\operatorname{ord}_{V,k}(\alpha,x)$ instead of $\operatorname{ord}_V(\alpha,x)$.)

REMARK. The reader may keep in mind that (α,x) is intended to be replaced by " αdx " after its definition in (28.13).

(28.8) LEMMA. <u>Let</u> x <u>be any separating transcendental of</u> K/k, <u>let</u> β <u>and</u> y <u>be any elements in</u> K, <u>and let</u> V <u>be any member of</u> $\mathfrak{X}(K,k)$. <u>Then</u>

(1) $$\operatorname{ord}_V(\beta,y) = \operatorname{ord}_V \beta\left(\frac{dy}{dx}, x\right)$$

and

(2) $$\operatorname{ord}_V(\beta,y) = \operatorname{ord}_V\left(\frac{\beta}{\alpha}\frac{dy}{dx}\right) + \operatorname{ord}_V(\alpha,x) \quad \text{for all} \quad 0 \neq \alpha \in K.$$

PROOF. If either $\beta = 0$ or y is not a separating transcendental of K/k then both sides of (1) as well as both sides of (2) are ∞. So henceforth suppose that $\beta \neq 0$ and y is a separating transcendental of K/k. If $x \in V$ and $y \in V$ then (1) follows directly from (28.5); if $x \notin V$ and $y \in V$ then (1) follows by applying (28.5) to the pair (x^{-1},y) and noting that $\frac{dy}{dx^{-1}} = -x^2\frac{dy}{dx}$; if $x \in V$ and $y \notin V$ then (1) follows by applying (28.5) to the pair (x,y^{-1}) and noting that $\frac{dy^{-1}}{dx} = -y^{-2}\frac{dy}{dx}$; and, finally, if $x \notin V$ and $y \notin V$ then (1) follows by applying (28.5) to the pair (x^{-1},y^{-1}) and noting that $\frac{dy^{-1}}{dx^{-1}} = y^{-2}x^2\frac{dy}{dx}$. This completes the proof of (1), and obviously (2) follows from (1).

(28.9) LEMMA. <u>If</u> x,y,α,β <u>are any elements in</u> K <u>such that</u> $\beta\frac{dy}{dz} = \alpha\frac{dx}{dz}$ <u>for some</u> (and hence every) <u>separating transcendental</u> z <u>of</u> K/k <u>then</u>

$$\operatorname{ord}_V(\beta,y) = \operatorname{ord}_V(\alpha,x) \quad \underline{\text{for all}} \quad V \in \mathfrak{X}(K,k).$$

PROOF. Follows by twice applying formula (1) of (28.8).

(28.10) LEMMA. If x and y are any separating transcendenta of K/k and α and β are any nonzero elements of K, then:

(28.10.1) $$0 \neq \frac{\beta}{\alpha}\frac{dy}{dx} \in K$$

(28.10.2) $$\operatorname{ord}_V(\beta,y) = \operatorname{ord}_V\left(\frac{\beta}{\alpha}\frac{dy}{dx}\right) + \operatorname{ord}_V(\alpha,x) \quad \text{for all} \quad V \in \mathfrak{X}(K,k)$$

and

$$(28.10.3)\quad \begin{cases} \sum_{V\in \mathfrak{X}(K,k)} [\mathrm{ord}_V(\beta,y)][V/M(V) : k] \\ \qquad = \sum_{V\in \mathfrak{X}(K,k)} [\mathrm{ord}_V(\alpha,x)][V/M(V) : k] \\ \qquad = \underline{\text{an integer}}. \end{cases}$$

PROOF. (28.10.1) is obvious. (28.10.2) follows from formula (2) of (28.8). In view of (28.10.1) and (28.10.2), (28.10.3) follows from (4.1).

(28.11) DEFINITION. By (28.10.3) there exists a unique rational number, to be denoted by genus(K,k), such that

$$\sum_{V\in \mathfrak{X}(K,k)} [\mathrm{ord}_V(\alpha,x)][V/M(V) : k] = 2\ \mathrm{genus}(K,k) - 2$$

for every separating transcendental x of K/k and every nonzero element α in K.

(28.12) THEOREM. If $K = k(x)$ for some $x \in K$ then $\mathrm{genus}(K,k) = 0$.

PROOF. Now obviously (or, say by (27.6)) $\mathrm{ord}_V \mathfrak{D}(k(x) \cap V,K) = 0$ for all $V \in \mathfrak{X}(K,k)$, and hence our assertion follows by taking $\alpha = 1$ in (28.11).

(28.13) DEFINITION. By a differential of K/k we mean an object of the form αdx with α and x in K, where these objects are to satisfy:

$$\beta dy = \alpha dx \Leftrightarrow \beta \frac{dy}{dz} = \alpha \frac{dx}{dz}$$ for some (and hence every) separating transcendetal z of K/k.

In other words, in the set of all pairs (α,x) of elements in

K we introduce the equivalence relation

$$(\beta,y) \sim (\alpha,x) \Leftrightarrow \beta \frac{dy}{dz} = \alpha \frac{dx}{dz} \text{ for some (and hence every) separating transcendental } z \text{ of } K/k$$

and then we let αdx stand for the equivalence class containing (α,x).

(When we want to make an explicit reference to K and k, we could write something like $\alpha d_{K/k}x$ instead of αdx.)

By dx we mean $1dx$.

We let 0 stand for $0d0$, and we observe that for any α and x in K we then have: $\alpha dx \neq 0 \Leftrightarrow \alpha \neq 0$ and x is a separating transcendental of K/k.

We convert the set of all differentials of K/k into a K-vector-space by defining

$$\alpha dx + \alpha^* dx^* = \left(\alpha \frac{dx}{dz} + \alpha^* \frac{dx^*}{dz}\right) dz \quad \text{and} \quad \gamma(\alpha dx) = (\gamma\alpha) dx$$

for all $\alpha,\alpha^*,\gamma,x,x^*,z$ in K with z a separating transcendental of K/k. Note that these definitions are "well-defined", i.e., if β,β^*,y,y^*,u are any other elements in K such that $\beta dy = \alpha dx$ $\beta^* dy^* = \alpha^* dx^*$, and u is a separating transcendental of K/k, then

$$\left(\beta \frac{dy}{du} + \beta^* \frac{dy^*}{du}\right) du = \left(\alpha \frac{dx}{dz} + \alpha^* \frac{dx^*}{dz}\right) dz \quad \text{and} \quad (\gamma\beta) dy = (\gamma\alpha) dx .$$

We observe that the K-vector-space-dimension of the said K-vector-space is one, and it is generated by any nonzero member of it, i.e., by any differential of the form αdx with $0 \neq \alpha \in K$ and x a separating transcendental of K/k.

We note that for any x and x^* in K we have

$$d(xx^*) = xdx^* + x^*dx .$$

We also note that for any $x \in K$, any $\eta(Y) \in k[Y]$, and any separating transcendental z of K/k we have

$$d\eta(x) = \eta'(x)dx = \left(\eta'(x)\frac{dx}{dz}\right)dz = \frac{d\eta(x)}{dx}\,dz\ .$$

For any differential αdx of K/k and any $V \in \mathfrak{X}(K,k)$ we <u>define</u>

$$\mathrm{ord}_V \alpha dx = \mathrm{ord}_V(\alpha,x)$$

and we note that, in view of (28.9), this definition is "well-defined' i.e., : $\beta dy = \alpha dx \Rightarrow \mathrm{ord}_V(\beta,y) = \mathrm{ord}_V(\alpha,x)$. We observe that:

$$\mathrm{ord}_V \alpha dx = \text{an integer or } \infty\ ,$$

$$\mathrm{ord}_V \alpha dx \neq \infty \Leftrightarrow \alpha dx \neq 0\ ,$$

and

$$\mathrm{ord}_V \gamma(\alpha dx) = \mathrm{ord}_V \gamma + \mathrm{ord}_V \alpha dx \quad \text{for all} \quad \gamma \in K\ ;$$

we also observe that if $\alpha^* dx^*$ is any other differential of K/k then:

$$\mathrm{ord}_V(\alpha dx + \alpha^* dx^*) \geq \min(\mathrm{ord}_V \alpha dx\ ,\ \mathrm{ord}_V \alpha^* dx^*)$$

with equality if $\mathrm{ord}_V \alpha dx \neq \mathrm{ord}_V \alpha^* dx^*$.

(28.14) THEOREM. <u>For any</u> α <u>and</u> x <u>in</u> K <u>with</u> $\alpha dx \neq 0$ (i.e., with $\alpha \neq 0$ and x a separating transcendental of K/k) <u>we have</u>

$$\sum_{V \in \mathfrak{X}(K,k)} (\mathrm{ord}_V \alpha dx)[V/M(V) : k] = 2\,\mathrm{genus}(K,k) - 2\ .$$

PROOF. This is simply a restatement of (28.11).

(28.15) DEDEKIND'S FORMULA FOR CONDUCTOR AND DIFFERENTIAL. <u>Let</u> x <u>and</u> y <u>be any elements in</u> K <u>such that</u> $k(x,y) = K$. <u>Assume that</u> x <u>is a separating transcendental of</u> K/k. <u>Let nonconstant irreducible</u> $\zeta(X,Y) \in k[X,Y]$ <u>be such that</u> $\zeta(x,y) = 0$. <u>Let</u> n <u>be the</u> (total) <u>degree of</u> $\zeta(X,Y)$. <u>Let</u>

$$W = \{V \in \mathfrak{X}(K,k) : x \in V \text{ and } y \in V\},$$

$$W' = \{V \in \mathfrak{X}(K,k) : x \notin V \text{ and } yx^{-1} \in V\}, \text{ and}$$

$$W^* = \{V \in \mathfrak{X}(K,k) : y \notin V \text{ and } xy^{-1} \in M(V)\} .$$

(Note that then clearly: $W = \mathfrak{Y}(k[x,y])$, $W' \cup W^* = \mathfrak{X}(K,k)\backslash W$, and $W' \cap W^* = \phi$.) <u>For any</u> $V \in \mathfrak{X}(K,k)$ <u>let</u>

$$S(V) = \begin{cases} k[x,y] & \underline{\text{if}}\ V \in W \\ k[x^{-1},yx^{-1}] & \underline{\text{if}}\ V \in W' \\ k[y^{-1},xy^{-1}] & \underline{\text{if}}\ V \in W^* \end{cases} \quad \underline{\text{and}}$$

$$R(V) = S(V)_{S(V) \cap M(V)} .$$

<u>Also let</u>

$R = k[x,y]$ <u>and</u> R^* = <u>the integral closure of</u> $k[x,y]$ <u>in</u> K.

<u>Then we have</u>

(28.15.1) $$\operatorname{ord}_V \mathfrak{C}(R(V)) + \operatorname{ord}_V dx - \operatorname{ord}_V \zeta_Y(x,y)$$

$$= \begin{cases} 0 & \underline{\text{for all}}\ V \in W \\ (n-3)\operatorname{ord}_V x^{-1} & \underline{\text{for all}}\ V \in W' \\ (n-3)\operatorname{ord}_V y^{-1} & \underline{\text{for all}}\ V \in W^* \end{cases}$$

(28.15.2) $$\begin{cases} 2 + \sum_{V \in \mathfrak{X}(K,k)\backslash \mathfrak{Y}(R)} [\operatorname{ord}_V dx - \operatorname{ord}_V \zeta_Y(x,y)][V/M(V) : k] \\ = 2\operatorname{genus}(K,k) + \sum_{V \in \mathfrak{Y}(R)} [\operatorname{ord} \mathfrak{C}(R(V))][V/M(V) : k] \\ = 2\operatorname{genus}(K,k) + [R^*/\mathfrak{C}(R) : k] \end{cases}$$

and

(28.15.3) $$2\operatorname{genus}(K,k) = (n-1)(n-2) - \sum_{V \in \mathfrak{X}(K,k)} [\operatorname{ord}_V \mathfrak{C}(R(V))][V/M(V) : k] .$$

(Note that since $K = k(x,y)$ is separably generated over k, by

Maclane's theorem either x or y is a separating transcendental of K/k; consequently, (28.15.3) would remain valid without the assumption of x being a separating transcendental of K/k.)

PROOF. By (28.6) we get

(1) $$\operatorname{ord}_V \mathfrak{C}(R(V)) + \operatorname{ord}_V dx - \operatorname{ord}_V \zeta_Y(x,y) = 0 \quad \text{for all } V \in W.$$

Now

$$\operatorname{ord}_V \zeta_Y(x,y)^{-1} dx = \operatorname{ord}_V dx - \operatorname{ord}_V \zeta_Y(x,y) \quad \text{for all } V \in \mathfrak{X}(K,k)$$

and hence by taking $\zeta_Y(x,y)^{-1}$ for α in (28.14) we get

(2) $$\begin{cases} 2\,\operatorname{genus}(K,k) = 2 + \sum_{V \in W} [\operatorname{ord}_V dx - \operatorname{ord}_V \zeta_Y(x,y)][V/M(V) : k] \\ \qquad + \sum_{V \in W' \cup W^*} [\operatorname{ord}_V dx - \operatorname{ord}_V \zeta_Y(x,y)][V/M(V) : k]. \end{cases}$$

In view of (3.1) and (5.5) we have

(3) $$[R^*/\mathfrak{C}(R) : k] = \sum_{V \in W} [\operatorname{ord}_V \mathfrak{C}(R(V))][V/M(V) : k].$$

Now (28.15.2) follows from (1), (2) and (3).

Given any $V \in W'$, upon letting

$$\xi(X,Y) = X^n \zeta(X^{-1}, YX^{-1})$$

we clearly have that nonconstant irreducible $\xi(X,Y) \in k[X,Y]$ with $\xi(x^{-1},yx^{-1}) = 0$, and hence by applying (28.6) to the pair (x^{-1},yx^{-1}) we get

(4_1) $$\operatorname{ord}_V \mathfrak{C}(R(V)) + \operatorname{ord}_V dx^{-1} = \operatorname{ord}_V \xi_Y(x^{-1},yx^{-1}) ;$$

now

$$dx^{-1} = -x^{-2} dx$$

and hence

$$(4_2) \qquad \mathrm{ord}_V dx^{-1} = \mathrm{ord}_V x^{-2} + \mathrm{ord}_V dx \ ;$$

by the chain rule we have

$$(4_3) \qquad \xi_Y(x^{-1}, yx^{-1}) = x^{1-n}\zeta_Y(x,y)$$

and hence

$$(4_4) \quad \mathrm{ord}_V \xi_Y(x^{-1}, yx^{-1}) = \mathrm{ord}_V x^{-2} + \mathrm{ord}_V x^{3-n} + \mathrm{ord}_V \zeta_Y(x,y) \ .$$

By (4_1), (4_2) and (4_4) we conclude that

$$(4) \quad \mathrm{ord}_V \mathfrak{S}(R(V)) + \mathrm{ord}_V dx - \mathrm{ord}_V \zeta_Y(x,y) = (n-3)\mathrm{ord}_V x^{-1} \quad \text{for all } V \in W' \ .$$

Now let $V \in W^*$ be given. Upon letting

$$\eta(X,Y) = Y^n \zeta(XY^{-1}, Y^{-1})$$

we clearly have that nonconstant irreducible $\eta(X,Y) \in k[X,Y]$ and $\eta(xy^{-1}, y^{-1}) = 0$. Now (by Maclane's theorem) either xy^{-1} is a separating transcendental of K/k or y^{-1} is a separating transcendental of K/k. If xy^{-1} is a separating transcendental of K/k, then by applying (28.6) to the pair (xy^{-1}, y^{-1}) we get

$$(*_1) \qquad \mathrm{ord}_V \mathfrak{S}(R(V)) + \mathrm{ord}_V d(xy^{-1}) = \mathrm{ord}_V \eta_Y(xy^{-1}, y^{-1}) \ ;$$

since

$$d(xy^{-1}) = \frac{d(xy^{-1})}{dx} dx = \left(y^{-1} - xy^{-2}\frac{dy}{dx}\right) dx$$

we get

$$(*_2) \qquad \mathrm{ord}_V d(xy^{-1}) = \mathrm{ord}_V \left(y^{-1} - xy^{-2}\frac{dy}{dx}\right) + \mathrm{ord}_V dx \ ;$$

by the chain rule we have

$$(*_3) \qquad \eta_Y(xy^{-1}, y^{-1}) = y^{-n}[-xy\zeta_X(x,y) - y^2\zeta_Y(x,y)]$$

and

$$(*_4) \qquad 0 = \frac{d\zeta(x,y)}{dx} = \zeta_X(x,y) + \zeta_Y(x,y)\frac{dy}{dx} ;$$

by $(*_3)$ and $(*_4)$ we get

$$(*_5) \qquad \eta_Y(xy^{-1},y^{-1}) = -y^{3-n}\left(y^{-1} - xy^{-2}\frac{dy}{dx}\right)\zeta_Y(x,y)$$

and hence

$$(*_6) \quad \mathrm{ord}_V\eta_Y(xy^{-1},y^{-1}) = \mathrm{ord}_V y^{3-n} + \mathrm{ord}_V\left(y^{-1} - xy^{-2}\frac{dy}{dx}\right) + \mathrm{ord}_V\zeta_Y(x,y) ;$$

now by $(*_1)$, $(*_2)$ and $(*_6)$ we get

$$(*) \qquad \mathrm{ord}_V\mathfrak{C}(R(V)) + \mathrm{ord}_V dx - \mathrm{ord}_V\zeta_Y(x,y) = (n-3)\mathrm{ord}_V y^{-1} .$$

If y^{-1} is a separating transcendental of K/k then: by applying (28.6) to the pair (y^{-1},xy^{-1}) we get

$$(**_1) \qquad \mathrm{ord}_V\mathfrak{C}(R(V)) + \mathrm{ord}_V dy^{-1} = \mathrm{ord}_V\eta_X(xy^{-1},y^{-1}) ;$$

since

$$dy^{-1} = \frac{dy^{-1}}{dy}\frac{dy}{dx}dx = -y^{-2}\frac{dy}{dx}dx$$

we get

$$(**_2) \qquad \mathrm{ord}_V dy^{-1} = \mathrm{ord}_V y^{-2} + \mathrm{ord}_V\frac{dy}{dx} + \mathrm{ord}_V dx ;$$

by the chain rule we have

$$(**_3) \qquad \eta_X(xy^{-1},y^{-1}) = y^{1-n}\zeta_X(x,y)$$

and

$$(**_4) \qquad 0 = \frac{d\zeta(x,y)}{dx} = \zeta_X(x,y) + \zeta_Y(x,y)\frac{dy}{dx} ;$$

by $(**_3)$ and $(**_4)$ we get

$$(**_5) \qquad \eta_X(xy^{-1}, y^{-1}) = - y^{1-n}\zeta_Y(x,y)\frac{dy}{dx}$$

and hence

$$(**_6) \qquad \mathrm{ord}_V\eta_X(xy^{-1},y^{-1}) = \mathrm{ord}_V y^{3-n} + \mathrm{ord}_V\left(y^{-2}\frac{dy}{dx}\right) + \mathrm{ord}_V\zeta_Y(x,y) \ ;$$

now by $(**_1)$, $(**_4)$ and $(**_6)$ we get

$$(**) \qquad \mathrm{ord}_V\mathfrak{C}(R(V)) + \mathrm{ord}_V dx - \mathrm{ord}_V\zeta_Y(x,y) = (n-3)\mathrm{ord}_V y^{-1}.$$

Thus in the above paragraph we have shown that

$$(5) \qquad \mathrm{ord}_V\mathfrak{C}(R(V)) + \mathrm{ord}_V dx - \mathrm{ord}_V\zeta_Y(x,y) = (n-3)\mathrm{ord}_V y^{-1} \quad \text{for all } V \in W^* .$$

By (1), (4) and (5) we get (28.15.1). By (1), (2), (4) and (5) we get

$$(6) \quad \begin{cases} 2\,\mathrm{genus}(K,k) = \\ = 2 + (n-3)\{\sum_{V\in W'}(\mathrm{ord}_V x^{-1})[V/M(V) : k] + \sum_{V\in W^*}(\mathrm{ord}_V y^{-1})[V/M(V):k]\}. \\ \quad - \sum_{V\in\mathfrak{X}(K,k)}[\mathrm{ord}_V\mathfrak{C}(R(V))][V/M(V) : k] . \end{cases}$$

Let $A = k[X,Y,Z]$ where we regard A to be a homogeneous domain in the obvious manner. We have a unique nonzero $\psi(X,Y,Z) \in H_n(A)$ such that $\psi(X,Y,1) = \zeta(X,Y)$. Upon letting $C = \psi(X,Y,Z)A$ we have $C \in \mathfrak{D}_1(A)$, and by (25.9) we get

$$(7) \qquad \mathrm{Deg}[A,C] = n.$$

Upon letting $\Phi = ZA$, we have $\Phi \in H^*(A)$ with $\mathrm{Deg}_A\Phi = 1$ and $\Phi \not\subset C$; consequently by Bezout's little theorem (23.9) we get

$$(8) \qquad \mu^*([A,C],\Phi) = \mathrm{Deg}[A,C] .$$

In view of (15.4) and (15.5) we easily see that

$$(9)\quad \begin{cases} \mu^*([A,C],\Phi) \\ = \sum_{V\in W'} (\mathrm{ord}_V x^{-1})[V/M(V) : k] + \sum_{V\in W^*} (\mathrm{ord}_V y^{-1})[V/M(V) : k] . \end{cases}$$

Now (28.15.3) follows from (6), (7), (8) and (9).

(28.16) REMARK. A slight variation of the above proof of (28.15.3) is thus. Firstly, with A, ψ and C as in the last paragraph we can "identify" $\mathfrak{R}([A,C])$ with K, and then we can either check that (28.15.1) is equivalent to

$$(28.16.1)\quad \begin{cases} \mathrm{ord}_V dx = \mathrm{ord}([A,C],\psi_Y(X,Y,Z),V) \\ \qquad - \mathrm{ord}_V([A,C],Z^2,V) - \mathrm{ord}_V \mathfrak{C}(R(V)) \\ \qquad\qquad \text{for all } V \in \mathfrak{X}(K,k) , \end{cases}$$

or, alternatively, we can also read off the above proof of (28.15.1) as a proof of (28.16.1); namely:

$$\psi_Y(x,y,1) = \zeta_Y(x,y) \qquad \begin{cases}\text{because by definition} \\ \psi(x,y),1) = \zeta(x,y) ;\end{cases}$$

$$x^2\psi_Y(1,yx^{-1},x^{-1}) = -\xi_Y(x^{-1},yx^{-1}) \frac{dx}{dx^{-1}} \qquad \begin{cases}\text{because clearly} \\ \psi(1,yx^{-1},x^{-1}) = \xi(x^{-1},yx^{-1}) \\ \text{by chain rule, see } (4_4) ;\end{cases}$$

$$= x^{3-n}\zeta_Y(x,y)$$

by Euler's formula

$$X\psi_X(X,Y,Z) + Y\psi_Y(X,Y,Z) + Z\psi_Z(X,Y,Z) = n\psi(X,Y,Z)$$

and hence, assuming $y \neq 0$,

$y^2\psi_Y(xy^{-1},1,y^{-1})$

$= -xy\eta_X(xy^{-1},y^{-1}) - y\eta_Y(xy^{-1},y^{-1})$ because clearly $\psi(xy^{-1},1,y^{-1}) = \eta(xy^{-1},y^{-1})$

$$= \begin{cases} \eta_Y(xy^{-1},y^{-1})\dfrac{dx}{dy^{-1}}, & \text{if } y^{-1} \text{ is a separating transcendental of } K/k \\ -\eta_X(xy^{-1},y^{-1})\dfrac{dx}{dxy^{-1}}, & \text{if } xy^{-1} \text{ is a separating transcendental of } K/k \end{cases} \quad \text{because } \eta(xy^{-1},y^{-1}) = 0$$

$= \quad y^{3-n}\zeta_Y(x,y)$ by chain rule, see (*5) and ($**_5$);

etc.

Secondly, by Bezout's Little theorem (23.9) we have

$$\sum_{V\in\mathfrak{X}(K,k)} \operatorname{ord}([A,C],\psi_Y(X,Y,Z),V)[V/M(V) : k] = (n-1)n$$

and

$$\sum_{V\in\mathfrak{X}(K,k)} \operatorname{ord}([A,C],Z^2,V)[V/M(V) : k] = 2n .$$

and by (28.14) we have

$$\sum_{V\in\mathfrak{X}(K,k)} [\operatorname{ord}_V dx][V/M(V) : k] = 2 \text{ genus}(K,k) - 2 ;$$

consequently, upon multiplying both sides of (28.16.1) by $[V/M(V):k]$ and then summing over all $V \in \mathfrak{X}(K,k)$ we get

$$2 \text{ genus}(K,k) - 2 = (n-1)n-2n - \sum_{V\in\mathfrak{X}(K,k)} [\operatorname{ord}_V \mathfrak{C}(R(V))][V/M(V) : k]$$

and hence

$$2 \text{ genus}(K,k) = (n-1)(n-2) - \sum_{V\in\mathfrak{X}(K,k)} [\operatorname{ord}_V \mathfrak{C}(R(V))][V/M(V) : k] .$$

(28.17) REMARK. Let the notation be as in (28.15). For any $V \in \mathfrak{Y}(R)$, by (28.15.1) we see that $\operatorname{ord}_V \zeta_Y(x,y)^{-1}dx$ depends only on R and not on its particular generators x and y; this can actually be shown directly (see §2 of [4]); also one can easily see that:

$\mathrm{ord}_V \zeta_Y(x,y)^{-1}dx = 0 \Leftrightarrow R(V)$ is regular $\Leftrightarrow \mathrm{ord}_V \mathfrak{C}(R(V)) = 0$; thus, one can guess at Dedekind's formula $\mathrm{ord}_V \zeta_Y(x,y)^{-1}dx = \mathrm{ord}_V \mathfrak{C}(R(V))$ and hence also guess at his formulas (27.2), (28.1), and (28.6). Let us also remark that by (28.15.2) we see that the first summation occurring in (28.15.2) depends only on R and not on its particular generators x and y; this too can be shown directly (see §2 of [4]).

In (28.18), (28.19) and (28.20) we shall give special attention to those members of $\mathfrak{X}(K,k)$ which are residually separable over k. These items (28.18), (28.19) and (28.20) are independent of the material from (28.4) to (28.17) except that in an <u>alternative</u> proof of (28.19.5) we shall use (28.5).

(28.18) DEFINITION. Given $V \in \mathfrak{X}(K,k)$, by a <u>uniformizing parameter</u> of V we mean an element of $M(V)\backslash M(V)^2$, and by a <u>uniformizing coordinate</u> of V/k we mean an element $x \in V$ such that $(k[x] \cap M(V))V = M(V)$; we note that every uniformizing parameter of V is clearly a uniformizing coordinate of V.

(28.19) LEMMA. <u>For any</u> $V \in \mathfrak{X}(K,k)$ <u>which is residually separable over</u> k, <u>we have the following</u>.

(28.19.1) <u>If</u> t <u>is any uniformizing coordinate of</u> V/k, <u>then</u> t <u>is a separating transcendental of</u> K/k, $\mathrm{ord}_V \mathfrak{D}(k(t) \cap V, K) = 0$, <u>and for every</u> $x \in V$ <u>we have</u> $\frac{dx}{dt} \in V$.

(28.19.2) <u>If</u> t <u>is any uniformizing coordinate of</u> V/k, <u>then for every</u> $0 \neq x \in K$ <u>we have</u>:

$$\mathrm{ord}_V \frac{dx}{dt} = (\mathrm{ord}_V x) - 1 \quad \underline{\text{if}} \quad \mathrm{ord}_V x \not\equiv 0 \ (\underline{\text{characteristic of}}\ k)$$

<u>and</u>

$$\mathrm{ord}_V \frac{dx}{dt} > (\mathrm{ord}_V x) - 1 \quad \underline{\text{if}} \quad \mathrm{ord}_V x \equiv 0 \ (\underline{\text{characteristic of}}\ k).$$

(28.19.3) <u>If</u> t <u>is any uniformizing coordinate of</u> V/k <u>and</u> u

is any element of K then:

$$u \text{ is a uniformizing coordinate of } V/k \Leftrightarrow u \in V \text{ and } \operatorname{ord}_V \frac{du}{dt} = 0.$$

(28.19.4) If t and u are any two uniformizing coordinates of V/k, then for every $x \in K$ we have $\operatorname{ord}_V \frac{dx}{du} = \operatorname{ord}_V \frac{dx}{dt}$.

(28.19.5) If t is any uniformizing coordinate of V/k and x is any separating transcendental of K/k, then

$$\operatorname{ord}_V \frac{dx}{dt} = \begin{cases} \operatorname{ord}_V \mathfrak{D}(k(x) \cap V) & \text{if } x \in V \\ \operatorname{ord}_V \mathfrak{D}(k(x) \cap V) + \operatorname{ord}_V x^2 & \text{if } x \notin V. \end{cases}$$

PROOF. (28.19.1) follows by taking $S = k[t]$ in (27.6).

If t and u are any two uniformizing coordinates of V/k then by (28.19.1) we have

$$\frac{du}{dt} \in V \text{ and } \frac{dt}{du} \in V$$

and by the chain rule we have

$$\frac{du}{dt}\frac{dt}{du} = 1 ;$$

consequently we must have

$$(1_1) \qquad \operatorname{ord}_V \frac{du}{dt} = 0 = \operatorname{ord}_V \frac{dt}{du} ;$$

moreover, for any $x \in K$, by the chain rule we have

$$\frac{dx}{dt} = \frac{dx}{du}\frac{du}{dt}$$

and hence by (1_1) we get

$$(1_2) \qquad \operatorname{ord}_V \frac{dx}{du} = \operatorname{ord}_V \frac{dx}{dt}$$

which proves (28.19.4).

In the above paragraph we have shown that:

(2) $\begin{cases} \text{if } t \text{ and } u \text{ are any two uniformizing coordinates of } V/k \text{ then} \\ \operatorname{ord}_V \frac{du}{dt} = 0 = \operatorname{ord}_V \frac{dt}{du} . \end{cases}$

To prove (28.19.2) let any uniformizing coordinate t of V/k and any $0 \neq x \in K$ be given; let $n = \operatorname{ord}_V x$ and fix any uniformizing parameter u of V; then $x = \delta u^n$ where n is an integer and

$$(3_1) \qquad \delta \in V\backslash M(V) ;$$

now

$$(3_2) \qquad \frac{dx}{dt} = \frac{dx}{du}\frac{du}{dt} = \left(u^n \frac{d\delta}{du} + \delta n u^{n-1}\right)\frac{du}{dt} ;$$

by (28.19.1) we have

$$(3_3) \qquad \frac{d\delta}{du} \in V$$

and by (2) we have

$$(3_4) \qquad \operatorname{ord}_V \frac{du}{dt} = 0 ;$$

by (3_1) to (3_4) we see that

$$\operatorname{ord}_V \frac{dx}{dt} = n - 1 \quad \text{if} \quad n \not\equiv 0 \quad (\text{characteristic of } k)$$

and

$$\operatorname{ord}_V \frac{dx}{dt} > n-1 \quad \text{if} \quad n \equiv 0 \ (\text{characteristic of } k) ;$$

this completes the proof of (28.19.2).

Since every uniformizing coordinate of V/k is, by definition, an element of V, the implication " $\Rightarrow$ " of (28.19.3) follows from (2). To prove the reverse implication, let any uniformizing coordinate t of V/k and any $u \in V$ with

$$(4_1) \qquad \operatorname{ord}_V \frac{du}{dt} = 0$$

be given; then in particular $\frac{du}{dt} \neq 0$ and hence u is a separating transcendental of K/k ; upon letting $W = k(u) \cap V$, we can take $x \in k[u]$ with

$$(4_2) \qquad \mathrm{ord}_W x = 1 \ ;$$

now clearly $W \in \mathfrak{X}(k(u),k)$ is residually separable over k and u is a uniformizing coordinate of W/k, and hence, in view of (4_2), by applying (28.19.2) to W we get

$$(4_3) \qquad \mathrm{ord}_W \frac{dx}{du} = 0 \ ;$$

by (4_2) we get

$$(4_4) \qquad \mathrm{ord}_V x > 0$$

and by (4_3) we get

$$(4_5) \qquad \mathrm{ord}_V \frac{dx}{du} = 0 \ ;$$

by the chain rule we have

$$\frac{dx}{dt} = \frac{dx}{du}\frac{du}{dt}$$

and hence by (4_1) and (4_5) we get

$$(4_6) \qquad \mathrm{ord}_V \frac{dx}{dt} = 0 \ ;$$

now, in view of (4_4) and (4_5), this time by applying (28.19.2) to V we deduce that

$$(4_7) \qquad \mathrm{ord}_V x = 1 \ ;$$

since $x \in k[u]$, by (4_7) we conclude that u is a uniformizing coordinate of V/k. This completes the proof of (28.19.3).

For any uniformizing parameter t of V/k and any separating transcendental x of K/k with $x \notin V$ we have that x^{-1} is a separatin transcendental of K/k, $x^{-1} \in V$ and

$$\mathrm{ord}_V \frac{dx}{dt} = \mathrm{ord}_V x^2 + \mathrm{ord}_V \frac{dx^{-1}}{dt} ;$$

consequently, to prove (28.19.5) it suffices to show that:

(*) for every uniforming coordinate t of V/k and every separating transcendental x of K/k with $x \in V$ we have
$$\mathrm{ord}_V \frac{dx}{dt} = \mathrm{ord}_V \mathfrak{D}(k(x) \cap V, K).$$

In view of (28.19.1), (*) follows from (28.5) by taking (t,x) for (x,y). Now, without using any material from (28.4) to (28.17), we shall give another proof of (*).

So let any uniformizing coordinate t of V and any separating transcendental x of K/k be given with $x \in V$. Let R be the integral closure of $k(x) \cap V$ in K. Let $f: R \to R/(M(V) \cap R)$ be the canonical epimorphism, and let Ω be the set of all maximal ideals in R different from $M(V) \cap R$. Since V is residually separable over k, by the Chinese remainder theorem we can find $\alpha \in R \backslash M(V)$ such that $\alpha \in P$ for all $P \in \Omega$ and $f(R) = f(k)(f(\alpha))$. Let $\xi(Y)$ be the unique monic polynomial in Y with coefficients in k such that upon applying f to the coefficients of $\xi(Y)$ we get the minimal monic polynomial of $f(\alpha)$ over $f(k)$. Now we have

(5) $$\xi(\alpha) \in M(V) \quad \text{and} \quad \xi'(\alpha) \notin M(V) .$$

Clearly

(6) $$\xi(Y+Z) = \xi(Y) + Z\xi'(Y) + Z^2 \text{ times an element in } k[Y,Z] .$$

By the Chinese remainder theorem we can find

(7) $$\gamma \in (R \cap M(V)) \backslash M(V)^2 \quad \text{with} \quad \gamma \in P \quad \text{for all} \quad P \in \Omega .$$

Upon letting

$$Y = \begin{cases} \alpha & \text{if } \xi(\alpha) \notin M(V)^2 \\ \alpha+\gamma & \text{if } \xi(\alpha) \in M(V)^2 \end{cases}$$

by (5), (6) and (7) we see that

(8) $\quad y \in R\backslash M(V)$, $f(R) = f(k)(f(y))$, $y \in P$ for all $P \in \Omega$

and

(9) $$\xi(y) \in (R \cap M(V))\backslash M(V)^2 .$$

In view of (8) and (9), by (27.4) we get

(10) $$k(x,y) = K .$$

By (9) we see that

(11) $\quad y$ is a uniformizing coordinate of V/k,

and hence by (28.19.1) we get that

(12) $\quad y$ is a separating transcendental of K/k.

We can find a nonconstant irreducible $\zeta(X,Y) \in k[X,Y]$ such tha $\zeta(x,y) = 0$. Let $Q = k[x,y] \cap M(V)$.

In view of (8) and (10), by (28.1) we get

(13) $$\text{ord}_V \mathfrak{C}(k[x,y]_Q) + \text{ord}_V \mathfrak{D}(k(x) \cap V, K) = \text{ord}_V \zeta_Y(\alpha,y) ;$$

(in view of (8) and (9), by (3.1) and (27.4) it can be seen that $\text{ord}_V \mathfrak{C}(k[x,y]_Q) = 0$; however, we shall not use this observation).

In view of (12), by applying (28.3) (with x and y interchanged) we can find a separating transcendental z of K/k and a nonconstant irreducible $\eta(X,Z) \in K[X,Z]$ with $\eta(x,z) = 0$ such that

(14) $\quad x$ is integral over $k[z]$

(15) $$k[x,z] = k[x,y]$$

(16) $$\mathrm{ord}_V \frac{dz}{dy} = 0$$

and

(17) $$\mathrm{ord}_V \eta_X(x,z) = \mathrm{ord}_V \zeta_X(x,y).$$

By (15) we have

(18) $$z \in V .$$

In view of (14), (15) and (18), by taking (z,x) for (x,y) in (28.1) we get

(19) $$\mathrm{ord}_V \mathfrak{C}(k[x,y]_Q) + \mathrm{ord}_V \mathfrak{D}(k(z) \cap V,K) = \mathrm{ord}_V \eta_X(x,z).$$

By the chain rule we have

$$0 = \frac{d\zeta(x,y)}{dy} = \zeta_X(x,y) \frac{dx}{dy} + \zeta_Y(x,y)$$

and hence

(20) $$\mathrm{ord}_V \zeta_Y(x,y) = \mathrm{ord}_V \frac{dx}{dy} + \mathrm{ord}_V \zeta_X(x,y).$$

By (13), (17), (19) and (20) we get that

(21) $$\mathrm{ord}_V \mathfrak{D}(k(x) \cap V,K) = \mathrm{ord}_V \frac{dx}{dy} + \mathrm{ord}_V \mathfrak{D}(k(z) \cap V,K).$$

In view of (11), (16) and (18), by (28.19.3) we see that z is a uniformizing coordinate of V/k and hence by (28.19.1) we conclude that

(22) $$\mathrm{ord}_V \mathfrak{D}(k(z) \cap V,K) = 0.$$

Now t is a uniformizing coordinate of V/k and by (11) so is y; consequently by (28.19.4) we get

$$\text{ord}_V \frac{dx}{dy} = \text{ord}_V \frac{dx}{dt} . \tag{23}$$

Thus, by (21), (22) and (23) we have

$$\text{ord}_V \mathfrak{D}(k(x) \cap V, K) = \text{ord}_V \frac{dx}{dt}$$

and this completes the proof of (*).

(28.20) EXAMPLE. The following example shows that the condition of V being residually separable over k is essential in (28.19) even when we restrict our attention to uniformizing parameters of V which are separating transcendentals of K/k.

Take k to be a nonperfect field of characteristic $p \neq 0$ and fix any $a \in k$ with $a^{1/p} \notin k$. Take $K = k(x)$ where x is an indeterminate. Let V be the unique member of $\mathfrak{X}(K,k)$ such that $\text{ord}_V(x^p+a) = 1$. For any positive integer q, upon letting

$$t(q) = x(x^p+a)^{p^q} + x^p(x^p+a) ,$$

we clearly have that $t(q)$ is a uniformizing parameter of V as well as a separating transcendental of K/k, and we have

$$\frac{dt(q)}{dx} = (x^p+a)^{p^q} \text{ and hence } \frac{dx}{dt(q)} = (x^p+a)^{-p^q} ;$$

consequently

$$\text{ord}_V \frac{dx}{dt(q)} = -p^q < 0$$

although clearly x is a separating transcendental of K/k with $x \in V$ and

$$\text{ord}_V \mathfrak{D}(k(x) \cap V, K) = 0 .$$

For any positive integers q and r we get

$$\frac{dt(q)}{dt(r)} = \frac{dt(q)}{dx}\,\frac{dx}{dt(r)} = (x^p+a)^{(p^q - p^r)}$$

and hence

$$\operatorname{ord}_V \frac{dt(q)}{dt(r)} = p^q - p^r \begin{cases} > 0 & \text{if } q > r \\ < 0 & \text{if } q < r \end{cases}$$

although both $t(q)$ and $t(r)$ are uniformizing parameter of V as well as separating transcendentals of K/k.

(28.21) REMARK. In view of (28.19.4), for any $V \in \mathfrak{X}(K,k)$ which is residually separable over k and any differential αdx of K/k we could have defined $\operatorname{ord}_V \alpha dx$ by the formula

$$\operatorname{ord}_V \alpha dx = \operatorname{ord}_V \alpha + \operatorname{ord}_V \frac{dx}{dt}$$

where t is a uniformizing coordinate of K/k. Thus, if k were perfect we could have proceeded without the intervention of (28.4) to (28.12). However, (28.20) shows that this procedure could not be followed for imperfect k. So, motivated by (28.19.5), we proceeded as we did.

(28.22) LEMMA. Let $V \in \mathfrak{X}(K,k)$ be residually separable over k. Then for any $t \in K$ we have:

t is a uniformizing coordinate of $V/K \Leftrightarrow t \in V$ and $\operatorname{ord}_V dt = 0$;

(whence, in particular, every uniformizing coordinate of V/k is a separating transcendental of K/k). Moreover, for any uniformizing coordinate t of V/k and any differential αdx of K/k (with α and x in K) we have

$$\operatorname{ord}_V \alpha dx = \operatorname{ord}_V \alpha \frac{dx}{dt} \begin{cases} = \operatorname{ord}_V \alpha + (\operatorname{ord}_V x) - 1 & \text{if } \alpha \neq 0 \neq x \text{ and } \operatorname{ord}_V x \not\equiv 0 \text{ (characteristic of } k) \\ > \operatorname{ord}_V \alpha + (\operatorname{ord}_V x) - 1 & \text{if } \alpha \neq 0 \neq x \text{ and } \operatorname{ord}_V x \equiv 0 \text{ (characteristic of } k). \end{cases}$$

PROOF. This is simply a reformulation of (28.19) in the language of differentials.

§29. Genus of an abstract curve.

Let R be a homogeneous domain such that $\operatorname{Dim} R = 1$. Assume that $\mathfrak{R}(R)$ is separably generated over $H_0(R)$.

Note that clearly (or, say in view of (27.6)) we have:

$H_0(A)$ is algebraically closed

$\Rightarrow H_0(A)$ is perfect

$\Rightarrow$ every member of $\mathfrak{Z}(R)$ is residually separable over $H_0(A)$

$\Rightarrow$ some member of $\mathfrak{Z}(R)$ is residually separable over $H_0(A)$

$\Rightarrow \mathfrak{R}(R)$ is separably generated over $H_0(A)$

and

$H_0(A)$ is algebraically closed

$\Rightarrow$ every member of $\mathfrak{Z}(R)$ is residually rational over $H_0(A)$

$\Rightarrow$ some member of $\mathfrak{Z}(R)$ is residually rational over $H_0(A)$

$\Rightarrow \begin{cases} \mathfrak{R}(R) \text{ is separably generated over } H_0(A)\text{, and} \\ H_0(A) \text{ is (relatively) algebraically cloed in } \mathfrak{R}(R). \end{cases}$

We <u>define</u>

$$\text{genus } R = \text{genus}(\mathfrak{R}(R), H_0(R)) .$$

Now let

$$n = \operatorname{Deg} R \text{ and } g = \text{genus } R .$$

In view of (15.4), (15.5) and (25.9), by (28.15.3) we get:

(29.1) GENUS FORMULA. *If* $\operatorname{Emdim} R \leq 2$, *then*

$$g = (1/2)(n-1)(n-2) - (1/2)\mu_{\mathfrak{C}}^{*}(R) .$$

Now we shall prove:

(29.2) THEOREM. Assume that Emdim $R \le 2$ and

(*) some $V \in \mathfrak{Z}(R)$ is residually rational over $H_0(A)$.

Also assume that

(**) for every $P \in \mathfrak{D}_0(R)$, upon letting $\mathfrak{R}(R,P)^*$ to be the integral closure of $\mathfrak{R}(R,P)$ in $\mathfrak{K}(R)$, we have:

$$[\mathfrak{R}(R,P)^*/\mathfrak{C}(\mathfrak{R}(R,P)) : \mathfrak{R}(R,P)] = 2[\mathfrak{R}(R,P)/\mathfrak{C}(\mathfrak{R}(R,P)) : \mathfrak{R}(R,P)] .$$

(Note that by (10.1.11): $\mu(R,P) \le 2$ for all $P \in \mathfrak{D}_0(R) \Rightarrow$ (**)).

Then we have

$$(1/2)\mu_{\mathfrak{C}}^{*}(R) = \sum_{P\in\mathfrak{D}_0(R)} [\mathfrak{R}(R,P)/\mathfrak{C}(\mathfrak{R}(R,P)) : H_0(A)]$$

and

$$\text{(29.2.1)} \quad \begin{cases} g = (1/2)(n-1)(n-2) - (1/2)\mu_{\mathfrak{C}}^{*}(R) \\ \quad = \text{a nonnegative integer} . \end{cases}$$

Moreover (as a converse of (28.12)) we have that:

$$\text{(29.2.2)} \quad \begin{cases} \text{if } g = 0 \text{ and } m \text{ is any integer such that} \\ m \ge \max(1,n-2), \text{ then there exist nonzero elements} \\ x \text{ and } y \text{ in } H_m(R) \text{ such that } \mathfrak{K}(R) = H_0(R)(x/y). \end{cases}$$

PROOF. By (17.4), (17.5) and (25.9) we see that for every nonnegative integer m we have

$$[H_m(R):H_0(R)] = \begin{cases} \binom{m+2}{2} & \text{if } m < n \\ \binom{m+2}{2} - \binom{m-n+2}{2} & \text{if } m \geq n \end{cases}$$

i.e.,

$$(1) \quad [H_m(R):H_0(R)] = \begin{cases} (1/2)(m+1)(m+2) & \text{if } m < n \\ mn + 1 - (1/2)(n-1)(n-2) & \text{if } m \geq n \end{cases}$$

In view of (**), by (5.4) and (5.5) we have

$$(2) \qquad (1/2)\mu^*_{\mathfrak{C}}(R) = \sum_{P \in \mathfrak{Q}_0(R)} [\mathfrak{R}(R,P)/\mathfrak{C}(\mathfrak{R}(R,P)) : H_0(A)]$$

and hence in particular

$$(3) \qquad (1/2)\mu^*_{\mathfrak{C}}(R) = \text{an integer} .$$

For all nonnegative integers m and e let

$$(4) \quad \begin{cases} E(m,e) = \{x \in H_m(R): \operatorname{ord}(R,x,V) \geq e + \operatorname{ord}_V \mathfrak{C}(\mathfrak{R}(R,\mathfrak{Z}^*(R,V))), \text{ and} \\ \qquad \operatorname{ord}(R,x,W) \geq \operatorname{ord}_W \mathfrak{C}(\mathfrak{R}(R,\mathfrak{Z}^*(R,W))) \\ \qquad\qquad \text{whenever } V \neq W \in \mathfrak{Z}(R)\}. \end{cases}$$

Then for any nonnegative integers m and e, in view of (2), by (24.20) we see that: $E(m,e)$ is an $H_0(R)$-vector-subspace of $H_m(R)$,

$$(5) \qquad [E(m,e) : H_0(R)] + e + (1/2)\mu^*_{\mathfrak{C}}(R) \geq [H_m(R) : H_0(R)] ,$$

$$(6) \qquad \mu^*(R,x) \geq e + \mu^*_{\mathfrak{C}}(R) \quad \text{for all } x \in E(m,e) ,$$

and

$$(7)\quad \begin{cases} \text{if } e > 0 \text{ then } E(m,e) \subset E(m,e-1) \text{ and} \\ [E(m,e) : H_0(R) \geq [E(m,e-1) : H_0(R)] - 1 . \end{cases}$$

For every nonnegative integer m upon letting

$$(8)\qquad e(m) = [H_m(R) : H_0(R)] - 1 - (1/2)\mu^*_{\mathfrak{C}}(R)$$

by (3) we see that $e(m)$ is an integer, and by (1) we see that $e(m) \geq 0$ for all large m. So for a moment we can fix any integer $m \geq n$ such that $e(m)$ is a nonnegative integer and then by (5) and (8) we have

$$[E(m,e(m)) : H_0(A)] > 0$$

and hence we can take

$$\Phi \in H^*_m(R) \quad \text{with} \quad \Phi = \varphi R \quad \text{for some} \quad \varphi \in E(m,e(m)) \ ;$$

now by (1), (6) and (8) we get

$$\mu^*(R,\Phi) \geq mn + (1/2)\mu^*_{\mathfrak{C}}(R) - (1/2)(n-1)(n-2)$$

and hence by Bezout's Little Theorem (24.9) we must have

$$(9)\qquad (1/2)\mu^*_{\mathfrak{C}}(R) - (1/2)(n-1)(n-2) \leq 0 \ ;$$

In view of (3) and (9), by (29.1) we get that

$$(10)\quad \begin{cases} g = (1/2)(n-1)(n-2) - (1/2)\mu^*_{\mathfrak{C}}(R) \\ \quad = \text{a nonnegative integer.} \end{cases}$$

Henceforth assume that $g = 0$ and let any integer m be given such that

$$(11)\qquad m \geq \max(1, n-2) .$$

Then by (10) we have

$$(12) \qquad \mu^*_{\mathfrak{C}}(R) = (n-1)(n-2)$$

and hence by (1), (8) and (11) we get that

$$(13) \qquad e(m) \geq 1$$

and

$$(14) \qquad e(m) + \mu^*_{\mathfrak{C}}(R) = mn \ .$$

If $[E(m,e(m)+1) : H_0(R)] > 0$, then we could take

$$\boldsymbol{\psi} \in H^*_m(R) \quad \text{with} \quad \boldsymbol{\psi} = \psi R \quad \text{for some} \quad \psi \in E(m,e(m+1))$$

and then by (6) and (14) we would have

$$\mu^*(R,\boldsymbol{\psi}) \geq 1 + mn$$

in contradiction to Bezout's Little Theorem (24.9).

Therefore we must have

$$(15) \qquad [E(m, e(m)+1) : H_0(R)] = 0 \ .$$

By (5) and (8) we have

$$(16) \qquad [E(m,e(m)) : H_0(R)] \geq 1 \ .$$

By (7), (15) and (16) we get

$$(17) \qquad [E(m, e(m)) : H_0(R)] = 1 \ .$$

Again, in view of (13),by (5) and (8) we have

$$(18) \qquad [E(m, e(m)-1) : H_0(R)] \geq 2$$

and hence in view of (7) we get

$$(19) \qquad E(m, e(m)-1) \backslash E(m, e(m)) \neq \emptyset$$

In view of (17) we can take

(20) $$0 \neq x \in E(\mathfrak{m}, e(\mathfrak{m}))$$

and in view of (19) we can take

(21) $$0 \neq y \in E(\mathfrak{m}, e(\mathfrak{m}) - 1) \setminus E(\mathfrak{m}, e(\mathfrak{m})) \ .$$

In view of (4), (14), (20) and (21), by Bezout's Little Theorem (24.9) we conclude that

(22) $$\operatorname{ord}(R,x,V) = e(\mathfrak{m}) + \operatorname{ord}_V \mathfrak{C}(\mathfrak{R}(R,\mathfrak{Z}^*(R,V))) = 1 + \operatorname{ord}(R,y,V)$$

and

(23) $$\begin{cases} \operatorname{ord}(R,x,W) = \operatorname{ord}_W \mathfrak{C}(\mathfrak{R}(R,\mathfrak{Z}^*(R,W))) \leq \operatorname{ord}(R,y,W) \\ \qquad\qquad \text{whenever } V \neq W \in \mathfrak{Z}(R). \end{cases}$$

In view of (22) and (23), by (19.10) we see that now $x/y \in \mathfrak{K}(R)$

and, in view of (22) and (23), by (19.10) we see that

$$\operatorname{ord}_V(x/y) = 1, \text{ and } \operatorname{ord}_W(x/y) \leq 0$$
$$\text{whenever } V \neq W \in \mathfrak{Z}(R) \ ;$$

consequently by (4.2) we must have $\mathfrak{K}(R) = H_0(R)(x/y)$.

(29.3) REMARK. The genesis of (29.2.2) as well as of its proof given above, and so also a good launching pad for much of birational geometry, is the elementary and ancient idea of parametrization of a conic. The latter can of course be deduced as a corollary of (29.2.2). But that would be a bit facetious! So here is the elementary version, not using any material after §25.

(29.4) PARAMETRIZATION OF A CONIC. _Let_ $\pi = zR$ _with any_ $0 \neq z \in H_1(R)$. _Let_ $k = H_0(R)$ _and_ $R' = k[H_1(R)z^{-1}]$. _Assume that_ $n = 2$. _Also assume that some_ $V \in \mathfrak{Z}(R,\pi)$ _is residually rational over_ k, *and let* $P = \mathfrak{Z}^*(R,V)$. *Then* Emdim $R = 2$, $g = 0$, *and* R' *is an euclidean domain*. *Moreover, there exist nonzero elements* x *and* y *in* $H_1(R)$ *with* $xk + yk + zk = H_1(R)$ *and* $(x,z)R = P$ *such that*:

$$(*) \qquad R' = k[x/z] \text{ and } zy = x^2 \text{ in case } \mathfrak{D}_0(R,\pi) = \{P\},$$

and

$$(**) \qquad R' = k[x/z, z/x] \text{ and } xy = z^2 \text{ in case } \mathfrak{D}_0(R,\pi) \neq \{P\}.$$

NOTE. Before writing down the proof let us draw suggestive figures for the two cases where, in both cases, the projecting center P is the point at ∞ along the y-axis, and so we are projecting the conic birationally onto (and hence parametrizing by) the x-axis along vertical (i.e., parallel to the y-axis) directions.

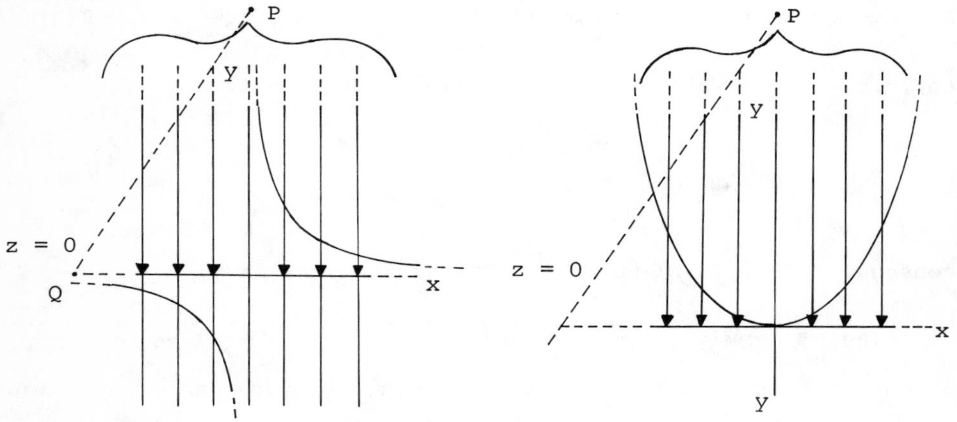

$z = 0$: line at ∞.

Hyperbola: $xy = z^2$.
$\mathfrak{D}_0(R,\pi) = \{P,Q\}$.
Projection from P is integral outside $z = 0$ *except* above the origin.

Parabola: $zy = x^2$.
$\mathfrak{D}_0(R,\pi) = \{P\}$.
Projection from P is integra everywhere outside $z = 0$.

PROOF. By (24.12) we get Emdim $R \geq 2$, and then by the Special Projection Formula (24.15) we see that the projection from P is birational and Emdim $R^P = 1$. So we must have Emdim $R = 2$. Now

$$\begin{aligned} \mu^*(R,P) &< \mu^*(R,T_1(R,V),P) \\ &\leq \mu^*(R,T_1(R,V)) \\ &= 2 \qquad \text{by Bezout's Little Theorem (24.9)} \end{aligned}$$

and hence

(1) $$\mu^*(R,P) = 1.$$

First suppose that $\mathfrak{D}_0(R,\pi) = \{P\}$. Then by Bezout's Little Theorem (24.9) we have $\mu^*(R,\pi,P) = 2$ and hence, in view of (1), by (24.18), (25.2), (25.3), (25.8), (25.9) and (25.10) we can find nonzero elements x and y in $H_1(R)$ such that $xk + yk + zk = H_1(R)$, $(x,z)R = P$, and $zy = x^2$. It follows that $R' = k[x/z]$.

Next suppose that $\mathfrak{D}_0(R,\pi) \neq \{P\}$. Since by Bezout's Little Theorem $\mu^*(R,\pi) = 2$, we must then have $\mathfrak{D}_0(R,\pi) = \{P,Q\}$ where $Q \neq P$ and

$$\mu^*(R,\pi,P) = 1 = \mu^*(R,\pi,Q) .$$

Consequently, in view of (24.18), we can find nonzero elements x and y' in $H_1(R)$ such that

$$xk + y'k + zk = H_1(R)$$

$$(x,z)R = P \quad \text{and} \quad T_1^*(R,P) = \{xR\}$$

and

$$(y',z)R = Q \quad \text{and} \quad T_1^*(R,Q) = \{y'R\} .$$

Now by (25.2), (25.3), (25.8), (25.9) and (25.10) we see that $axy' = z^2$ for some $0 \neq a \in k$. Let $y = ay'$. Then $xk + yk + zk = H_1(R)$ and $xy = z^2$. It follows that $R' = k[x/z, z/x]$.

Obviously R' is an euclidean domain in both the cases. By (28.12) we also see that $g = 0$ in both the cases.

Let us record that, in view of (24.12), by (28.12) (or alternatively by (29.2.1)) we have:

(29.5) PARAMETRIZATION OF A LINE. <u>Assume that</u> $n = 1$. <u>Then</u> Emdim $R = 1$ <u>and</u> $g = 0$. <u>Moreover, upon taking any free</u> $H_0(R)$-basis (x,z) <u>of</u> $H_1(R)$ <u>and upon letting</u> $k = H_0(R)$ <u>and</u> $R' = k[H_1(R)z^{-1}]$; <u>we have</u> $R' = k[x/z]$; <u>whence in particular</u> R' <u>is an euclidean domain</u>

Let us round off the above discussion by calculating:

(29.6) GENUS OF A CUBIC. <u>Assume that</u> $n = 3$. <u>Also assume that some</u> $V \in \mathfrak{Z}(R)$ <u>is residually rational over</u> $H_0(R)$. <u>Then either</u>: $g = 1$ <u>and</u> Emdim $R = 2$, <u>or</u>: $g = 0$ <u>and</u> Emdim $R = 2$, <u>or</u>: $g = 0$ <u>and</u> Emdim $R = 3$.

PROOF. Let $P = \mathfrak{Z}^*(R,V)$, $e =$ Emdim R, and $\pi = T_{e-1}(R,P)$.

Then

$$
\begin{aligned}
2 &\le e && \text{by (24.12)}\\
&\le e - 1 + \mu(R,P)\\
&\le e - 1 + \mu^*(R,P)\\
&\le \mu^*(R,\pi,P) && \text{by (22.1)}\\
&\le \mu^*(R,\pi)\\
&= 3 && \text{by Bezout's Little Theorem (24.9)}
\end{aligned}
$$

and hence we must be in one of the following three (mutually exclusive cases:

Case (1). $e = 2 = \mu^*(R,P) = \mu(R,P)$.

Case (2). $e = 2, \mu^*(R,P) = \mu(R,P) = 1, \mathfrak{Z}(R,P) = [V]$, and $T_1^*(R,P) = \{\pi\}$.

Case (3) $e = 3$, $\mu^*(R,P) = \mu(R,P) = 1, \mathfrak{Z}(R,P) = \{V\}$, and $\mu(R,\pi,P) = 3$

In Case (1), by the Special Projection Formula (24.15) we see that the projection from P is birational and hence $g = \text{genus } R^P$; now $\text{Deg } R^P = 1$ and hence $\text{genus } R^P = 0$ by (29.5); therefore $g = 0$.

In Case (2): First note that $\mu^*(R,\pi) = 3$ by Bezout's Little Theorem (24.9), and hence we clearly get

$$(*) \qquad \mu^*(R,Q) = \mu(R,Q) = 1 \quad \text{for all} \quad Q \in \mathfrak{Q}_0(R,\pi) .$$

Next, in view of (25.2), (25.3), (25.8), (25.9) and (25.10), upon letting $k = H_0(R)$, we can find a free k-basis (x,y,z) of $H_1(R)$ and a nonconstant irreducible $\varphi(X,Y,Z) \in k[X,Y,Z]$ such that: $zR = \pi$, $(x,z)R = P$, $\varphi(x,y,z) = 0$, and

$$\varphi(X,Y,Z) = ZY^2 + \varphi_2(X,Z)Y + \varphi_3(X,Z)$$

where

nonzero homogeneous $\varphi_3(X,Z) \in k[X,Z]$ is of degree 3

and where either $0 = \varphi_2(X,Z) \in k[X,Z]$ or

nonzero homogeneous $\varphi_2(X,Z) \in k[X,Z]$ is of degree 2 .

Upon letting

$$\xi = x/z \ , \ \eta = y/z \ , \ S' = k[\xi] \ , \ R' = S'[\eta] \ , \text{ and}$$

$$\psi(Y) = Y^2 + \varphi_2(\xi,1)Y + \varphi_3(\xi,1) \in k[\xi][Y] \ ,$$

we clearly have that ξ is transcendental over k, $R' = k[H_1(R)z^{-1}]$, $\psi(Y)$ is irreducible in $k(\xi)[Y]$, and $\psi(\eta) = 0$; now in view of (4.2) we immediately see that

$$\lambda(E) \leq 2 \quad \text{for all} \quad E \in \mathfrak{B}(R') \ ,$$

and by (15.4) and (15.5) we have

$$\mathfrak{B}(R') = \{\mathfrak{R}(R,Q) \ : \ Q \in \mathfrak{Q}_0(R,\backslash\pi)\} \ ;$$

therefore

(**) $\mu(R,Q) \leq 2$ for all $Q \in \mathfrak{Q}_0(R,\backslash\pi)$.

Thus by (*) and (**) we have

$$\mu(R,Q) \leq 2 \text{ for all } Q \in \mathfrak{Q}_0(R) .$$

and hence by (29.2.1) we get $g = 0$ or 1.

In Case (3), by the Special Projection Formula we see that the projection from P is birational and $\text{Deg } R^P = 2$, and hence by (29.4) we get $g = 0$.

(29.7) THEOREM. Assume that some member of $\mathfrak{Z}(R)$ is residually rational over $H_0(R)$. Also assume that $\text{Emdim } R \leq 2$, $4 \leq n \leq 5$, $g \leq 1$,

$$\mu(R,P) \leq 2 \text{ for all } P \in \mathfrak{Q}_0(R) .$$

and

$$\mu(R,P_0) = 2 \text{ for some } P_0 \in \mathfrak{Q}_0^1(R) .$$

Let $m = n - 3$. Then there exists

$$\Theta \in \text{Tradj}(R,\backslash P_0) \cap H_m^*(R)$$

such that Θ is irreducible in the sense that:

$$\Theta = \Theta_1\Theta_2 \text{ with } \Theta_1 \text{ and } \Theta_2 \text{ in } H^*(R) \Rightarrow \text{either } \Theta_1 = R \text{ or } \Theta_2 = R .$$

PROOF. In view of (17.4), (17.5), (25.9), (10.1.11) and (29.2.1) our assertion follows from (24.21) by taking $\pi = P_0$ and $s(P_0)$ and $s(P_0) = 1 - g$.

§30. Genus of an embedded curve.

Let A be a homogeneous domain and let $C \in \mathfrak{Q}_1(A)$ be such that $\mathfrak{K}([A,C])$ is separably generated over $H_0(A/C)$.

We define

$$\text{genus}[A,C] = \text{genus } A/C .$$

Now let

$$n = \text{Deg}[A,C] \quad \text{and} \quad g = \text{genus}[A,C] .$$

By (29.1) we get:

(30.1) GENUS FORMULA. If $\text{Emdim}[A,C] \leq 2$, then

$$g = (1/2)(n-1)(n-2) - \mu_{\mathfrak{C}}^{*}([A,C]) .$$

By (29.2) we get:

(30.2) THEOREM. Assume that $\text{Emdim}[A,C] \leq 2$ and

(*) some $V \in \mathfrak{Z}([A,C])$ is residually rational over $H_0(A/C)$.

Also assume that

(**) for every $P \in \mathfrak{Q}_0([A,C]$, upon letting $\mathfrak{R}([A,C],P)^{*}$ to be the integral closure of $\mathfrak{R}([A,C],P)$ in $\mathfrak{K}([A,C])$, we have:

$$[\mathfrak{R}([A,C],P)^{*}/\mathfrak{C}(\mathfrak{R}([A,C],P)) : \mathfrak{R}([A,C],P)]$$
$$= 2[\mathfrak{R}([A,C],P)/\mathfrak{C}(\mathfrak{R}([A,C],P)) : \mathfrak{R}([A,C],P)] .$$

(Note that by (10.1.11): $\mu([A,C],P) \leq 2$ for all $P \in \mathfrak{Q}_0([A,C]) \Rightarrow$ (**).)

Then we have

$$(1/2)\mu_{\mathfrak{C}}^{*}([A,C]) = \sum_{P \in \mathfrak{Q}_0([A,C])} [\mathfrak{R}([A,C],P)/\mathfrak{C}(\mathfrak{R}([A,C],P)) : H_0(A/C)]$$

and

$$(30.2.1)\quad \begin{cases} g = (1/2)(n-1)(n-2) - (1/2)\mu^{*}_{\mathfrak{C}}([A,C]) \\ \quad = \underline{\text{a nonnegative integer}}. \end{cases}$$

Moreover (as a converse of (28.12)) we have that:

(30.2.2) { if $g = 0$ and m is any integer such that $m \geq \max(1,n-2)$, then there exist elements x and y in $H_m(A)\backslash C$ such that $\mathfrak{R}([A,C] = H_0(A\backslash C)(f(x)/f(y))$ where $f: A \to C$ is the canonical epimorphism.

Finally, in view of (25.9), (29.4), (29.5) and (29.7), Theorem (26.12) yields its augmented version as stated below:

(30.3) THEOREM ON 2-EQUIMULTIPLE PROJECTIONS OF PROJECTIVE SPACE QUNITICS. Assume that $\operatorname{Emdim} A = \operatorname{Dim} A = 3$, $H_0(A)$ is algebraically closed, $n \leq 5$, $g \leq 1$, and let $\pi = zA$ with $z \in H_1(A)\backslash C$. Assume that

$$\mu([A,C],P) = 1 \quad \text{for all} \quad P \in \mathfrak{D}_0([A,C],\backslash\pi).$$

(Note that by (29.4), (29.5) and (29.6), $g \leq 1$ is automatic if $n \leq 3$.)

Assume that there does not exist any $\Phi \in H_2^{*}(A)$ such that: Φ is a π-quasiplane and $C \in \mathfrak{D}_1(A,\Phi)$. Also assume that there does not exist any $N \in \mathfrak{D}_0(A,\pi)$ such that: the projection of C from N is birational, the projection of C from N is π-integral, $\mu([A^N,C^N],Q)$ $= 1$ for all $Q \in \mathfrak{D}_0([A^N,C^N],\backslash\pi^N)$, and $\min(4,n) \leq \operatorname{Deg}[A^N,C^N] \leq n$.

Then $4 \leq n \leq 5$ and there exists $N \in \mathfrak{D}_0(A,\pi)$ such that: the projection of C from N is birational, the projection of C from N is π-integral, $\mu([A^N,C^N],Q) \leq 2$ for all $Q \in \mathfrak{D}_0(A^N)$, $\mu([A^N,C^N],Q_0)$ $= 2$ for some $Q_0 \in \mathfrak{D}_0([A^N,C^N],\backslash\pi^N)$, $\mu([A^N,C^N],Q_1) = 2$ for some $Q_1 \in \mathfrak{D}_0([A^N,C^N],\pi^N)$, and $\min(4,n) \leq \operatorname{Deg}[A^N,C^N] \leq n$. Moreover, for _any such_ N, _upon letting_ $m = n - 3$, _there exists_ $0 \neq \theta \in H_m(A^N)$ _such that_

$$\Theta \in \mathrm{Tradj}([A^N, C^N], \backslash\pi^N) \cap H_m^*(A^N) \cap \mathfrak{Q}_1(A^N, \backslash\pi^N) \quad \underline{\text{where}} \quad \Theta = \theta A^N.$$

Finally, for any such N _and_ θ, _upon letting_

$$B' = H_0(A)[H_1(A^N)z^{-1}] \quad \underline{\text{and}} \quad \Theta' = (\theta/z^m)B'$$

we have that B'/Θ' _is an euclidean domain; in greater detail, there exist elements_ x, y, t _in_ $H_1(A)$ _with_ $(x,y,t,z)A = \Delta(A)$ _and_ $(x,y,z)A = P$ _such that, upon letting_

$$\sigma : B' \to B'/\Theta' \quad \underline{\text{to be the canonical epimorphism}}$$

and

$$k = \sigma(H_0(A)) \quad \underline{\text{and}} \quad t = \sigma(x/z) ,$$

we have that t _is transcendental over the field_ k _and_

$$B'/\Theta' = k[t] \quad \text{or} \quad k[t, t^{-1}] .$$

CHAPTER IV. AFFINE GEOMETRY OR FILTERED DOMAINS

§31. Filtered domains

DEFINITION. Let A be a domain. A _filtration_ on A is a sequence $(G_n)_{0\le n<\infty}$ of additive subgroups of A such that:

(i) $A = \bigcup_{n=0}^{\infty} G_n$,

(ii) $G_0 \subset G_1 \subset G_2 \ldots$,

(iii) $G_m G_n \subset G_{m+n}$, for all $0 \le m,n < \infty$,

(iv) G_0 is a subfield of A,

(v) $[G_1:G_0] < \infty$, and

(vi) $A = G_0[G_1]$.

The pair $(A,(G_n)_{0\le n<\infty})$ is said to be a _filtered domain_. If there is no confusion, we simply say that A is a filtered domain and we denote G_n by $F_n(A)$. In the rest of the chapter, A will denote a filtered domain, unless said otherwise. Note that, by definition it follows that, $F_m(A)F_n(A) = F_{m+n}(A), F_n(A) = F_1(A)^n$ and $0 < [F_n(A):F_0(A)] < \infty$ for all nonnegative integers m,n. Also note that A is clearly a noetherian domain.

$F_0(A)$ will be said to be the _ground field_ of A and we will also say that A is a _filtered domain over the ground field_ $F_0(A)$.

Example. Let R be an affine domain over a field k (i.e. a finitely generated domain extension of k). Let any finite generating set S of R over k (or in fact the finite dimensional k vector space Sk) be fixed with $k \subset Sk$. Then $(R,(G(Sk,n))_{0\le n<\infty})$ is a filtered domain over the ground field k, where

$$G(Sk,n) = \begin{cases} k & , \text{ if } n = 0 \\ S^n k & , \text{ if } 0 < n < \infty . \end{cases}$$

$(G(Sk,n))_{0 \le n < \infty}$ shall be called the natural filtration induced by S (or Sk). It is easy to see that in fact every filtered domain is obtained this way.

For a filtered domain A we define

$$\text{emdim } A = [F_1(A):F_0(A)] - 1 .$$

We note that then

$$\text{emdim } A \geq \dim A = \text{trdeg}_{F_0(A)} \mathfrak{F}(A) .$$

In case dim A = 0, by deg A we denote $[\mathfrak{F}(A):F_0(A)]$.

Now let f: A → B be a homomorphism where A is a filtered domain. In case B is also a filtered domain, f is said to be a filtered homomorphism, if $f(F_n(A)) \subset F_n(B)$ for all n. If f(A) is a domain, it is clear that, upon taking $F_n(f(A)) = f(F_n(A))$, f(A) becomes a filtered domain and f: A → f(A) becomes a filtered homomorphism. $(f(F_n(A)))_{0 \le n < \infty}$ may be called the natural filtration induced by f on f(A). In particular, for any $P \in \mathfrak{P}(A)$, if we let f: A → A/p to be the canonical homomorphism, we get that A/p is a filtered domain by the natural filtration induced by f and, unless otherwise mentioned, A/p will be assumed to be a filtered domain with this filtration. In view of this filtration, we further define $F_n([A,P]) = F_n(A/p)$ and emdim[A,P] = emdim(A/p).

If B is a subdomain of A, and if B is a filtered domain, then we say that B is a filtered subdomain of A provided that the inclusion map of B into A is a filtered homomorphism.

Note that every filtered subdomain B of A is obtained by selecting some subfield k of $F_0(A)$ and some finite subset S containing 1 of $F_1(A)$ (or in fact the finite dimensional vector space Sk) and taking B = k[Sk] with the natural filtration induced by S (or Sk).

Finally for any filtered domain A, we shall write λ^* instead of $\lambda^{F_0(A)}$ as in various definitions in §5 and §6. Thus for example $\lambda^*(A,I,J) = \lambda^{F_0(A)}(A,I,J)$ etc.

§32. Homogenization.

(32.1) DEGREE. Let A be a filtered domain. For any $x \in A$ we <u>define</u>

$$\deg_A x = \begin{cases} -\infty, & \text{if } x = 0, \text{ and} \\ \min\{n: x \in F_n(A)\}, & \text{if } x \neq 0. \end{cases}$$

If $\Phi \in F^*(A)$, we <u>define</u>,

$$\deg_A \Phi = \min\{\deg_A \varphi: 0 \neq \varphi \in \Phi\}.$$

Note that as a consequence we have, for any $g,h \in F^*(A)$ or $g,h \in A$, $\deg_A gh \leq \deg_A g + \deg_A h$. Also note that $\deg_A \Phi$ is a non-negative integer for all $\Phi \in F^*(A)$. We <u>define</u> $F^*_n(A) = \{\Phi \in F^*(A): \deg_A \Phi = n\}$.

(32.2) DEFINITION. Let A be a filtered domain and let z be an element in an overdomain of A such that z is transcendental over A. For any $x \in A$, we <u>define</u>

$$H_{(A,z)}(x) = \begin{cases} s, & \text{if } x \in F_0(A), \text{ and} \\ xz^n, & \text{if } x \in F_n(A)\backslash F_{n-1}(A), \text{ for some } n > 0. \end{cases}$$

For any $I \subset A$ and $0 \leq n < \infty$, we <u>define</u>

$H_{(A,z,n)}(I)$ = The additive subgroup of $A[z]$ generated by $\{xz^n: x \in I \cap F_n(A)\}$,

and

$H_{(A,z)}(I)$ = The additive subgroup of $A[z]$ generated by $\bigcup_{n=0}^{\infty} H_{(A,z,n)}(I)$.

We may drop " A " from the above notation and simply write $H_z(x)$, $H_{(z,n)}(I)$ and $H_z(I)$ respectively, if the reference to A is clear from the context.

(32.3) Observe that, upon defining $H_n(H_z(A)) = H_{(z,n)}(A)$ for $0 \le n < \infty$, $H_z(A)$ becomes a homogeneous domain. We say that the homogeneous domain $H_z(A)$ is the <u>natural</u> z-<u>homogenization of</u> A. Note that, if z and z' are two transcendental elements over A, then $H_z(A)$ and $H_{z'}(A)$ are canonically isomorphic, i.e., there is a unique isomorphism between $H_z(A)$ and $H_{z'}(A)$ which makes az^n and az'^n to correspond to each other for every $a \in F_n(A)$.

(32.4) It is clear from the definition that $\mathfrak{R}(H_z(A)) = \mathfrak{F}(A)$. Thus it follows that $\dim A = \text{Dim } H_z(A)$.

(32.5) Also clearly $F_1(A)$ is isomorphic to $H_1(H_z(A))$ as an $F_0(A) = H_0(H_z(A))$ vector space and hence $\text{emdim } A = \text{Emdim } H_z(A)$.

(32.6) For $x \in A$, clearly we have

$$\text{Deg}_{H_z(A)} H_z(x) = \deg_A x ,$$

and in particular $x \to H_z(x)$ gives an injective map $H_z\colon A \to H(H_z(A))$.

It also follows that for any $\Phi \in F_n^*(A)$, $H_z(\Phi) \in H_n^*(H_z(A))$ and $\Phi \to H_z(\Phi)$ gives an injective map $H_z\colon \ F_n^*(A) \to H_n^*(H_z(A))$.

(32.7) For $I \subset A$ an ideal in A and S a generating set for I, we have that $H_z(I)$ is a homogeneous ideal in $H_z(A)$ generated by $\{H_z(x)\colon x \in S\}$. Further it is clear that $H_z(I) \cap H(H_z(A)) = \{xz^n : x \in I \text{ and } \deg_A x \le n < \infty\}$. It follows that for any $x \in A$, we have

$$H_z(x) \in H_z(I) \Leftrightarrow x \in I.$$

In particular we have, for any two ideals I, I' in A,

$$H_z(I) = H_z(I') \Leftrightarrow I = I' .$$

It also follows that, $I = A, \Leftrightarrow 1 \in I \Leftrightarrow z \in H_z(I)$, so that if $I \neq A$ then $z \notin H_z(I)$.

(32.8). Now let $P \in \mathfrak{P}(A)$ and let $f: A \to A/p$ be the canonical homomorphism. If z' is any transcendental over A/p, then f has a unique extension $f': A[z] \to (A/p)[z']$ such that $f'(z) = z'$ and $f'(a) = f(a)$ for every $a \in A$. It is then clear that f' induces an isomorphism of the homogeneous domains $H_z(A)/H_z(P)$ and $H_{z'}(A/P)$. In particular, we may take $\bar{z} = z$ modulo $H_z(P)$ and then we have $H_z(A)/H_z(P) = H_{\bar{z}}(A/P)$.

In particular it follows that $P \to H_z(P)$ gives a map H_z: $\mathfrak{P}(A) \to \mathfrak{Q}(H_z(A))$. In view of (32.4), it follows that $H_z(\mathfrak{P}_i(A)) \subset \mathfrak{Q}_i(H_z(A))$ and by (32.7), it follows that H_z: $\mathfrak{P}_i(A) \to \mathfrak{Q}_i(H_z(A))$ is an injective map.

§33. Dehomogenization.

(33.1) DEFINITION. Let B be a homogeneous domain. Let $0 \neq z \in H_1(B)$. For any $h \in H(B)$, we <u>define</u>

$$F_{(B,z)}(h) = \begin{cases} h \quad , \text{ if } h \in H_0(B), \text{ and} \\ hz^{-n}, \text{ if } h \in H_n(B) \text{ for } 0 < n < \infty . \end{cases}$$

Also for $I \subset B$ and $0 \leq n < \infty$, we <u>define</u>

$$F_{(B,z,n)}(I) = \text{The additive subgroup of } \mathfrak{R}(B) \text{ generated by } \{hz^{-n}: h \in I \cap H_n(B)\} ;$$

and

$$F_{(B,z)}(I) = \text{The additive subgroup of } \mathfrak{R}(B) \text{ generated by } \bigcup_{n=0}^{\infty} F_{(B,z,n)}(I).$$

Just as before we may drop " B " from the above notation and simply write $F_z(h)$, $F_{(z,n)}(I)$ and $F_z(I)$ respectively, if the reference to B is clear from the context.

Observe that, upon defining $F_n(F_z(B)) = F_{(z,n)}(B)$, for $0 \le n < \infty$, $F_z(B)$ becomes a filtered domain. $F_z(B)$ is said to be the <u>(natural)</u> z-<u>dehomogenization</u> of B. Note that $zB = z'B \Rightarrow F_n(F_z(B)) = F_n(F_{z'}(B))$, for all $0 \le n < \infty$, and thus $F_z(B)$ and $F_{z'}(B)$ are naturally isomorphic filtered domains. Thus, if $zB = \pi$, we may write $F_\pi(B)$ for $F_z(B)$ and call it the <u>(natural)</u> π-<u>dehomogenization</u> of B.

(33.2) Let $h, h' \in H(B)$. Then we have,

$F_z(h) = F_z(h') \Leftrightarrow h = z^r h'$ or $h' = z^r h$,

for some $0 \le r < \infty$.

Further, it is clear that,

$\deg_{F_z(B)} F_z(h) \le \mathrm{Deg}_B h$ and

$\deg_{F_z(B)} F_z(h) < \mathrm{Deg}_B h \Leftrightarrow F_z(h) = F_z(h')$ for some $h' \in H(B)$ with $\mathrm{Deg}_B h' < \mathrm{Deg}_B h$.

Now given any $x \in F_z(B)$, consider $\{h \in H(B): F_z(h) = x\}$. From the above observations it follows that there exists a unique $h_0 \in H(B)$ with $F_z(h_0) = x$, such that h_0 satisfies one of the three equivalent conditions

(1) $\deg_{F_z(B)} x = \mathrm{Deg}_B h_0$

(2) $h_0 \in H(B) \backslash zB$.

(3) $F_z(h) = x \Leftrightarrow h = h_0 z^r$ for some $0 \le r < \infty$.

In particular, it follows that $H_z(x) = h_0$ and hence $F_z\colon H(B)\backslash zB \to F_z(B)$ is a bijective map with the inverse map being given by H_z.

It also follows that $F_z : H_n^*(B)\backslash (zB)H^*(B) \to F_n^*(B)$ is also a bijective map with the inverse being given by H_z.

Note that for every $h \in H(B)$, we have,

$h = F_z(h) z^{\mathrm{Deg}_B h} \in H_z(F_z(B))$ and hence $H_z(F_z(B)) = B$.

(33.3) If $h_i \in H(B)$ for $i = 1,2,\ldots,t$, then for any n such that $n \geq \mathrm{Deg}_B h_i$, for all $i = 1,2,\ldots,t$, we have,

$$h = \sum_{i=1}^{t} h_i z^{n-\mathrm{Deg}_B h_i} \in H_n(B) \quad \text{and} \quad F_z(h) = \sum_{i=1}^{t} F_z(h_i) .$$

From this observation, it is easy to see that if $I \subset B$ is a homogeneous ideal, then

$$F_{(z,n)}(I) = \{F_z(h): h \in I \cap H_n(B)\} \quad \text{and}$$

$$F_z(I) = \{F_z(h): h \in I \cap H(B)\} .$$

(33.4) By (33.2) and (33.3), for any homogeneous ideal I and $h' \in H(B)$, we have

(1) $\qquad F_z(h') \in F_z(I) \Leftrightarrow z^r h' \in I$ for some $0 \leq r < \infty$.

Now let us call a homogeneous ideal I <u>prime to</u> z, if I satisfies

(*) $\qquad z^r h \in I$ with $0 \leq r < \infty$ and $h \in B \Rightarrow h \in I$.

Note that we clearly have,

(2) $\qquad I \in \mathfrak{D}(B,\backslash z) \Rightarrow I$ is prime to z .

Now we shall prove that:

(3) If J is any ideal in $F_z(B)$, then $H_z(J)$ is a homogeneous ideal in B prime in z.

Let $z^r h' \in H_z(J)$, for some $0 \leq r < \infty$.

We must have

$z^r h' = \mathbf{x} z^n$ where $0 \leq n = r + \mathrm{Deg}_B h'$ and $x \in J$.

Hence $h' = xz^{\mathrm{Deg}_B h'}$. Since $x \in F_z(B)$, it follows that $\deg_{F_z(B)}(x) \leq \mathrm{Deg}_B h'$ and hence $h' \in H_z(J)$.

Now for a homogeneous ideal I prime to z, in view of (1), we get:

(4) $$F_z(h') \in F_z(I) \Leftrightarrow h' \in I, \text{ for any } h' \in H(B).$$

We get the following consequences of (4).

(5) If I and I' are homogeneous ideals prime to z, then

$$F_z(I) = F_z(I') \Leftrightarrow I = I' .$$

(6) If I is a homogeneous ideal prime to z, then by (4) and (3) we get,

$$H_z(F_z(I)) = I .$$

(7) If $P \in \mathfrak{Q}(B,\backslash z)$, then, in view of (2), by (4) we can see that $F_z(P) \in \mathfrak{P}(F_z(B))$.

Further in view of (5), (6) and (32.8) we get that $F_z\colon \mathfrak{Q}(B,\backslash z) \to \mathfrak{P}(F_z(B))$ is a bijective map, the inverse map being given by H_z. In fact we also have that $F_z\colon \mathfrak{Q}_i(B,\backslash z) \to \mathfrak{P}_i(F_z(B))$ is a bijective map with the inverse given by H_z.

§34. Relation between homogenization and dehomogenization.

Let A be a filtered domain and let B be a homogeneous domain. Assume that there exists $z \in H_1(B)$ such that one of the following holds:

(*) $$B = H_z(A)$$

(**) $$A = F_z(B) .$$

Note that as indicated in (32.3.1), (**) ⇒ (*). On the other hand from the definition it is clear that (*) ⇒ (**). Thus we assume that both (*) and (**) hold.

We have the following relations between B and A.

(34.1) $H_z: A \to H(B)\setminus(zB)H^*(B)$ is a bijection map with the inverse being given by F_z.

Both are restatements from (33.2).

(34.2) $H_z : \mathfrak{P}_i(A) \to \mathfrak{Q}_i(B,\setminus z)$ is a bijection map with the inverse being given by F_z. (This is a restatement from (33.4), (7).)

(34.3) Let $P \in \mathfrak{P}(A)$ and $\bar{z} = z$ modulo $H_z(P)$ and $Q \in \mathfrak{Q}(B,\setminus zB)$ and $t = z$ modulo Q. Then we have

$$H_{\bar{z}}(A/P) = B/H_z(P) \quad \text{and} \quad F_t(B/Q) = A/F_z(Q).$$

In view of (34.2), this follows from (32.8).

(34.4) $\mathfrak{R}(B) = \mathfrak{J}(A)$ and $\mathfrak{R}([B,Q]) = \mathfrak{J}(A/P) = \mathfrak{J}([A,P])$, whenever $Q \in \mathfrak{Q}(B,\setminus zB)$, $P \in \mathfrak{P}(A)$ and $Q = H_z(P)$ or equivalently $P = F_z(Q)$.

This follows from (32.4) and (34.3).

Also we clearly have $H_0(B) = F_0(A)$.

(34.5) emdim A = Emdim B and emdim(A/P) = Emdim(B/Q) = Emdim$[B,Q]$, whenever $Q \in \mathfrak{Q}(B,\setminus zB)$, $P \in \mathfrak{P}(A)$ and $Q = H_z(P)$ or equivalently $P = F_z(Q)$. This follows from (32.5) together with (34.3) Emdim(B/Q) = Emdim$[B,Q]$ follows from definitions, as remarked in §13.

(34.6) Let $P = F_z(Q)$ or equivalently $Q = H_z(P)$, for some $P \in \mathfrak{P}(A)$ and $Q \in \mathfrak{Q}(B,\setminus z)$. Then clearly we have $\mathfrak{R}(B,Q) = A_P$ (as already observed in (15.4)). Further we have $\mathfrak{R}(B,I,Q)A_P = F_z(I)A_P$, for any $I \in H(B)$ or $I \subset B$; this follows by (33.3) in view of (15.1) and the definition of $\mathfrak{R}(B,I,Q)$.

Further, from definitions and the above result it is easily checked that we have the following relations, where we assume $\dim A = \operatorname{Dim} B = 1$, $P \in \mathfrak{P}_0(A)$ and $Q \in \mathfrak{Q}_0(B)$.

(34.6.1) $\mu^*(B,I,Q) = \lambda^*(A,F_z(I),P)$, and

$\mu^*(B,H_z(J),Q) = \lambda^*(A,J,P)$.

(34.6.2) $\mu^*_{\mathfrak{C}}(B,Q) = \lambda^*_{\mathfrak{C}}(A,P)$, and

$\mu^*_{\mathfrak{C}}(B,\backslash z) = \lambda^*_{\mathfrak{C}}(A)$.

(34.6.3) $F_z(\operatorname{Adj}(B,Q)) = \operatorname{adj}(A,P)$, $H_z(\operatorname{adj}(A,P)) = \operatorname{Adj}(B,Q)$;

$F_z(\operatorname{Tradj}(B,Q)) = \operatorname{tradj}(A,P)$, $H_z(\operatorname{tradj}(A,P)) = \operatorname{Tradj}(B,Q)$;

$F_z(\operatorname{Adj}(B,\backslash z)) = \operatorname{adj}(A)$, $H_z(\operatorname{adj}(A)) = \operatorname{Adj}(B,\backslash z)$;

and

$F_z(\operatorname{Tradj}(B,\backslash z)) = \operatorname{tradj}(A)$, $H_z(\operatorname{tradj}(A)) = \operatorname{Tradj}(B,\backslash z)$.

(34.7) LEMMA. <u>Assume that</u> $\operatorname{emdim} A = \dim A = r$ <u>and hence</u> (in view of (34.4) and (34.5)) <u>equivalently</u> $\operatorname{Emdim} B = \operatorname{Dim} B = r$.

<u>Let</u> $zB = \pi$ <u>and</u> $\Phi \in H_n^*(B)$ <u>a</u> <u>π-quasihyperplane</u>. <u>Let</u> $\varphi = F_z(\Phi)$. <u>Then there exist elements</u> $y_1,\dots,y_r \in A$ <u>such that</u> $\varphi = y_rA$ <u>and</u> $A = F_0(A)[y_1,\dots,y_r]$. <u>In other words</u>, <u>there is another filtration on</u> A (the one induced by $\{1\ y_1,\dots,y_r\}$ <u>in which</u> deg φ <u>becomes</u> 1.

PROOF. Observe that by (25.5) we can write $\Phi = (X_0^{n-1}X_r + \varphi_n)B$, where $B = H_0(B)[X_0,\dots,X_r]$, $X_i \in H_1(B)$, $X_0 = z$ and φ_n is a homogeneous polynomial of degree n in $X_0,\dots,X_{r-1}$ with coefficients in $H_0(B)$. Clearly we have, upon taking $y_i = F_z(X_i) = X_i/X_0$ for $1 \le i \le r-1$ and $y_r = X_r/X_0 + F_z(\varphi_n)$,

$$F_z(\Phi) = y_rA \quad\text{and}\quad A = F_0(A)[X_1/X_0,\dots,X_r/X_0]$$
$$= F_0(A)[y_1,\dots,y_r].$$

REMARK. The above Lemma is saying that a π-quasihyperplane is, in some sense, "essentially a hyperplane outside π ". This is the motivation behind the term π-quasihyperplane.

§35. Projection of a filtered domain.

(35.1) DEFINITION. Let A be a filtered domain. A filtered subdomain A' of A is said to be a <u>projection</u> of A, if $F_0(A) = F_0(A')$. Note that then $F_1(A')$ is an $F_0(A)$ subspace of $F_1(A)$ containing $F_0(A)$. We <u>define</u> $F_1(A')$ to be the center of the projection and we say that A' is the projection of A from $F_1(A')$.

We <u>define</u>

$$\mathfrak{L}^{\infty}(A) = \{L \supset F_0(A) : L \text{ is an } F_0(A) \text{ subspace of } F_1(A)\} \text{ and}$$

$$\mathfrak{L}_i^{\infty}(A) = \{L \in \mathfrak{L}^{\infty}(A) : [L : F_0(A)] = \text{emdim } A - i\}$$

Note that then $L \to F_0(A)[L]$ gives a bijection of $\mathfrak{L}^{\infty}(A)$ onto the set of all projections of A, where the filtration of $F_0(A)[L]$ is given by $F_0(F_0(A)[L]) = F_0(A)$ and $F_n(F_0(A)[L]) = L^n$, for $0 < n < \infty$. Clearly the inverse bijection is given by $A' \to F_1(A')$.

Also note that $L \to H_{(z,1)}(L)$ gives a bijection of $\mathfrak{L}_i^{\infty}$ onto $\{M \in \mathfrak{M}_i(H_z(A)) : z \in M\}$, the inverse mapping being given by $M \to F_{(z,1)}(M)$.

Now let A' be the projection of A from $L \in \mathfrak{L}^{\infty}(A)$. We <u>say that the projection from</u> L <u>is birational</u> (<u>respectively integral</u>), <u>if</u> $\mathfrak{J}(A) = \mathfrak{J}(A')$ (<u>respectively</u> A/A' <u>is integral</u>).

For any $I \subset A$, the set $I \cap A'$ is denoted by $I^{L,A}$ or by I^L if reference to A is clear; and we say that I^L is the projection of I from L (in A). Note that then,

$$A^{L,A} = A^L = A' .$$

If $C \in \mathfrak{P}(A)$, then clearly A^L/C^L is the projection of A/C from $\bar{L} = L$ modulo C. We <u>say that the projection of</u> C <u>from</u> L <u>is birational</u> (respectively <u>integral</u>) <u>whenever the projection of</u> A/C <u>from</u> $\bar{L}$ <u>is birational</u> (<u>respectively integral</u>).

(35.2) RELATION WITH HOMOGENIZATION AND DEHOMOGENIZATION.

Let z be transcendental over A. Then the following relations are obvious in view of (33.4), (34.4) and the definitions.

(35.2.1) $H_z(A)^{H_z(L)} = H_z(A^L)$, for $L \in \mathfrak{L}^\infty(A)$.

(35.2.2) $(F_z(Q))^L = F_z(Q^{H_z(L)})$, for every homogeneous ideal Q in $H_z(A)$ prime to z; and hence in particular $A^L = F_z(H_z(A)^{H_z(L)})$.

(35.2.3) The projection from L (in A) is birational if and only if the projection from $H_z(L)$ in $H_z(A)$ is birational.

(35.3) LEMMA. Let A be a filtered domain, $C \in \mathfrak{P}_1(A)$, and $L \in \mathfrak{L}^\infty(A)$. Let $A^* = H_z(A)$, $C^* = H_z(C)$, and $L^* = H_z(L)$. Then the following conditions are equivalent:

(*) The projection of C from L is integral.

(**) The projection of C^* from L^* is π-integral, where $\pi = zH_z(A)$.

(***) $C^L \in \mathfrak{P}_1(A^L)$ and for all $V \in \mathfrak{X}(A/C, F_0(A/C))$ we have

$$\mathrm{ord}_V(F_1(A/C)) < 0 \Rightarrow \mathrm{ord}_V(\bar{L}) < 0 \text{ , where } \bar{L} = L \text{ modulo } C$$

PROOF. First note that for any filtered domain B, (1') to (5') are equivalent, where $W \in \mathfrak{X}(B, F_0(B))$.

(1') $\mathrm{ord}_W B < 0$

(2') $\mathrm{ord}_W F_1(B) < 0$

(3') For some $u \in F_1(B)$, $\mathrm{ord}_W u < 0$.

(4') For some $x \in H_1(H_z(B))$, $\mathrm{ord}_W(x/z) < 0$; i.e.,

$$\mathrm{ord}(H_z(B), x, W) < \mathrm{ord}(H_z(B), z, W) \text{ .}$$

(5') $W \in \mathfrak{Z}(H_z(B), zH_z(B))$.

It is easy to see that, we also get that each of (1), (2), (3) is equivalent to (*).

(1) A/C is integral over A^L/C^L.

(2) $C^L \in \mathfrak{P}_1(A^L)$ and for every $V \in \mathfrak{X}(A/C, F_0(A/C))$ we have,

$\mathrm{ord}_V(A/C) < 0 \Rightarrow \mathrm{ord}_{V \cap \mathfrak{J}(A^L/C^L)}(A^L/C^L) < 0$.

(3) $C^L \in \mathfrak{P}_1(A^L)$ and for every $V \in \mathfrak{X}'(A/C, F_0(A/C))$ we have,

$V \in \mathfrak{Z}(H_{\bar{z}}(A/C), \bar{z}H_{\bar{z}}(A/C)) \Rightarrow V \cap \mathfrak{J}(A^L/C^L) \in \mathfrak{Z}(H_{\bar{z}}(A^L/C^L), \bar{z}H_{\bar{z}}(A^L/C^L))$

where $\bar{z} = z$ modulo C^*.

(we use the equivalence of (1') and (5').

In view of (19.13) and (19.14), (*) $\Leftrightarrow$ (**) follows from the equivalence of (*) and (3); while (*) $\Leftrightarrow$ (***) can be deduced from (*) $\Leftrightarrow$ (2) $\Leftrightarrow$ (1) and (1') $\Leftrightarrow$ (2').

(35.4) DEFINITION. Let $C \in \mathfrak{P}_1(A)$. Then we <u>define</u> $\deg[A,C] = \mathrm{Deg}\, H_z(A/C)$ for any indeterminate z over A/C; and note that by (32.3) $\deg[A,C]$ is well defined. In particular, when $\dim A = 1$, we <u>define</u> $\deg A = \deg[A,\{0\}]$.

Now let $F_0(A)$ be algebraically closed. Using §34, we can get an alternative description of $\deg[A,C]$ (when $C \in \mathfrak{P}_1(A)$) as in (24.16):

$$\begin{aligned}\deg[A,C] &= \max\{[A/C : A^L/C^L] : L \in \mathfrak{L}^\infty(A),\ [L : F_0(A)] = 2\} \\ &= \max\{[A/C : F_0(A)(x)] : x \in F_1(A/C)\setminus F_0(A/C)\}\ .\end{aligned}$$

We further <u>define</u>

$\mathrm{genus}[A,C] = \mathrm{genus}[H_z(A), H_z(C)]$,

and,

$\mathrm{genus}\, A = \mathrm{genus}\, H_z(A)$, whenever $\dim A = 1$.

(35.5) THEOREM. <u>Let</u> A <u>be a filtered domain with</u> $\mathrm{emdim}\, A = \dim A = 3$. <u>Let</u> $A^* = H_z(A)$ <u>and note that then</u> $\mathrm{Emdim}\, A^* = 3$. <u>Assume</u>

that $H_0(A^*) = F_0(A)$ is algebraically closed. Let $C \in \mathfrak{P}_1(A)$ and $C^* = H_z(C)$. Let $zA^* = \pi$. Clearly $C^* \in \mathfrak{P}_1(A^*,\backslash\pi)$. Let $n = \mathrm{Deg}[A^*,C^*]$ and $g = \mathrm{genus}[A^*,C^*] = \mathrm{genus}[A,C]$.

Applying the Theorem (30.3) and (26.12) to A^* and C^*, and systematically applying " F_z " to various quantities involved we get the following theorem.

THEOREM ON 2-EQUIMULTIPLE PROJECTIONS OF AFFINE SPACE QUNITICS. Let $n \leq 5$ and let $\lambda([A,C],P) = 1$ for all $P \in \mathfrak{P}_0([A,C])$.

Then at least one of the following situations prevails.

(1) There exists $\varphi \in C \cap F_2(A)$ such that $\Phi = H_{(z,2)}(\varphi)A^* \in H_2^*(A^*)$ and Φ is a π-quasiplane, and hence (by (34.7)) we have that there exist $y_1, y_2, y_3 = \varphi$ in A such that $A = F_0(A)[y_1,y_2,y_3]$

(2) $3 \leq n \leq 5$ and there exists $L \in \mathfrak{L}_0^\infty(A)$ such that the projection of C from L (in A) is birational and integral; further, $\lambda([A^L,C^L],P) = 1$, for all $P \in \mathfrak{P}_0([A^L,C^L])$: and $\min(4,n) \leq \deg[A^L,C^L] \leq n$.

(3) $4 \leq n \leq 5$ and there exists $L \in \mathfrak{L}_0^\infty(A)$ such that upon letting $B = H_z(A^L)$. $D = H_z(C^L)$ and $\Delta = zB$ we have

(i) the projection of C from L (in A) is birational and integral.

(ii) $\max\{\mu([B,D],Q) : Q \in \mathfrak{Q}_0([B,D],\backslash\Delta)\}$
$= \max\{\mu([B,D],Q) : Q \in \mathfrak{Q}_0([B,D]],\Delta)\} = 2$;
in particular, we have
$\max\{\lambda([A^L,C^L],P) : P \in \mathfrak{P}_0([A^L,C^L])\} = 2$, and

(iii) $\min(4,n) \leq \deg[A^L,C^L] \leq n$.

Further assume that $g \leq 1$. Then we have in case (3) above:

with L *as above, upon putting* $m = n - 3$, *there exists* $\theta \in F_m(A^L) \setminus F_0(A^L)$ *with* $\theta A^L = \Theta \in \mathrm{tradj}([A^L, C^L]) \cap \mathfrak{P}_1(A^L)$. (Note that $m = 1$ or 2.) *Further we have that* A^L/Θ *is an euclidean domain*. Specifically we have that, there exist $x, y \in F_1(A)$, with $xF_0(A) + yF_0(A) + F_0(A) = L$, *such that* $A^L/\Theta = k[t]$ *or* $k[t, t^{-1}]$ *where* $t =$ image of x modulo Θ *and* $k =$ image of $F_0(A) = F_0(A^L)$ modulo Θ.

§36. Complete intersections.

(36.1) DEFINITION. Let A be a domain and $P \in \mathfrak{P}(A)$. We say that P is a *complete intersection* (in A) if P can be generated by $\dim A - \dim[A, P]$ generators.

Now let $(F_n(A))_{0 \le n \le \infty} = \mathfrak{F}$, say, be a filtration on A and let $\dim A = \mathrm{emdim}(A, \mathfrak{F}) = r$, say. We say that P is *hyper-planar relative to* $\mathfrak{F}$ if $P \cap F_1(A) \neq \{0\}$. We say that P is *essentially hyper-planar* (in A), if, for some filtration $\mathfrak{F}$ with $\dim A = \mathrm{emdim}(A, \mathfrak{F})$, we have that P is hyperplanar relative to $\mathfrak{F}$. We drop the word "hyper" if $r = 3$.

(36.2) LEMMA. *Let* $\dim A = 3$ *and let* $C \in \mathfrak{P}_1(A)$ *be essentially planar. Then* C *is a complete intersection*.

PROOF. Let C be planar relative to a filtration $(F_n(A))_{0 \le n \le \infty}$. Then there exist $u, v, w \in F_1(A)$, such that $A = F_0(A)[u, v, w]$ and $w \in C$. Then we have, $A/C \approx F_0(A)[u, v]/C \cap F_0(A)[u, v]$. Hence $C \cap F_0(A)[u, v] \in \mathfrak{P}_1(F_0(A)[u, v]$ and taking γ to be any generator of $C \cap F_0(A)[u, v]$ we easily see that $C = (\gamma, w)A$. Hence the proof.

(36.3) COROLLARY. *In particular, if in Theorem* (35.5), *the situation* (1) *prevails, then* C *is a complete intersection*.

(36.4) LEMMA. *Let* A, C *be as in Theorem* (35.5) *and assume that situation* (2) *prevails. Then* C *is a complete intersection*.

PROOF. We have $\mathrm{emdim}\, A^L = \dim A^L = 2$, and $C^L \in \mathfrak{P}_1(A^L)$. Let

$\gamma \in A^L$ such that $\gamma A^L = C^L$. Now by hypothesis we clearly have that, A^L/C^L is normal, $\mathfrak{I}(A/C) = \mathfrak{I}(A^L/C^L)$ and A/C is integral over A^L/C^L. It follows that $A/C = A^L/C^L$. Let $w \in F_1(A)$ such that $L + wF_0(A) = F_1(A)$. Then we have $A = A^L[w]$. In view of $A/C = A^L/C^L$, we see that $w - h \in C$ for some $h \in A^L$. It is now obvious that $C = (\gamma, w-h)A$.

(36.5) A SUFFICIENT CONDITION FOR COMPLETE INTERSECTION.

The following two Theorems ((36.5) and (36.7)) are, respectively, a restatement of Theorem 2' in [3] in our present terminology and a generalization of Theorem 3' in [3]. We include both proofs for the sake of completeness. For our final Theorem (36.9), however, the old Theorem 3' in [3.] is enough. This generalization (36.7) was needed for an older version of the proof of (36.9) and is related to the generalization of euclidean domains as would be carried out in §39.

(36.5) THEOREM. Let A be a filtered domain and B a projection of A such that $F_1(B) \in \mathfrak{L}_0^{\infty}(A)$. Let emdim A = dim $A = 3$, and hence emdim B = dim $B = 2$. Let $Z \in F_1(A)$ such that $A = B[Z]$. Assume that the projection of C from $F_1(B)$ is birational and integral. Let $C \in \mathfrak{P}_1(A)$ such that $\lambda([A,C],P) = 1$, for all $P \in \mathfrak{P}_0(A,C)$. Further assume that $C \cap B = \gamma B$ for some $\gamma \in B$ and hence $\gamma B \in \mathfrak{P}_1(B)$. Let $\alpha, \beta, \rho, \pi \in B$ such that $\alpha Z - \beta \in C$ and $(\rho,\pi)B = B$. Further assume that (1'), (2') and (3') are satisfied.

(1') $\rho \notin \gamma B$.

(2') for some integer $n > 1$, upon putting $\alpha' = \alpha\rho$ and $\beta' = \beta\rho + \pi\gamma$, we have $\gamma\rho^{n-1} \in (\{\alpha',\beta'\}B)^n$.

(3') $\gamma\rho^{n-1} \notin Q(\{\alpha',\beta\}B)^n$ for any $Q \in \mathfrak{P}_0(B,\{\alpha',\beta'\})$.

Then there exist $a_0, a_1, \ldots, a_n \in B$ such that $C = \{\alpha'Z-\beta', a_0Z^N + a_1Z^{n-1}+\ldots+a_n\}A$, and thus C is a complete intersection.

PROOF. By (2') we can write

(1) $\gamma\rho^{n-1} = a_0(\beta')^n + a_1(\beta')^{n-1}\alpha' + \ldots + a_n(\alpha')^n$, where $a_0, \ldots a_n \in B$.

Put

(2) $f = a_0 Z^n + a_1 Z^{n-1} + \ldots + a_n$ and $g = \alpha' Z - \beta'$.

Now we claim that

(3) $\{a_0, a_1, \ldots, a_n, \alpha', \beta'\}B = B$.

For proof, observe that, in the contrary case there will exist $Q \in \mathfrak{P}_0(A)$ such that $a_0, a_1, \ldots, a_n, \alpha', \beta' \in Q$, and hence $\gamma\rho^{n-1} \in Q(\{\alpha', \beta'\}B^n)$, in contradiction with (3').

Now let $h \in A$ such that $f, g \in hA$. Then by (1) we have $\gamma\rho^{n-1} \in hA$. Hence $h \in B$ and consequently $a_0, a_1, \ldots, a_n, \alpha', \beta' \in hB$. In view of (3) we get $1 \in hB$. Thus we have proved,

(4) the greatest common divisor of f, g is 1.

In view of (4), by Macaulay's unmixedness theorem we get that,

(5) if $(f,g)A = \bigcap_{i=1}^{r} Q_i$ is an irredundent primary decomposition of $(f,g)A$, where Q_i are primary for P_i ; then $\dim[A, P_i] = 1$ for all i and all P_i and Q_i are unique (being isolated components); in particular, $f, g \in P \in \mathfrak{P}_1(A) \Rightarrow P \in \{P_1, \ldots, P_r\}$.

We shall prove that $r = 1$ and $Q_1 = P_1 = C$.

First we claim that $P_i \cap B \in \mathfrak{P}_1(B)$ for all i. Because in the contrary case, we have $P_i \cap B \in \mathfrak{P}_0(B)$, $P_i = (P_i \cap B)A$ and hence $a_0, a_1, \ldots, a_n, \alpha', \beta' \in P_i \cap B$, in contradiction with (3).

Now we claim that $P_i \cap B = \gamma B$ for all i. Note that, since both γB and $P_i \cap B \in \mathfrak{P}_1(B)$, we only have to prove $\gamma \in P_i \cap B$. Assuming, if possible, $r \notin P_i \cap B$, we get by (1), $\rho \in P_i \cap B$ and hen using $g \in P_i$ and $(\rho, \pi)B = B$, we conclude $\gamma \in P_i$, a contradiction.

Now we claim that $C \in \{P_1, \ldots, P_r\}$. It is enough to show (by (5 that $f, g \in C$. Now $g \in C$, and in view of (1), $(\alpha')^n f \in (\gamma\rho^{n-1}, g)A \subset$ Now $\alpha' \notin C$, for otherwise $\alpha' \in C \cap B = \gamma B$ and hence $\beta' \in C \cap B =$

and (1), (1') and the fact that $n > 1$ yield a contradiction. It follows that $f \in C$.

Applying Proposition A1[3, P.97], with $B = R$, $A = S$, $Z = t$, $\gamma\rho^{n-1} = u$ and $\gamma B = Q$, we get

(6) $$1 = \mathrm{ord}_Q(\gamma\rho^{n-1}) = \mathrm{ord}_Q(\text{Z-resultant of } f,g)$$

$= \sum [A_P/PA_P : B_Q/QB_Q][A_P/(f,g)A_P : A_P]$, the summation being extended over all $P \in \mathfrak{P}_1(A)$ with $f,g \in P$ and $P \cap B = Q$

$$= \sum_{i=1}^{r} [A_{P_i}/P_iA_{P_i} : B_Q/QB_Q][A_{P_i}/Q_iA_{P_i} : A_{P_i}].$$

It follows from (6) that, $r = 1$ and $Q_1 = P_1 = C$, as claimed.

(36.6) ELEMENTARY TRANSFORMATIONS.

Let R be an commutative ring with 1. An $n \times n$ matrix M over R is said to be an <u>elementary</u> n-<u>step</u> <u>in</u> R if there exist distinct $i,j \in \{1,\dots,n\}$ and $t \in R$ such that, for each n-tuple $(r_1,\dots,r_n)$ over R we have.

$$(r_1,\dots,r_n)M = (s_1,\dots,s_n), \text{ where}$$

$$s_\alpha = \begin{cases} r_\alpha & \text{if } \alpha \neq j \\ r_j + tr_i, & \text{if } \alpha = j. \end{cases}$$

We shall denote M by $\mathfrak{E}_R(i,j;t)$. A finite product of elementary n-steps in R is said to be an <u>elementary</u> n-<u>transformation</u> in R. Whenever the number n is clear from the context we may just use the words <u>elementary</u> <u>step</u> and <u>elementary</u> <u>transformation</u> respectively. Also we may write $\mathfrak{E}(i,j;t)$ for $\mathfrak{E}_R(i,j;t)$, if reference to R is clear. The following results about elementary transformations are easy to prove by induction on the number of elementary steps in an elementary transformation, and are well known (in the substance).

(36.6.1) If M is an elementary n-transformation in R, then the determinant of M is $+1$ and hence elements in any row or column of M generate the unit ideal in R.

(36.6.2) If $(r_1,\ldots,r_n)M = (s_1,\ldots,s_n)$, where M is an elementary transformation, then

$$\{r_1,\ldots,r_n\}R = \{s_1,\ldots,s_n\}R .$$

(36.6.3) All elementary n-transformations in R form a group where $\mathfrak{E}(i,j,t)^{-1} = \mathfrak{E}(i,j,-t)$.

(36.6.4) Let $\psi: R \to S$ be an epimorphism. Assume that $S = S_1\oplus\cdots\oplus S_m$ and let $\pi_i^* : S \to S_i$ be the projection onto the ith component. For an n-tuple $r = (r_1,\ldots,r_n)$ over R we shall denote by $\psi(r)$ the n-tuple $(\psi(r_1),\ldots,\psi(r_n))$ over S. Let $N_1,\ldots,N_m$ be elementary n-transformations in $S_1,\ldots,S_m$.

Then there exists an elementary n-transformation N in R such that $\pi_i^*(\psi(rN)) = \pi_i^*(\psi(r)N_i$ for every n-tuple r over R and $i = 1,2,\ldots,m$.

For constructing N first let N_i be written as $N_i = \theta_1^{(i)}\theta_2^{(i)} \cdots \theta_{s_i}^{(i)}$, where $\theta_j^{(i)} = \mathfrak{E}_{S_i}(p(i,j),q(i,j)\ ;\ t(i,j))$. Then upon taking $N = \sigma_1^{(1)}\sigma_2^{(1)} \cdots \sigma_{s_1}^{(1)} \cdots\cdots \sigma_1^{(m)} \cdots\cdots \sigma_{s_m}^{(m)}$, N will clearly have the required property, where we take

$\sigma_j^{(i)} = \mathfrak{E}_R(p(i,j),q(i,j)\ ;\ t^*(i,j))$, where $t^*(i,j) \in R$ is chosen such that

$$\pi_s(\psi(t^*(i,j)) = \begin{cases} 0 & \text{, if } s \neq i \\ t(i,j) & \text{, if } s = i . \end{cases}$$

(36.6.5) Let S be an euclidean domain. Then for any n-tuple $s = (s_1,\ldots,s_n)$ over S, there exists an elementary n-transformation M (depending on s) such that

$$sM = (s_1^*,\ldots,s_n^*) \quad \text{with} \quad s_i^* = 0 \quad \text{for} \quad 2 \le i \le n .$$

Hence by (36.6.2) we have

$$s_1^* S = \{s_1, \ldots, s_n\} S$$

For constructing M we simply take the obvious n-steps as elementary n-steps of M which perform single stages of the euclidean algorithm on $(s_1, \ldots, s_n)$.

(36.6.6) Let $\psi: R \to S$ be an epimorphism and assume that for an n-tuple $r = (r_1, \ldots, r_n)$ over R there exists an elementary n-transformation M in S such that

$$\psi(r)M = (s_1^*, \ldots, s_n^*) \quad \text{with} \quad s_i^* = 0 \quad \text{for} \quad 2 \le i \le n \,.$$

As a special case we may assume that S is an euclidean domain.

Then there exists an n-tuple $(m_1, \ldots, m_n)$ over R such that

$$\operatorname{Ker} \psi + \{r_1, \ldots, r_n\} R = \operatorname{Ker} \psi + (m_1 r_1 + \ldots + m_n r_n) R$$

and $\{m_1, \ldots, m_n\} R = R$.

For proof, we first take the n-transformation N in R as obtained in (36.6.4). Then we must have

$$rN = (r_1^*, \ldots, r_n^*) \quad \text{where} \quad r_i^* \in \operatorname{Ker} \psi, \text{ for } \quad 2 \le i \le n \,.$$

Then $r_1^* = m_1 r_1 + \ldots + m_n r_n$ for some $m_i \in R$ where $(m_1, \ldots, m_n)$ is the transpose of the first column of N. The rest of the proof follows from (36.6.1) and (36.6.2).

(36.7) THEOREM. Assume that A, B, C, r are as in Theorem (36.5). Further assume that α is a true adjoint of γ in B. It follows that there exists $\beta \in B$ such that $\alpha Z - \beta \in C$. Assume that α has the property,

(*) for some β with $\alpha Z - \beta \in C$, there exist $\rho', \pi' \in B$ such that $\{\rho', \pi'\} B = B$ and $\{\beta\rho' + \pi'\gamma, \alpha\} B = \{\beta, \gamma, \alpha\} B$.

Further assume that,

(**) $\{\lambda^*([B,\gamma],P) : P \in \mathfrak{P}_0([B,\gamma])\} = \{1,2\}$.

Then there exist $\rho,\pi \in B$ such that all the conditions of Theorem (36.5) are satisfied, when a suitable β is chosen and α is as above, a true adjoint with property (*).

Further, α has the property (*), if $B/\alpha B$ is an euclidean domain.

PROOF. Choose $\beta,\rho',\pi' \in B$, such that,

(1) $\alpha Z - \beta \in C$,

(2) $\{\rho',\pi'\}B = B$, and

(3) $\{\beta\rho' + \pi'\gamma,\alpha\}B = \{\beta,\gamma,\alpha\}B$.

Let

(4) $\Omega = \{P \in \mathfrak{P}_0(B) : \rho', \beta\rho' + \pi'\gamma \in P\}$.

We claim that,

(5) $P \in \Omega \Rightarrow \alpha \notin P$.

For proof, assume if possible, that $\alpha \in P$. By (*) we have

(5_1) $\gamma \in P$ and $\beta \in \{\beta\rho' + \pi'\gamma,\alpha\}B_P$

Since α is a true adjoint, in view of (**), (4), (5_1) and (10.1.11), we get

(5_2) $\beta \in P(\{\beta\}B_P) + \alpha B_P$ and consequently $\beta \in \alpha B_P$.

Let $\varphi: A \to A/C$ be the canonical homomorphism. Then we have that $\varphi(B)[\varphi(Z)]$ is the integral closure of $\varphi(B)$. In view of (1) and (5_2) we get

(5_3) $\varphi(Z) \in \varphi(B)_{\varphi(P)}$ and in particular $\varphi(B)_{\varphi(P)}$ is normal.

Hence we must have $\lambda_{\mathfrak{C}}([B,\gamma],P) = 0$ (by the last result in (6.1)

and hence α being a true adjoint $\alpha \notin P$; a contradiction. This concludes the proof of (5).

From definition of Ω, by (1) and (5) it follows that

(6) $P \in \Omega \rightarrow \rho', \gamma \in P$ and $\lambda([B,\gamma],P) = 1$, and hence by (9.1), $\gamma \notin P^2$.

Clearly Ω is a finite set and we can choose $c \in B$ such that

(7) $\rho' + c\alpha\pi'\gamma \in P \backslash P^2$, for all $P \in \Omega$.

Now put

(8) $\rho = \rho' + c\alpha\pi'\gamma$, $\pi = \pi'$, $\beta' = \beta\rho + \pi\gamma = \beta\rho' + \pi'\gamma + c\alpha\pi'\gamma$.

It is clear that (*) implies

(9) $\{\rho,\pi\}B = B$ and $\{\beta',\alpha\}B = \{\beta,\gamma,\alpha\}B$.

In view of (6) and (**) we get that

(10) there exists $P_0 \in \mathfrak{P}_0([B,\gamma])$ with $\lambda([B,\gamma],P_0) = 2$ and hence $\rho' \notin P_0$; in particular $\rho' \notin \gamma B$ and hence $\rho \notin \gamma B$.

Now let

$\Omega_1 = \{P \in \mathfrak{P}_0(B) : \alpha,\beta' \in P\}$, $\Omega_2 = \{P \in \mathfrak{P}_0(B) : \rho,\beta' \in P\}$ and $\Omega^* = \{P \in \mathfrak{P}_0(B) : \alpha',\beta' \in P\}$.

Then we have, using (5)

(11) $\Omega_2 = \Omega$, $\Omega_1 \cap \Omega_2 = \phi$ and $\Omega^* = \Omega_1 \cup \Omega_2$.

Let $P \in \Omega_1$. Then by (**) we must have $\lambda([B,\gamma],P) = 2$. Further by (11) $\rho \notin P$, $\alpha' \in \mathrm{tradj}([B_P,\gamma],P)$, and by applying (10.1.11) we get,

(12) $P \in \Omega_1 \Rightarrow \gamma\rho \in ((\alpha',\beta')B_P)^2$ and $\gamma\rho \notin PB_P(\{\alpha',\beta'\}B_P)^2$.

Now let $P \in \Omega_2$. Then from (11), (6) and (7) we get,

(13) $P \notin \Omega_2 \Rightarrow \rho,\gamma \in P$; $\alpha,\pi \notin P$; $\{\alpha',\beta'\}B_P = \{\rho,\gamma\}B_P$ and $\gamma\rho \in (\{\alpha',\beta'\}B_P)^2$, $\gamma\rho \notin PB_P(\{\alpha',\beta'\}B_P)^2$

(14) $P \notin \Omega \Rightarrow \{\alpha',\beta'\}B_P = B_P$ (by (11) and hence)

$$\gamma_\rho \in \left(\{\alpha',\beta'\}B_P\right)^2 .$$

From (10), (12), (13) and (14), it follows that the conditions (1'), (2') and (3') for Theorem (36.5) are satisfied.

The last statement readily follows from (36.6.6).

(36.8) COROLLARY. In particular, if in Theorem (35.5), the situation (3) prevails, then as a consequence of Theorem (36.5) and (36.7) we get that C is a complete intersection.

(36.9) THEOREM. Let A be a filtered domain over an algebraically closed ground field and with $\dim A = \operatorname{emdim} A = 3$.

Let $C \in \mathfrak{P}_1(A)$ with $\deg[A,C] \leq 5$ and $\operatorname{genus}[A,C] \leq 1$. Further assume that $\lambda([A,C],P) = 1$ for all $P \in \mathfrak{P}_0([A,C])$. Then C is a complete intersection.

PROOF. In view of Theorem (35.5), the proof is finished by (36.3), (36.4) and (36.8).

CHAPTER V. APPENDIX

§37. Double points of algebroid curves.

In §10 we proved certain properties of an abstract local domain R of dimension one and multiplicity two such that the integral closure R^* of R in $\mathfrak{J}(R)$ is a finite R-module. We gave a direct proof. One can also give a proof by the so-called method of "going to completion", i.e., by replacing R by $\mathcal{O}$, the completion of R. The advantage of working with $\mathcal{O}$ is that we can get a much more concrete and standard-looking description of $\mathcal{O}$. The disadvantage is that we need a lot of background material for the "completion proof". In this section we shall illustrate how the completion proof can be carried out.

The proof of (10.1.1), (10.1.2) and (10.1.3) shall be the same as in (10.1). The proofs of (10.1.10) to (10.1.14) shall be also the same as in §10. (10.1.9) can be proved alternatively (not by a "completion method") as follows:

(37.1) LEMMA. _With_ R _as in Theorem_ (10.1), _we have,_

$$\text{emdim } R = 2.$$

PROOF. We have, $\lambda(R) = e(R)$, where $e(R)$ is the multiplicity of R in the sense of Zariski-Samuel, by using Corollary 1, page 299, or better still by formula (8"), page 300 [18]. Thus $e(R) = \lambda(R) = e(\mathcal{O}) = \lambda(\mathcal{O}) = 2$.

If $R/M(R)$ is an infinite field, then we have Abhyankar's result (1) [2],

$$\text{emdim } R \leq \dim (R) + e(R) - 1 \text{ ;}$$

and since R is not regular (otherwise $\lambda(R) = 1$ by (5.10)) we must have emdim $R = 2 =$ emdim $\mathcal{O}$.

In the general case we proceed as in §9; namely, we replace R by $R' = R[Y]_{M(R)R[Y]}$ where Y is an indeterminate. Then R' is also a one dimensional local domain with $\lambda(R') = \lambda(R)$, $M(R)R' = M(R')$

and an infinite residue field $R'/M(R') \approx R/M(R)(Y)$. Hence emdim $R' = 2 =$ emdim R as required.

(37.2) REDUCTION TO THE COMPLETE CASE. We will now show that it is enough to prove (10.1.4) to (10.1.8) by replacing R by its completion $\mathcal{O}$. Let $\{V_1,\ldots,V_r\} = \mathfrak{Y}(R)$ and let $P_i = M(V_i) \cap R^*$. Let $\mathcal{O}$ be the completion of R and let $\mathcal{O}^*$ be the integral closure of $\mathcal{O}$ in $\mathfrak{J}(\mathcal{O})$. Using that R^* is a finite R-module and that R has dimension 1, we can easily deduce the following description of $\mathcal{O}$ and $\mathcal{O}^*$ using any standard treatment of completions, say Zariski-Samuel [18] Chapter VIII. (In particular, we refer to Theorem 11, Theorem 18, Theorem 24, Theorem 27 and their corollaries.)

$\mathcal{O}^*$ coincides with a completion of R^* which in turn coincides with the direct sum of completions of V_i, say $\mathcal{O}_i^*$. In particular $\mathcal{O}^*$ is a finite $\mathcal{O}$-module. Further, $\mathcal{O}$ is a one dimensional local ring and emdim $\mathcal{O}$ = emdim R.

$\mathfrak{J}(R)$ is canonically a subring of $\mathfrak{J}(\mathcal{O})$. Since every nonzero ideal in R is primary for $M(R)$, $I \to I\mathcal{O}$ and $J \to J \cap R$ are inverse bijections between nonzero ideals of R and $\mathcal{O}$. In particular for nonzero corresponding ideals I, J in $R,\mathcal{O}$ respectively we have

$$[R/I : R] = [\mathcal{O}/J : \mathcal{O}].$$

Further we have, $\mathfrak{C}(R)\mathcal{O} = \mathfrak{C}(\mathcal{O})$ and hence $[R/\mathfrak{C}(R) : R] = [\mathcal{O}/\mathfrak{C}(\mathcal{O}) : \mathcal{O}]$.

Similarly for any $I \in R$ or $I \subset R$ we have $[R^*/IR^* : R] = [\mathcal{O}^*/I\mathcal{O}^* : \mathcal{O}]$ and hence

$$\lambda(R,I) = \lambda(\mathcal{O},I) = \lambda(\mathcal{O},I\mathcal{O}) \quad \text{(by (5.5))}.$$

In view of these remarks, it is clear that to prove (10.1.4) to (10.1.8) we may replace everywhere, R by $\mathcal{O}$, R^* by $\mathcal{O}^*$, V by its completion, say $\mathcal{V}$, and W by its completion, say $\mathcal{W}$.

This gives the following description for the three cases:

Case (1^*). $\mathcal{O}$ is not a domain and $\mathcal{O}^* = \mathcal{V} \oplus \mathcal{W}$ where $\mathcal{V}$ and $\mathcal{W}$ are complete one dimensional regular local domains. Further $\mathrm{ord}_{\mathcal{V}} M(\mathcal{O}) = \mathrm{ord}_{\mathcal{W}}(M(\mathcal{O})) = 1$ and $\mathcal{V}$ and $\mathcal{W}$ are residually rational over $\mathcal{O}$.

Case (2^*) $\mathcal{O}$ is a domain and $\mathcal{O}^* = \mathcal{V}$, a complete one dimensional regular local domain. Further, $\mathrm{ord}_{\mathcal{V}}(M(\mathcal{O})) = 2$ and $\mathcal{V}$ is residually rational over $\mathcal{O}$.

Case (3^*) $\mathcal{O}$ is a domain and $\mathcal{O}^* = \mathcal{V}$, a complete one dimensional regular local domain. Further, $\mathrm{ord}_{\mathcal{V}}(M(\mathcal{O})) = 1$ and $[\mathcal{V}/M(\mathcal{V}) : \mathcal{O}/M(\mathcal{O}))] = 2$.

(37.3) DESCRIPTION WITH AN EXTRA ASSUMPTION. Note that the above description is yet no simpler than the one in §10. To be able to use the completeness of $\mathcal{O}$ effectively we shall make an extra assumption:

(*) $\mathcal{O}$ and $\mathcal{O}/M(\mathcal{O})$ have the same characteristic.

Note that the assumption (*) holds good for the case of points of algebraic curves over a ground field, since the ground field is contained in both $\mathcal{O}$ and $\mathcal{O}/M(\mathcal{O})$.

Assuming (*) we get by Cohen's Structure (Theorem 9 [8]):

There exists a coefficient field for $\mathcal{O}$, i.e. there is a subfield of $\mathcal{O}$ which maps isomorphically onto $\mathcal{O}/M(\mathcal{O})$ under the canonical map; further if K is such a coefficient field then there exists an epimorphism $\varphi: B \to \mathcal{O}$ where B is a formal power series ring in emdim $\mathcal{O}$ variables over K and φ restricts to identity on K.

Thus, in our particular case we get that, if x,y is any basis of $M(\mathcal{O})$, then $\mathcal{O} = K[[x,y]]$. The map $\varphi : K[[X,Y]] = B \to \mathcal{O}$ may be defined by $\varphi(X) = x$ and $\varphi(Y) = y$. Also $\mathrm{Ker}\,\varphi$ is principal, say $\mathrm{Ker}\,\varphi = \gamma' B$. In view of (9.1) we get from $\lambda(\mathcal{O}) = 2$ that

$ord_{M(B)}\gamma' = 2$.

Further, if $x \in \mathcal{O}$ with $\lambda(\mathcal{O},x) = 2$, then in view of (8.2) it is easy to see that we can write

$$\gamma' = r_1Y^2 + r_2XY + r_3X^2 \text{ , where } r_i \in B \text{ and } r_1 \text{ is a unit.}$$

By Weierstrass Preparation Theorem we may write:

$$\gamma'\eta = Y^2 + p^*Y + q^* \text{ , where } p^*,q^* \in XK[[X]] \text{ and } \eta \text{ is a unit.}$$

Thus we have proved the following statement:

If x,y is a basis of $M(\mathcal{O}), \varphi(X) = x, \varphi(Y) = Y$, and $\lambda(\mathcal{O},x) = 2$, then there exists $\gamma \in B$ such that $Ker\ \varphi = \gamma B$ and

(1) $$\gamma = Y^2 + p^*Y + q^* \text{ , where } p^*,q^* \in XK[[X]].$$

We shall now redescribe the three cases. First we introduce the following notation:

NOTATION. Let $K[[t]]$ be any formal power series ring over K. If we simply say $h \in K[[t]]$, we mean

$$h = h_0 + h_1t+\ldots+h_nt^n+\ldots \in K[[t]].$$

and we may write $h = h(t)$. Also if X is any other variable, then we write $h(X)$ to mean

$$h(x) = h_0 + h_1X+\ldots+h_nX^n+\ldots \in K[[X]].$$

Further in $K[[t]]$ by $ord_t h$ we mean as usual $\min\{i: h_i \neq 0\}$.

Case (1^*). Since $\mathcal{O}$ is not a domain, γ is reducible in $K[[X]][X]$. Then we may write:

(2) $$\gamma = (Y-p)(Y-q) \text{ , } p,q \in K[[X]], \text{ and}$$

(3) $$e = ord_X(p-q).$$

We shall prove later that $e = d = [\mathcal{O}/\mathfrak{C}(\mathcal{O}) : \mathcal{O}]$.

(4) $\mathcal{O}^* = K[[\tau]] \oplus K[[\sigma]]$, where σ, τ are independent variables, $\mathcal{V} = K[[\tau]]$, $\mathcal{W} = K[[\sigma]]$, $\mathrm{ord}_{\mathcal{V}}(h,h') = \mathrm{ord}_{\tau}(h(\tau))$ and $\mathrm{ord}_{\mathcal{W}}(h,h') = \mathrm{ord}_{\sigma} h'(\sigma)$.

(5) $\varphi: B \to \mathcal{O}$ is defined by
$\varphi(h(X,Y)) = (h(\tau, p(\tau)), h(\sigma, q(\sigma)))$.

From the definition of φ, it is easy to check that

(6) $\mathcal{O} = \{(h,h') \in \mathcal{O}^* : \mathrm{ord}_X(h(X) - h'(X)) \geq e\}$.

Case (2^*). In this case γ is irreducible in $K[[X]][Y]$ and we may write:

(7) $\mathcal{O}^* = \mathcal{V} = K[[\tau]]$ and $\mathrm{ord}_{\mathcal{V}} = \mathrm{ord}_{\tau}$.

(8) $\mathcal{O} = K[[x,y]]$ where x,y is any basis of $M(\mathcal{O})$ and $x = x(\tau)$, $y = y(\tau) \in K[[\tau]]$. Further by our choice of x we have

$$\mathrm{ord}_{\tau} x(\tau) = 2.$$

Case (3^*). In this case γ is irreducible and we may write:

(9) $\mathcal{O}^* = \mathcal{V} = K'[[\tau]]$ where $K' = \mathcal{V}/M(\mathcal{V})$ is a quadratic extension of K. In particular we have
$\sigma \in K' \setminus K \Rightarrow K(\sigma) = K'$.

(10) $\mathcal{O} = K[[x,y]]$ where x,y is any basis of $M(\mathcal{O})$ and $x = x(\tau)$, $y = y(\tau) \in K'[[\tau]]$. Further by our choice of x we have

$$\mathrm{ord}_{\tau} x(\tau) = 1 .$$

(11) If $g: \mathcal{V} \to \mathcal{V}/M(\mathcal{V}) = K'$ denotes the canonical residue map, then we have $g(h) = h_0$, for any
$h = h(\tau) = h_0 + h_1\tau + \ldots + h_n\tau^n + \ldots \in K'[[\tau]]$.

We shall now give proofs of (10.1.4) to (10.1.8) (after going to completion) in each of the three cases above with the assumption (*).

(37.4) CASE OF HIGH NODES. Assume the description of case (1^*) in (37.3) i.e., assume (1) to (6) in (37.3).

First we claim that

(1) $\mathfrak{C}(\mathcal{O}) = M(\mathcal{V})^e \cap M(\mathcal{W})^e$

$= \{(h,h') \in \mathcal{O}^* : \text{ord}_\tau h(\tau) \geq e \text{ and } \text{ord}_\sigma h'(\sigma) \geq e\}$.

For proof let $\mathfrak{A}$ be the ideal on the right side.

From (6) in (37.3) it is clear that $\mathfrak{A}$ is an ideal in $\mathcal{O}$ as wel as $\mathcal{O}^*$. Hence clearly $\mathfrak{A} \subset \mathfrak{C}(\mathcal{O})$.

On the other hand if $(h,h') \in \mathfrak{C}(\mathcal{O})$ then we must have that $(h(\tau),h'(\sigma))(1,0)$ and $(h(\tau), h'(\sigma))(0,1)$ belong to $\mathcal{O}$, and hence from (6) in (37.3) we clearly have $\text{ord}_\tau h(\tau) \geq e$ and $\text{ord}_\sigma h'(\sigma) \geq e$, i.e., $\mathfrak{C}(\mathcal{O}) \subset \mathfrak{A}$. Hence the claim.

Now let $x = (a,b) \in \mathcal{O}$ such that $\lambda(\mathcal{O},x) = 2$. Then it is easy to see that we must have

(2) $a_i = b_i$ for $0 \leq 1 \leq e - 1$.

(3) $a_0 = b_0 = 0$ and $a_1 = b_1 \neq 0$.

In particular we see that

(4) $K[[\tau]] = K[[a]]$ and $K[[\sigma]] = K[[b]]$.

In view of (6) in (37.3) again we see that given any $(h,h') \in \mathcal{O}$ we can write

(5) $(h,h') = \psi(x) + (0,f)$ where $\text{ord}_\sigma f(\sigma) \geq e$, and $\psi \in K[[a]]$ such that $\psi(a) = h(\tau)$.

Thus $(h,h') = \psi(x)$ (modulo $\mathfrak{C}(\mathcal{O})$).

(6) Since $x^e \in \mathfrak{C}(\mathcal{O})$ by (1), we clearly see that the only ideals of $\mathcal{O}/\mathfrak{C}(\mathcal{O})$ are those generated by x^i modulo $\mathfrak{C}(\mathcal{O})$ for $i = 0, 1,2,\ldots,e-1$.

In particular,

(7) $[\mathcal{O}/\mathfrak{C}(\mathcal{O}) : \mathcal{O}] = e$ and hence $e = d$.

In view of (6) in (37.3) and (1), (6) and (7) it is clear that the proof of (10.1.4), (10.1.6), (10.1.7) and (10.1.8) is now complete.

It only remains to prove (10.1.5). Thus, let $y = (a',b') \in \mathcal{O}$ with $\operatorname{ord}_\tau a'(\tau) = d < \operatorname{ord}_\sigma b'(\sigma)$.

Then we have

(8) $a_i' = 0$ for $i = 0,1,\ldots,d-1$, $a_d' \neq 0$, and

(9) $b_i' = 0$ for $i = 0,1,\ldots,d$.

Also let $y^* = y - a_d' a_1^{-d} x^d = (a^*,b^*)$ say. Then $\operatorname{ord}_\sigma b^*(\sigma) = d < \operatorname{ord}_\tau a^*(\tau)$ and we have:

(10) $a_i^* = 0$ for $i = 0,1,\ldots,d$,

(11) $b_i^* = 0$ for $i = 0,1,\ldots,d-1$, and $b_d^* \neq 0$.

Now $\{x,y\}\mathcal{O}$ is clearly primary for $M(\mathcal{O})$ and hence we have that the number

(12) $m = \max\{\min\{\operatorname{ord}_\tau h(\tau), \operatorname{ord}_\sigma h'(\sigma)\} : (h,h') \in \mathcal{O}\setminus\{x,y\}\mathcal{O}\}$
is a finite nonnegative integer.

We claim that

(13) $m = 0$.

For let, if possible, $m > 0$ and

(13_1) $(h,h') \in \mathcal{O}\setminus\{x,y\}$ with $\min\{\operatorname{ord}_\tau h(\tau), \operatorname{ord}_\sigma h'(\sigma)\} = m > 0$.

Note that $m \geq d$, since otherwise for $(f,f') = (h,h') - (h_m a_1^{-m})x^m$ we clearly have

$$\min\{\operatorname{ord}_\tau f(\tau), \operatorname{ord}_\sigma f'(\sigma)\} > m$$

(in view of the fact that $h_m = h_m'$ and $a_1 = b_1$).

If $m \geq d$, then clearly we can choose $\eta_1, \eta_2, \eta_3 \in K$ such that

(13_2) $h_m = \eta_1 a_1^m + \eta_2 a'_d a_1^{m-d}$, and

$$h'_m = \eta_1 b_1^m + \eta_3 b_d^* b_1^{m-d} .$$

Then it is easy to see that in view of (3), (8), (9), (10), (11) and (13_1) we have, upon letting $(f,f') = (h,h') - (\eta_1 x^m + \eta_2 y x^{m-d} + \eta_3 y^* x^{m-d})$, that:

(13_3) $\min\{\mathrm{ord}_\tau f(\tau), \mathrm{ord}_\sigma f'(\sigma)\} > m$ and

$$(f,f') \in \mathcal{O}\backslash\{x,y\}\mathcal{O} .$$

Since this is a contradiction to the maximality of m, (13) is proved.

From (12) and (13), it follows that $M(\mathcal{O}) = \{x,y\}\mathcal{O}$, thus proving (10.1.5).

(37.5) CASE OF HIGH CUSPS. Assume the description of case (2^*) in (37.3), i.e., assume (7) and (8) in (37.3).

Since $\mathrm{ord}_\tau x(\tau) = 2$, it is clear that every element $h(\tau) \in K[[\tau]]$ can uniquely written as

(1) $h(\tau) = h^{(1)}(x) + \tau\, h^{(2)}(x)$, where $h^{(1)}(x), h^{(2)}(x) \in K[[x]]$.

Let

(2) $e = \min\{\mathrm{ord}_x h^{(2)}(x) : h^{(1)}(x) + \tau\, h^{(2)}(x) \in \mathcal{O}$
for some $h^{(1)}(x)\}$.

Since $K[[x]] \subset \mathcal{O}$ we clearly have

(3) $e = \min\{\mathrm{ord}_x h(x) : \tau\, h(x) \in \mathcal{O}\}$.

If $\mathfrak{J}(K[[x]]) = \mathfrak{J}(\mathcal{O})$; we would get that $\mathrm{ord}_{\mathcal{V}}(\theta) \equiv 0 \ (2)$ for every $0 \neq \theta \in \mathfrak{J}(\mathcal{O})$; and since this is a contradiction we have $\mathcal{O} \neq K[[x]]$. Also clearly $\tau \notin K[[x]]$ and hence we see that e is a positive integer.

Also note that in (3), we have $\mathrm{ord}_\tau \tau h(x) = 1 + 2\, \mathrm{ord}_x h(x)$ and hence

(4) $e = \min\{s : \operatorname{ord}_\tau h(\tau) = 2s + 1$ for some $h \in \mathcal{O}\}$.

Note that in view of (1) and (3), we have

(5) $\mathcal{O} = \{h^{(1)}(x) + \tau h^{(2)}(x) : \operatorname{ord}_x h^{(2)}(x) \geq e\}$.

Now we claim that

(6) for $h \in \mathcal{O}$, $h \in \mathfrak{C}(\mathcal{O}) \Leftrightarrow \operatorname{ord}_x h^{(1)}(x) \geq e$.

Namely, in view of (5) and (1), $\Leftarrow$ is obvious; while for $\Rightarrow$ observe that

(6_1) $h \in \mathfrak{C}(\mathcal{O}) \Rightarrow \tau h \in \mathcal{O}$.

and if we write

(6_2) $\tau^2 = a(x) + \tau b(x)$, $a(x), b(x) \in K[[x]]$,

then from (6_1) and (5) we get

(6_3) $\operatorname{ord}_x(h^{(1)}(x) + b(x)h^{(2)}(x) \geq e$, $\operatorname{ord}_x h^{(2)}(x) \geq e$.

and thus $\operatorname{ord}_x h^{(1)}(x) \geq e$.

From (5) and (6), it follows that

(7) $\mathfrak{C}(\mathcal{O}) = M(\mathcal{V})^{2e}$, and

(8) every $h \in \mathcal{O}$ can be written as $h = h^*(x) + h^{**}$ where $h^{**} \in \mathfrak{C}(\mathcal{O})$ and $h^*(x)$ is a polynomial in x of degree at most $(e-1)$.

(9) From (8) it follows that the only ideals of $\mathcal{O}/\mathfrak{C}(\mathcal{O})$ are generated by the images of x^i modulo $\mathfrak{C}(\mathcal{O})$, for $i = 0, 1, \ldots, e-1$.

In particular we have

(10) $[\mathcal{O}/\mathfrak{C}(\mathcal{O})] = e$, and hence $e = d$.

Clearly the proof of (10.1.4), (10.1.6), (10.1.7) and (10.1.8) is now complete in view of (10) by (5), (9), (7) and (4) respectively.

It remains to prove (10.1.5). So let $y \in \mathcal{O}$ with

(11) $y(\tau) = y^{(1)}(x) + \tau y^{(2)}(x)$ with $ord_\tau y(\tau) = 2d + 1$.

Hence we must have

(12) $ord_x y^{(1)}(x) > d$, $ord_x y^{(2)}(x) = d$.

Upon taking $y^*(\tau) = \tau y^{(2)}(x)$ we have (since $y^{(1)}(x) \in x\mathcal{O}$) ,

(13) $\{x,y\}\mathcal{O} = \{x,y^*\}\mathcal{O}$.

But from (5) and (10) we see that

(14) $h = h^{(1)}(x) + \tau h^{(2)}(x) \in \mathcal{O} \Rightarrow \tau h^{(2)}(x) = y^* h^*(x)$ where $h^*(x) \in K[[x]]$.

Thus clearly $\mathcal{O} = K[[x]] + y^* K[[x]]$ and in particular, $M(\mathcal{O}) = \{x,y\}\mathcal{O}$.

Thus (10.1.5) is proved.

(37.6) CASE OF NONRATIONAL CUSPS. Assume the description of case (3^*) in (37.3), i.e., assume (9), (10) and (11) in (37.3).

Choose any field generator σ of K' over K. (Any element of $K'\setminus K$ would do.)

First note that $K'[[\tau]] = K'[[x]]$, since $ord_\tau x(\tau) = 1$. Now we have:

(1) every $h \in K'[[x]]$ can be uniquely written as $h = h^{(1)}(x) + \sigma h^{(2)}(x)$, where $h^{(1)}(x), h^{(2)}(x) \in K[[x]]$.

Let

(2) $e = \min\{ord_x h^{(2)}(x) : h^{(1)}(x) + \sigma h^{(2)}(x) \in \mathcal{O}$ for some $h^{(1)}(x)\}$

Since $K[[x]] \subset \mathcal{O}$ we clearly have

(3) $e = \min\{ord_x h(x) \in K[[x]] : \sigma h(x) \in \mathcal{O}\}$.

Since $\mathcal{O} \neq K'[[x]]$, $\sigma \notin \mathcal{O}$ and $e \neq 0$.

Also $\sigma \in \mathfrak{J}(\mathcal{O})$ and hence $e < \infty$. Thus e is a positive integer

Note that in view of (1) and (3), we have:

(4) $\mathcal{O} = \{h^{(1)}(x) + \sigma h^{(2)}(x) : ord_x h^{(2)}(x) \geq e\}$.

Now we claim that

(5) for $h \in \mathcal{O}$, $h \in \mathfrak{C}(\mathcal{O}) \Leftrightarrow \text{ord}_x h^{(1)}(x) \geq e$.

Namely in view of (5) and (1), $\Leftarrow$ is obvious, while for $\Rightarrow$ observe that

(5_1) $h \in \mathfrak{C}(\mathcal{O}) \Rightarrow \sigma h \in \mathcal{O}$,

and if we let

(5_2) $\sigma^2 = a + \sigma b$, $a,b \in K$,

Then from (5_1) and (4) we get

(5_3) $\text{ord}_x(bh^{(2)}(x) + h^{(1)}(x)) \geq e$, $\text{ord}_x h^{(2)}(x) \geq e$,

and thus $\text{ord}_x h^{(1)}(x) \geq e$.

From (4) and (5) it follows that

(6) $\mathfrak{C}(\mathcal{O}) = M(\mathcal{V})^e$, and

(7) every $h \in \mathcal{O}$ can be written as
$h = h^*(x) + h^{**}$ where $h^{**} \in \mathfrak{C}(\mathcal{O})$ and $h^*(x)$ is a polynomial in x of degree at most $(e-1)$.

(8) From (7) it follows that the only ideals of $\mathcal{O}/\mathfrak{C}(\mathcal{O})$ are generated by the images of x^i modulo $\mathfrak{C}(\mathcal{O})$, for $i = 0,1,\ldots,e-1$.

In particular we have

(9) $[\mathcal{O}/\mathfrak{C}(\mathcal{O}) : \mathcal{O}] = e$, and hence $e = d$.

Clearly the proof of (10.1.4), (10.1.6) and (10.1.7) follows, in view of (9), from (3), (8) and (6).

For proving (10.1.8), note that if $y,\theta \in \mathcal{O}$ with $\text{ord}_x y = \text{ord}_x \theta = s$ and if $h(x) \in K[[x]]$ with $\text{ord}_x h(x) = s$, then

$$g(y/\theta) \notin K \Leftrightarrow g(y/h(x) \notin K \text{ or } g(\theta/h(x)) \notin K.$$

Thus for (10.1.8) it is enough to prove that

$$d = \min\{s : \text{ord}_x y = s = \text{ord}_x h(x) \text{ and } g(y/h(x)) \notin K, \text{ for some } y \in \mathcal{O} \text{ and } h(x) \in K[[x]]\}.$$

This readily follows from (9) and (3).

It remains to prove (10.1.5). So let $y \in \mathcal{O}$ such that

(10) $y(x) = y^{(1)}(x) + \sigma y^{(2)}(x)$ and $\text{ord}_x y = d$ and $g(y/x^d) \notin K$.

Hence we must have

(11) $\text{ord}_x y^{(1)}(x) \geq d$, $\text{ord}_x y^{(2)}(x) = d$.

Upon taking $y^*(x) = \sigma y^{(2)}(x)$ we clearly have (since $d > 0$)

(12) $\{x,y\}\mathcal{O} = \{x,y^*\}\mathcal{O}$.

But from (4) and (9) we see that

(13) $h = h^{(1)}(x) + \sigma h^{(2)}(x) \in \mathcal{O} \Rightarrow \sigma h^{(2)}(x) = y^* h^*(x)$ where

$$h^*(x) \in K[[x]].$$

Thus clearly $\mathcal{O} = K[[x]] + y^* K[[x]]$ and in particular

$$M(\mathcal{O}) = \{x,y\}\mathcal{O} .$$

Thus (10.1.5) is proved.

§38. Bezout's theorem for two hypersurfaces.

Let A be a homogeneous domain.

We shall now give (essentially) Bezout's own proof of (25.14) which uses the resultant and which does not assume $r = 2$.

(38.1) DEFINITION. Assume that $\text{Emdim } A = \text{Dim } A = r \geq 2$. Let $P \in \mathfrak{D}_{r-2}(A)$ and $N \in \mathfrak{M}_0^*(A)$ be such that $P \not\subset N$. Let $B = A^N$ and $Q = P^N$. Now $\text{rad}_A(P+N) = H_1(A)A$ and hence (for instance see [1 : (12.1.5)]) it follows that $[\mathfrak{K}([A,P]) : \mathfrak{K}([B,Q])] < \infty$. Consequently $Q \in H^*(B) \cap \mathfrak{D}(B)$, (i.e., $Q \in \mathfrak{D}_{r-1}(B)$), and so $\text{Deg}_B Q$ and $[\mathfrak{K}([A,P]) : \mathfrak{K}([B,Q])]$ are positive integers. We <u>define</u>

$$\mathfrak{g}(A,P,N) = [\mathfrak{K}([A,P]) : \mathfrak{K}([B,Q])]\text{Deg}_B Q$$

and we note that $\mathfrak{g}(A,P,N)$ is then a positive integer.

(38.2) LEMMA. <u>Assume that</u> $\text{Emdim } A = \text{Dim } A = r = 2$ <u>or</u> 3. <u>Let</u>

$P \in \mathfrak{D}_{r-2}(A)$ <u>and</u> $N \in \mathfrak{M}_0^*(A)$ <u>be such that</u> $P \not\subset N$. <u>Then</u> $\mathfrak{g}(A,P,N) = \text{Deg}[A,P]$.

PROOF. If $r = 2$ then the assertion is obvious. If $r = 3$ then the assertion follows from (23.15) and (25.9).

We shall now deduce the following lemma from its affinte version given in [3: Proposition A.1 on page 67].

(38.3) BEZOUT'S LEMMA. <u>Assume that</u> $\text{Emdim}\, A = \text{Dim}\, A = r \geq 2$. <u>Let</u> $\Phi \in H_m(A)$, $\Psi \in H_n(A)$, <u>and</u> $N \in \mathfrak{M}_0^*(A)$ <u>be such that</u> $\mathfrak{D}_{r-1}(A,\Phi) \cap \mathfrak{D}_{r-1}(A,\Psi) = \emptyset$ <u>and</u> $N \notin \mathfrak{D}(A,\Phi) \cup \mathfrak{D}(A,\Psi)$. <u>Let</u> $\mathfrak{U} = \mathfrak{D}_{r-2}(A,\Phi) \cap \mathfrak{D}_{r-2}(A,\Psi)$. Then

$$\sum_{P \in \mathfrak{U}} \mu([A,\Phi,\Psi],P)\, \mathfrak{g}(A,P,N) = mn .$$

PROOF. If $mn = 0$ then $\mathfrak{U} = \emptyset$ and we have nothing more to show. So suppose that $mn \neq 0$, i.e., $m > 0$ and $n > 0$. Let $k = H_0(A)$ and let $Y_0, Y_1, \ldots, Y_r$ be indeterminates over k. We can take a homogeneous coordinate system $(X_0, X_1, \ldots, X_r)$ in A such that $N = (X_0, X_1, \ldots, X_{r-1})A$. Since $N \notin \mathfrak{D}(A,\Phi) \cup \mathfrak{D}(A,\Psi)$, we now have

$$\Phi = \varphi(X_0,\ldots,X_r)A \quad \text{with} \quad \varphi(Y_0,\ldots,Y_r) = Y_r^m + \sum_{1 \leq i \leq m} \varphi_i(Y_0,\ldots,Y_{r-1})Y_r^{m-i}$$

where $\varphi_i(Y_0,\ldots,Y_{r-1}) \in K[Y_0,\ldots,Y_{r-1}]$ is either zero or a nonzero homogeneous polynomial of degree i, and similarly

$$\Psi = \psi(X_0,\ldots,X_r)A \quad \text{with} \quad \psi(Y_0,\ldots,Y_r) = Y_r^n + \sum_{1 \leq i \leq n} \psi_i(Y_0,\ldots,Y_{r-1})Y_r^{n-i}$$

where $\psi_i(Y_0,\ldots,Y_{r-1}) \in K[Y_0,\ldots,Y_{r-1}]$ is either zero or a nonzero homogeneous polynomial of degree i. By assumption $\varphi(Y_0,\ldots,Y_r)$ and $\psi(Y_0,\ldots,Y_r)$ have no nonconstant common factor, and hence upon letting

(1) $\theta(Y_0,\ldots,Y_{r-1}) = \mathrm{Res}_{Y_r}(\varphi(Y_0,\ldots,Y_r),\ \psi(Y_0,\ldots,Y_r))$

where Res_{Y_r} denotes resultant with respect to Y_r, we have that

(2) $0 \neq \theta(Y_0,\ldots,Y_{r-1}) \in k[Y_0,\ldots,Y_{r-1}]$ is homogeneous of degree mn.

Let

$$B = A^N, \quad \mathfrak{B} = \mathfrak{D}_{r-2}(B), \quad \text{and} \quad \Theta = \theta(X_0,\ldots,X_{r-1})A .$$

Then in view of (2) we clearly have

(3) $$\sum_{Q\in\mathfrak{B}} \mu'([B,\Theta],Q)\mathrm{Deg}_B Q = mn.$$

For every $Q \in \mathfrak{B}$ let

$$\Omega(Q) = \{P \in \mathfrak{D}_{r-2}(A) : P^N = Q\}.$$

Then clearly

(4) $$\begin{cases} \sum_{P\in\mathfrak{A}} \mu([A,\Phi,\Psi],P)\,\mathfrak{g}(A,P,N) \\ = \sum_{Q\in\mathfrak{B}} \mathrm{Deg}_B Q \left(\sum_{P\in\Omega(Q)} \mu([A,\Phi,\Psi],P)[\mathfrak{K}([A,P]) : \mathfrak{K}([B,Q])] \right). \end{cases}$$

We shall now prove that for every $Q \in \mathfrak{B}$ we have

(5) $$\sum_{P\in\Omega(Q)} \mu([A,\Phi,\Psi],P)[\mathfrak{K}([A,P]) : \mathfrak{K}([B,Q])] = \mu'([B,\Theta],Q)$$

and, in view of (3) and (4), this will complete the proof.

So let any $Q \in \mathfrak{B}$ be given. Upon relabelling $X_0, X_1, \ldots, X_{r-1}$ suitably, we may suppose that $X_0 \notin Q$. We note that the elements $X_1/X_0,\ldots,X_r/X_0$ are algebraically independent over k, and we let:

$$B' = k[X_1/X_0,\ldots,X_{r-1}/X_0]$$
$$t = X_r/X_0$$
$$A' = B'[t]$$

$$\varphi'(t) = \varphi(X_0,\dots,X_r)/X_0^m = t^m + \sum_{1\le i\le m} \varphi_i(1,X_1/X_0,\dots,X_{r-1}/X_0)t^{m-i}$$

$$\psi'(t) = \psi(X_0,\dots,X_r)/X_0^n = t^n + \sum_{1\le i\le m} \psi_i(1,X_1/X_0,\dots,X_{r-1}/X_0)t^{n-i}$$

$$\theta' = \theta(X_0,\dots,X_{r-1})/X_0^{mn} = \theta(1,X_1/X_0,\dots,X_{r-1}/X_0)$$

$$Q' = \bigcup_{0\le i\le\infty} \{\zeta/X_0^i : \zeta \in Q \cap H_i(B)\}$$

$$\Omega' = \{P' \in \mathfrak{P}_{r-2}(A') : P' \cap B' = Q'\}$$

$$\Omega^* = \{P \in \Omega(Q) : X_0 \notin P\}$$

and

$$P' = \bigcup_{0\le i\le\infty} \{\zeta/X_0^i : \zeta \in P \cap H_i(A)\} \quad \text{for every } P \in \Omega^* .$$

Now, in view of (15.4) and (15.5), we have that

$$Q' \in F^*(B') \cap \mathfrak{P}(B') ,$$

$B'_{Q'} = \mathfrak{R}(B,Q)$ is a one dimensional regular local ring ,

$P \to P'$ gives a bijection of Ω^* onto Ω' ,

$$(6)\quad \left.\begin{cases} \mu([A,\Phi,\Psi],P) = \lambda([A',\varphi'(t)A',\psi'(t)A'],P') \\ [\mathfrak{R}([A,P]) : \mathfrak{R}([B,Q])] = [A'_{P'}/P'A'_{P'} : B'_{Q'}/Q'B'_{Q'}] \end{cases}\right\} \text{for all } P \in \Omega^*$$

and

$$(7)\quad \mu'([B,\Theta],Q) = \lambda'([B',\theta'B'],Q') .$$

Since $N \notin \mathfrak{D}(A,\Phi) \cup \mathfrak{D}(A,\Psi)$ we clearly have

$$(8)\quad \mu([A,\Phi,\Psi],P) = 0 \quad \text{for all } P \in \Omega(Q)\backslash\Omega^* .$$

By (6) and (8) we get

$$(9)\quad \begin{cases} \sum_{P\in\Omega(Q)} \mu([A,\Phi,\Psi],P)[\mathfrak{K}([A,P]) : \mathfrak{K}([B,Q])] \\ \\ = \sum_{P'\in\Omega'} \lambda([A',\varphi'(t)A',\psi'(t)A'],P')[A'_{P'}/P'A'_{P'} : B'_{Q'}/Q'B'_{Q'}] . \end{cases}$$

In view of (1) we have

$$\theta' = \mathrm{Res}_t(\varphi'(t),\psi'(t))$$

and hence by [3: Proposition A.1 on page 67] we get

$$(10)\quad \begin{cases} \sum_{P'\in\Omega'} \lambda([A',\varphi'(t)A',\psi'(t)A'],P')[A'_{P'}/P'A'_{P'} : B'_{Q'}/Q'B'_{Q'}] \\ \\ = \lambda'([B',\theta'B'],Q'] \end{cases}$$

Now (5) follows from (7), (9) and (10).

(38.4) BEZOUT'S THEOREM FOR PLANE CURVES AND FOR SURFACES IN THREE SPACE. <u>Assume that</u> Emdim A = Dim A = r = 2 or 3. <u>Let</u> $\Phi \in H^*(A)$ <u>and</u> $\Psi \in H^*(A)$ <u>be such that</u> $\mathfrak{D}_{r-1}(A,\Phi) \cap \mathfrak{D}_{r-1}(A,\Psi) = \phi$. <u>Let</u> $\mathfrak{U} = \mathfrak{D}_{r-2}(A,\Phi) \cap \mathfrak{D}_{r-2}(A,\Psi)$. <u>Then</u>

$$\sum_{P\in\mathfrak{U}} \mu([A,\Phi,\Psi],P)\mathrm{Deg}[A,P] = (\mathrm{Deg}_A\Phi)(\mathrm{Deg}_A\Psi) .$$

PROOF. If $H_0(A)$ is infinite then our assertion follows from (38.2) and (38.3) by taking $N \in \mathfrak{M}_0^*(A)$ with $N \notin \mathfrak{D}(A,\Phi) \cup \mathfrak{D}(A,\Psi)$. If $H_0(A)$ is finite then such an N may not exist and we proceed thus. Let Y be an indeterminate over A. Let $A' = (H_0(A)(Y))[A]$, and for all n let $H_n(A')$ be the $((H_0(A)(Y))$-submodule of A' generated by $H_n(A)$. Then A' is a homogeneous domain with Emdim A' = Dim A' = r. Let $\Phi' = \Phi A'$ and $\Psi' = \Psi A'$. Then clearly Φ' and Ψ' are members of $H^*(A)$ with $\mathrm{Deg}_{A'}\Phi' = \mathrm{Deg}_A\Phi$, $\mathrm{Deg}_{A'}\Psi' = \mathrm{Deg}_A\Psi$, and $\mathfrak{D}_{r-1}(A',\Phi') \cap \mathfrak{D}_{r-1}(A',\Psi') = \phi$. Also $P \to PA'$ gives a bijection of $\mathfrak{D}_{r-2}(A,\Phi) \cap \mathfrak{D}_{r-2}(A,\psi)$ onto $\mathfrak{D}_{r-2}(A',\Phi') \cap \mathfrak{D}_{r-2}(A',\psi')$, and it is easily seen that for every $P \in \mathfrak{D}_{r-2}(A,\Phi) \cap \mathfrak{D}_{r-2}(A,\psi)$ we have

$$\mu([A',\Phi',\psi'],PA') = \mu([A,\Phi,\psi],P)$$

$$\text{and} \quad \mathrm{Deg}[A',PA'] = \mathrm{Deg}[A,P] .$$

Thus, since $H_0(A') = H_0(A)(Y)$ is infinite, we are reduced to the previous case.

(38.5) DEFINITION. For any homogeneous ideal Q in A with $H_1(A) \not\subset \mathrm{rad}_A Q$, as in [1: (12.13)], we define the positive integer $\mathfrak{g}(A,Q)$ by means of the Hilbert polynomial; we also define $\mathfrak{g}(A) = \mathfrak{g}(A,\{0\})$. The following three facts are easily seen:

(38.5.1) $\mathfrak{g}(A',QA') = \mathfrak{g}(A,Q)$ where A' is the homogeneous domain obtained by taking an indeterminate Y over A and putting $A' = (H_0(A)(Y))[A]$ and $H_n(A')$ = the $(H_0(A)(Y))$-submodule of A' generated by $H_n(A)$.

(38.5.2) If $Q \in \mathfrak{D}_0(A)$, then $\mathfrak{g}(A,Q) = \mathrm{Deg}[A,Q]$. In particular, if $\mathrm{Dim}\,A = 0$ then $\mathfrak{g}(A) = \mathrm{Deg}\,A$.

(38.5.3) If $\mathrm{Emdim}\,A = \mathrm{Dim}\,A$ and $Q \in H^*(A)$, then $\mathfrak{g}(A,Q) = \mathrm{Deg}_A Q$.

In [1: (12.3.4)(1)] one finds:

(38.5.4) PROJECTION FORMULA. If $Q \in \mathfrak{D}_i(A)$ and $N \in \mathfrak{M}_0^*(A)$ with $Q \not\subset N$, then $Q^N \in \mathfrak{D}_i(A^N)$ and

$$\mathfrak{g}(A,Q) = [\mathfrak{K}([A,Q]) : \mathfrak{K}([A^N,Q^N])]\,\mathfrak{g}(A^N,Q^N) .$$

In view of (38.5.1) and (38.5.4) we easily get

(38.5.5) $\mathfrak{D}^1(A) = \{Q \in \mathfrak{D}(A) : \mathfrak{g}(A,Q) = 1\}$; (this explains our notation $\mathfrak{D}^1$). In particular, $\mathfrak{g}(A) = 1 \Leftrightarrow \mathrm{Emdim}\,A = \mathrm{Dim}\,A$.

In view of (38.5.2) and (38.5.4) we get

(38.5.6). If Emdim A = Dim A = r ≥ 2, then for any $P \in \mathfrak{Q}_{r-2}(A)$ and $N \in \mathfrak{M}_0^*(A)$ and $P \not\subset N$ we have $\mathfrak{g}(A,P,N) = \mathfrak{g}(A,P)$.

It is easily seen that:

(38.5.7) If $Q \in \mathfrak{Q}_1(A)$ and A' is as in (38.5.1), then $QA' \in \mathfrak{Q}_1(A')$ and Deg[A' ,QA'] = Deg[A,Q]. In particular, if Dim A = 1 and Q' is as in (38.5.1), then Dim A' = 1 and Deg A' = Deg A.

Finally, in view of (38.5.1), (38.5.6) and (38.5.7), by the Projection Formulas (23.15) and (38.5.4) we get:

(38.5.8) If $Q \in \mathfrak{Q}_1(A)$, then $\mathfrak{g}(A,Q)$ = Deg[A,Q]. In particular, if Dim A = 1, then $\mathfrak{g}(A)$ = Deg A.

(38.6) BEZOUT'S THEOREM FOR TWO HYPERSURFACES IN A PROJECTIVE SPACE. <u>Assume that</u> Emdim A = Dim A = r ≥ 2. <u>Let</u> $\Phi \in H^*(A)$ <u>and</u> $\Psi \in H^*(A)$ <u>be such that</u> $\mathfrak{Q}_{r-1}(A,\Phi) \cap \mathfrak{Q}_{r-1}(A,\Psi) = \phi$. <u>Let</u> $\mathfrak{A} = \mathfrak{Q}_{r-2}(A,\Phi) \cap \mathfrak{Q}_{r-2}(A,\Psi)$. <u>Then</u>

$$\sum_{P \in \mathfrak{A}} \mu([A,\Phi,\Psi],P)\,\mathfrak{g}(A,P) = (\mathrm{Deg}_A\Phi)(\mathrm{Deg}_A\Psi).$$

PROOF. If $H_0(A)$ is infinite, then our assertion follows from (38.5.4) and (38.3) by taking $N \in \mathfrak{M}_0^*(A)$ with $N \notin \mathfrak{Q}(A,\Phi) \cup \mathfrak{Q}(A,\Psi)$. If $H_0(A)$ is finite, then, in view of (38.5.1), we can proceed as in the proof of (38.4).

§39. Chains of euclidean curves.

The proof of Theorem (36.7) was based on condition (*), which in turn was satisfied when the true adjoint α was such that $B/\alpha B$ was an euclidean domain ((36.6).) In actual application we arrange this by what may be geometrically described as finding an α which represents either a line or a nonsingular conic in the affine plane. (α

was actually constructed in (24.21)). In an older version of the proof of Theorem (36.9) we also had to work with the case when α represented a pair of distinct lines with their common point outside the curve represented by γ. This needed a generalization of Theorem (36.7), which we present here in (39.3).

(39.1) DEFINITION. Let $E_1, E_2, \ldots, E_n$ be a sequence of euclidean domains. Let $E = E_1 \oplus \ldots \oplus E_n$ and let $\pi_i : E \to E_i$ denote the projection onto the ith component.

Let $\Lambda \subset \{1,2,\ldots,n-1\}$ and assume that there are fields k_i, epimorphisms $\varphi_i : E_i \to k_i$, $\psi_i : E_{i+1} \to k_i$ and monomorphisms $\tau_i : k_i \to E_i$ and $\sigma_i : k_i \to E_{i+1}$ for every $i \in \Lambda$, such that:

(1) if i, $i-1 \in \Lambda$, then $\operatorname{Ker} \varphi_i \neq \operatorname{Ker} \psi_{i-1}$, and

(2) $\varphi_i \circ \tau_i = \psi_i \circ \sigma_i =$ identity in k_i.

We define the ring D together with $(\Lambda, \{\varphi_i\}, \{\psi_i\}, \{\tau_i\}, \{\sigma_i\})$ to be a *chain of euclidean domains*, if $D \subset E$ is defined by

$$D = \{d \in E : \varphi_i \circ \pi_i(D) = \psi_i \circ \pi_{i+1}(d)$$

If $i \in \Lambda$, *we say that* D *is connected at the* i^{th} *component with the connecting relation* (φ_i, ψ_i).

D *is said to be defined over a field* k *if* $k_i = k$ *for all* $i \in \Lambda$. D *is said to be connected if* $\Lambda = \{1,2,\ldots,n-1\}$.

Let A be an affine domain over a field k, and $I \subset A$ be an ideal in A, such that, I is an intersection of finitely many members of $\mathfrak{P}_1(A)$. *We shall say that* I *is a chain* (*connected chain*) *of euclidean curves over* k *if* A/I *is isomorphic to a chain* (*connected chain*) *of euclidean domains defined over* k.

Note that, if Λ is empty, then $D = E$ is a chain of euclidean domains. Further, in that case, D is connected if and only if $n = 1$.

EXAMPLES. (1) In particular, upon taking $B = k[X,Y]$, a polynominal ring over a field k and $I = (f_1 f_2 \dots f_n)B$ where $f_i \in B\backslash k$ are linear expressions in X,Y with $f_iB + f_jB = B$ whenever $i \neq j$, we easily see that B/I is a chain of euclidean domains where $E_i = A/f_iB$ and Λ is empty. Thus I is a chain of euclidean curves over k and I is connected if and only if $n = 1$.

(2) Let g_1, g_2 be linear expressions in X,Y, such that $g_1B + g_2B \neq B$. Then $g_1B + g_2B$ is a maximal ideal in B with $B/g_1B + g_2B \approx k$. Take $h_1: B \to B/g_1B$ and $h_2: B \to B/g_2B$ to be canonical epimorphisms and let $\varphi_1: B/g_1B \to k$ and $\psi_1: b/g_2B \to k$ be the unique epimorphisms such that $\varphi_1 \circ h_1 = \psi_1 \circ h_2 =$ the canonical epimorphism $B \to k$ whose kernel is $g_1B + g_2B$. Upon taking $\tau_1 = h_1$ restricted to k and $\sigma_1 = h_2$ restricted to k, it is easy to see that $B/g_1g_2B \subset B/g_1B \oplus B/g_2B$ is a connected chain of euclidean domains connected at the first component by (φ_1, ψ_1).

(39.2) LEMMA. <u>Let</u> $B = k[X,Y]$, <u>a polynomial ring in</u> X,Y <u>over</u> k. <u>Let</u> $\alpha \in B$ <u>such that</u> αB <u>is a chain of euclidean curves over</u> k. <u>Let</u> $u: B \to D$ <u>be the composition of the canonical homomorphism</u> $B \to B/\alpha B$ <u>and the isomorphism</u> $B/\alpha B \to D$ <u>where</u> D <u>is a chain of euclidean domains</u>. <u>With exactly the same notation as in</u> (39.1) <u>assume that</u> $\beta, \gamma \in B$, <u>such that, for every</u> $i \in \Lambda$, <u>we have</u>

$$(\varphi_i \circ \pi_i(u(\beta)), \varphi_i \circ \pi_i(u(\gamma)) \neq (0,0).$$

<u>Then there exists an elementary</u> 2-<u>transformation</u> N <u>in</u> B <u>with</u> $(\beta,\gamma)N = (\beta', \gamma')$ where $\gamma' \in \alpha B$.

PROOF. We use the notation in (39.1) for D. First write $\{1,2,\dots,n\}$ as a disjoint union $\Omega_1 \cup \dots \cup \Omega_m$ such that each Ω_i is either of the form $\{s\}$ with $s, s-1 \notin \Lambda$, or of the form $\{s, s+1, \dots, s+t\}$ with $\{s, \dots, s+t-1\} \subset \Lambda$, $s+t \notin \Lambda$.

Let $E^{(i)} = \bigoplus_{i\epsilon\Omega_i} E_i$ and $\Lambda^{(i)} = \Lambda \cap \Omega_i$. Let $D^{(i)}$ be the euclidean domain contained in $E^{(i)}$ with connecting relations $(\varphi_j, \psi_j)_{j\epsilon\Lambda^{(i)}}$. It is clear that each $D^{(i)}$ is a connected chain of euclidean domains and $D = \bigoplus_{i=1}^{m} D^{(i)}$.

In view of (36.6.4) it is enough to prove that there exist elementary 2-transformations N_i in $D^{(i)}$ such that

$$(\pi_i^*(u(\beta)),\ \pi_i^*(u(\gamma))N_i = (\theta_i, 0) \quad \text{for some} \quad \theta_i \in D^{(i)}$$

where π_i^* denotes the projection of D onto $D^{(i)}$.

Thus we may replace D by $D^{(i)}$, u by $\pi_i^* \circ u$ and hence assume, that,

(1) D is a connected chain of euclidean domains.

We are then reduced to proving the following:

Let $f = u(\beta)$ $g = u(\gamma)$. Then we have $f, g \in D$ such that $(\varphi_i(\pi_i(f)),\ \varphi_i(\pi_i(g))) \neq (0,0)$ and $(\psi_i(\pi_{i+1}(f)),\ \psi_i(\pi_{i+1}(g))) \neq (0,0)$ for any $i \in \Lambda$. Then we want to prove the existance of an elementary 2-transformation N in D such that $(f,g)N = (h,0)$ for some $h \in D$.

We will prove the lemma by induction on n.

If $n = 1$, then D is an euclidean domain and the proof is well known (or as explained in (36.6.5)).

Now assume $n > 1$. Let $D^* \subset E^* = E_1 \oplus \ldots \oplus E_{n-1}$ be defined by $D^* = \{(d_1, \ldots, d_{n-1}) \in E^*$: there exists $(d_1, \ldots, d_n) \in D$ for some $d_n \in E_n\}$. Clearly D^* is a connected chain of euclidean domains with connecting relations $(\varphi_j, \psi_j)_{1 \leq j < n-1}$. Let $\eta = D \to D^*$ be the epimorphism.

$$\eta((d_1, \ldots, d_n)) = (d_1, \ldots, d_{n-1}).$$

Let $\eta(f) = f^*$ and $\eta(g) = g^*$. By induction hypothesis, there exists an elementary 2-transformation M^* in D^* such that

$$(f^*,g^*)M^* = (d^*,0).$$

By (36.6.4) applied to $\eta: D \to D^*$, there exists an elementary 2-transformation M in D such that

$$(f,g)M = (f',g')$$

where $\eta(g') = 0$.

(2) Clearly (f',g') satisfy the same conditions as (f,g) and we may thus assume without loss, that, upon writing $f = (f_1,\ldots,f_n)$ and $g = (g_1,\ldots,g_n)$, we have $g_1 = g_2=\ldots=g_{n-1} = 0$.

Now let $N^* = \theta_1'\theta_2'\ldots\theta_s'$ be an elementary 2-transformation in the euclidean domain E_n such that $\theta_1' = \mathfrak{E}_{E_n}(p(i),q(i) ; t'(i))$. Define $\theta_i = \mathfrak{E}_D(p(i),q(i);t(i))$, where $t(i) \in D$ is defined by

$$\pi_j(t(i)) = \begin{cases} \tau_j \circ \psi_j \circ \tau_{j+1} \circ \psi_{j+1} \circ \cdots \circ \tau_{n-1} \circ \psi_{n-1}(t'(i)) & , \text{if } 1 \le j \le n-1 , \\ t'(i) & , \text{if } j = n . \end{cases}$$

Let $N = \theta_1\theta_2\ldots\theta_n$. Then it is easy to check that if $(f,g)N = (f'',g'')$, then

(3) $\pi_j(f'') = \mu_j f_j$, $\pi_j(g'') = \nu_j f_j$ for some $\mu_j, \nu_j \in \tau_j(k)$, if $1 \le j < n$, and

(4) $$(\pi_n(f''), \pi_n(g'')) = (f_n,g_n)N^* .$$

In particular if we take N^* such that $(f_n,g_n)N^* = (d,0)$, which certainly exists as explained in (36.6.5), we have, in view of (4),

(5) $$\pi_n(g'') = 0.$$

(6) Now assume, if possible, $g'' \neq 0$.

Clearly, if $\nu_j = 0$ for $j = 1,2,\ldots,n-1$. Then $g'' = 0$. Hence assume that for some $1 \leq j \leq n-1$ we have

(7) $\nu_j \neq 0$; and $j = n-1$ or $\nu_i = 0$ for $i = j+1,\ldots,n-1$.

Then we have, since $g'' \in D$,

$$\varphi_j(\nu_j f_j) = \begin{cases} \psi_j(\nu_{j+1}f_{j+1}) = 0 & , \quad \text{if } j < n-1 \\ \psi_{n-1}(\pi_n(g'')) = 0 & , \quad \text{if } j = n-1. \end{cases}$$

Thus we have $\varphi_j(\nu_j)\varphi_j(f_j) = 0$. Since $0 \neq \nu_j \in \tau_j(k)$ and $\varphi_j \circ \tau_j =$ identity on **k**, we must have

(8) $\varphi_j(f_j) = 0$ and hence $\varphi_j(\pi_j(f'')) = \varphi_j(\pi_j(g'')) = 0$.

But clearly (or by (36.6.2)) we have

$$\{\pi_j(f),\pi_j(g)\}E_j = \{\pi_j(f''),\pi_j(g'')\}E_j$$

and hence (8) implies

(9) $\varphi_j(\pi_j(f)) = \varphi_j(\pi_j(g)) = 0$.

Thus (9) is a contradiction to the hypothesis on f,g and hence contrary to the assumption in (6) we must have $g'' = 0$, and hence $(f,g)N = (f'',0)$ as required.

(39.3) COROLLARY. <u>Let</u> α <u>be as in the statement of Theorem</u> (36.7), <u>without assuming the property</u> (*). <u>Let</u> α <u>be a chain of euclidean curves and</u> $D = B/\alpha B$ <u>be a chain of euclidean domains as described in</u> (39.1). <u>Let</u> $\{M_1,\ldots,M_n\}$ <u>be the set of maximal ideals in</u> B <u>such that for each</u> M_j <u>there exists some</u> $i \in \Lambda$ <u>such that</u>: <u>if</u> $u(M_j)$ <u>denotes the image of</u> M_j <u>in</u> D, <u>then either</u> $\text{Ker}\,\varphi_i = (\pi_i(u(M_j))$ <u>or</u> $\text{Ker}\,\psi_i = \pi_{i+1}(u(M_j))$.

If $(\beta,\gamma)B \not\subset M_j$ *for all* $j = 1,2,\ldots,r$; *then* α *has the property of* (*)

REMARK. In particular, if, as in Examples (39.1), $\alpha = g_1g_2$ (i.e. α represents a pair of parallel or intersecting lines) and if either $g_1B = g_2B = B$ or $\gamma \notin g_1B + g_2B$ (i.e. γ does not pass through the common point, if any, of the lines represented by α), then α has the property (*).

§40. Other treatments of differentials.

The study of differentials in §28 was devoted to the problem of defining the divisor of a differential of a separably generated function field K/k of transcendence degree one; namely, to defining $\mathrm{ord}_V(\alpha dx)$ for every $V \in \mathfrak{X}(K,k)$ and every differential αdx of K/k The definition should of course be such that

$$\mathrm{ord}_V \alpha dx = \mathrm{ord}_V \alpha + \mathrm{ord}_V dx ,$$

$$\alpha dx = \beta dy \Rightarrow \mathrm{ord}_V \alpha dx = \mathrm{ord}_V \beta dy .$$

In case V is residually separable over k, the problem is easy to solve and the answer is

$\mathrm{ord}_V \alpha dx = \mathrm{ord}_V \alpha \dfrac{dx}{dt}$, where t is any uniformizing coordinate of V/k. (28.10).

In particular, when the ground field k is perfect, the problem is completely solved.

In the general case one uses different "tricks" to get at the "correct" answer.

The correct answer is of course the one given in (28.13)

$$\mathrm{ord}_V dx = \mathrm{ord}_V \mathfrak{D}(k(x) \cap V, K) , \text{ if } x \in V ,$$

and in general if $x \notin V$

$$\mathrm{ord}_V dx = \mathrm{ord}_V x^2 + \mathrm{ord}_V d(1/x) .$$

Chevalley starts with an ad hoc definition of differentials - certain linear maps on repartitions, and then finally shows that they behave like the usual ones (Chapter II, III and VI, [9]).

Other treatments by Zarisk-Falb [17], Kunz [10], Nastold [13], start with the "usual" formal definition of differentials, but make use of derivations of the ground field k.

We point out that our definition of $\text{ord}_V dx$ and of genus(K,k) coincides with the usual genus of K/k as in Chevalley (when k is relatively algebraically closed in K). (See Theorem 2 p.106 [9].) In particular, genus (K,k) ≥ 0, at least when k is relative algebraically closed in K. Another proof of nonnegativity of genus, when some $V \in \mathfrak{X}(K,k)$ **is** residually rational, is given in §42.

§41. Generalized Dedekind's formula for conductor and different.

The formula (*) in (28.1) can be generalized to complete intersections by using results of Kunz, Berger and Roquette. In this section we shall indicate how the generalization is obtained. Except for Berger's result, we do not repeat any proofs or definitions which are not essential. Berger's proof is repeated because it is short and the notation in it needs clearification.

NOTATION. We shall use the following notation.

S is a discreet valuation ring with $\mathfrak{Z}(S) = L$.

K is a finite separable algebraic extension of L.

T is an integral ring extension of S with $\mathfrak{Z}(T) = K$.

$\bar{T}$ is the integral closure of T in K.

M is a prime nonzero ideal in T.

$R = T_M$.

$B = \bar{T}_M = \{a/b : a \in \bar{T}, b \in T \setminus M\}$ = the integral closure of R in K.

For any $\bar{T} \supset A \supset S$ we <u>define</u>

$A^* = \{x \in K : \text{Trace}_{K/L} xy \in S$ for all $y \in A\}$, and

$\mathfrak{D}(A/S) = \mathfrak{D}_D(A/S) = \{x \in K : xy \in A$ for all $y \in A^*\}$.

Further we <u>define</u>,

$R^* = R\cdot T^*$,

$\mathfrak{D}(R/S) = \mathfrak{D}_D(R/S) = R\cdot\mathfrak{D}_D(T/S)$,

$B^* = B\cdot\bar{T}^*$, and

$\mathfrak{D}(B/S) = \mathfrak{D}_D(B/S) = B\cdot\mathfrak{D}_D(\bar{T}/S)$.

REMARK. Note that for A^* as defined above (where we may have $A = R$ or $A = B$) $A^* \supset A$ is a finite A-module and $\mathfrak{D}_D(A/S)$ is an ideal in A. The suffix D stands for the Dedekind different as opposed to the Kähler different, which we shall use later. Also note that $\bar{T}^* = \mathfrak{C}^*(S,K)$ and $\mathfrak{D}(\bar{T}/S) = \mathfrak{D}(S,K)$ in our old notation in §27.

It is easy to check that,

$\mathfrak{D}(R/S) = \{x \in K : xy \in R$ for all $y \in R^*\}$, and

$\mathfrak{D}(B/S) = \{x \in K : xy \in B$ for all $y \in B^*\}$.

(41.1) LEMMA. (Berger). <u>With the notation as above we have</u>

$$B\cdot\mathfrak{D}(R/S) \subsetneq \mathfrak{C}(R)\mathfrak{D}(B/S) ;$$

<u>and further, if</u> R^* <u>is invertible</u>, i.e. $\mathfrak{D}(R/S)\cdot R^* = R$, <u>then we have equality in the above relation</u>.

PROOF. We clearly have $B^* \subset R^*$ and hence

(1) $B^*\cdot\mathfrak{D}(R/S) \subset R$.

Since B^* is a B-module, it follows from (1) that

(2) $B^*\cdot\mathfrak{D}(R/S) \subset \mathfrak{C}(R)$.

Now clearly B is a Dedekind domain, and then it is well known that every fractionary ideal of B and in particular,

(3) B^* is invertible, i.e., $B^*\cdot\mathfrak{D}(B/S) = B$.

Multiplying both sides of (2) by $\mathfrak{D}(B/S)$, we have by (3),

(4) $B\cdot\mathfrak{D}(R/S) \subset \mathfrak{C}(R)\cdot\mathfrak{D}(B/S)$.

Now we claim that

(5) $\mathfrak{C}(T)T^* \subset \bar{T}^*$.

For proof, we simply have to note that if $x \in \mathfrak{C}(T)$, $y \in T^*$ and $z \in \bar{T}$, we have $xz \in T$ and hence $\mathrm{Trace}_{K/L}(xy)z = \mathrm{Trace}_{K/L}(xz)y \in S$ by definition of T^*.

From (5), it follows that

(6) $\mathfrak{C}(R)R^* \subset B^*$;

since it is easy to check that $\mathfrak{C}(R) = R\cdot\mathfrak{C}(T)$.

Now if we further assume that $R^*\cdot\mathfrak{D}(R/S) = R$, then by multiplying both sides of (6) by $\mathfrak{D}(R/S)$ we get

(7) $\mathfrak{C}(R) \subset B^*\mathfrak{D}(R/S)$,

and again by (3) we get

(8) $\mathfrak{C}(R)\mathfrak{D}(B/S) \subset \mathfrak{D}(R/S)$.

The proof is complete by (4) and (8).

REMARK. The above proof is essentially repeated from Berger [7], §2, Lemma 9.5, except that his definitions of R^* and B^* are ambiguous. We have simply chosen as definitions, their descriptions, as he has proved earlier in his Lemma 9.

(41.2) LEMMA (Roquette). _Assume in the above notation, that_ T _is a complete intersection over_ S, i.e., _for some_ $n \geq 0$ _and indeterminates_ $X_1,\ldots,X_n$ _over_ S, _we have an_ S-_epimorphism_ $\psi: S[X_1,\ldots,X_n] \to T$ _such that_ Ker ψ _is generated by_ n _elements_ $g_1,\ldots,g_n$.

Then R^* is invertible, i.e. $\mathfrak{D}(R/S)\cdot R^* = R$.

PROOF. See [16].

(41.3) LEMMA (Kunz). _Using the above notation and assuming that_ T _is a complete intersection over_ S, _let_ J _denote the image by_ ψ _of the jacobian of_ $g_1,\ldots,g_n$ _with respect to_ $X_1,\ldots,X_n$. _Then it is well known that the ideal_ JT _is independent of the choice of_ ψ _and_

$g_1,\ldots,g_n$. Then we have

$$JB = B\cdot\mathfrak{D}(R/S) = \mathfrak{C}(R)\mathfrak{D}(B/S).$$

PROOF. In view of (41.1) and (41.2) the proof is reduced to showing $JR = \mathfrak{D}(R/S)$. In [11], the discussion after Satz 1, proves $\mathfrak{D}(R/S)$ = the Kähler different of R over S; and the following "reduction of the proof" establishes that the Kähler different of $R/S = JR$.

(41.4) COROLLARY. DEDEKIND'S FORMULA FOR CONDUCTOR AND DIFFERENT. Let K be a function field of one variable over k. Let $x \in K$ be a separating transcendental. Let $y_1,\ldots,y_n \in K$ and $Y_1,\ldots,Y_n$ be indeterminates over $k[x]$. Let $\psi\colon k[x][Y_1,\ldots,Y_n] \to k[x,y_1,\ldots,y_n]$ be the unique epimorphism with $\psi(x) = x$ and $\psi(Y_i)$ y_i. Assume that $\operatorname{Ker}\psi$ has n generators $g_1,\ldots,g_n$ and J = the image by ψ of the jacobian of $g_1,\ldots,g_n$ with respect to $Y_i,\ldots,Y_n$

Let $V \in \mathfrak{X}(K,k)$ with $k[x,y_1,\ldots,y_n] \subset V$. Further assume that $k[x,y_1,\ldots,y_n]$ is the integral over $k[x]$ and let $Q = k[x,y_1,\ldots,y_n] \cap M(V)$. Then we have

$$(*)\qquad \operatorname{ord}_V\mathfrak{C}(k[x,y_1,\ldots,y_n]_Q) + \operatorname{ord}_V\mathfrak{D}(k(x)\cap V,K) = \operatorname{ord}_V J\ .$$

PROOF. Let $S = k(x)\cap V$ and $T = S[y_1,\ldots,y_n]$. Then clearly $k[x,y_1,\ldots,y_n]_Q = T_{T\cap M(V)} = R$, say. Not it is easy to see that all the above notation applies and further we have, $B \subset V$ and $V = B_{B\cap M(V)}$. It follows that

$\operatorname{ord}_V(\mathfrak{D}(k(x)\cap V),K)$

$= \operatorname{ord}_V\mathfrak{D}(\bar{T}/S)$.

$= \operatorname{ord}_V B\ \mathfrak{D}(\bar{T}/S)$

$= \operatorname{ord}_V\mathfrak{D}(B/S)$.

The proof now follows from (41.3), by taking ord_V of both sides

§42. The general adjoint condition.

The assertion (10.1.11) (and hence also the assertion (10.1.10) is valid without assuming $\lambda(R) = 2$. Indeed we have the following Theorem, which may be regarded as a version of Max Noether's Famous $Af + B\varphi$ Theorem (see [14] and [15]), and whose proofs (under more or less stringent hypothesis) really abound in the literature. For a list and survey, see [8]. For a new proof in the spirit of Noether's Theorem, see [5].

(42.1) THEOREM ON ADJOINT CONDITIONS. Let R be any local ring such that: $\dim R = 1$, $\operatorname{emdim} R \le 2$, $M(R)$ contains a nonzero-divisor, and the integral closure R^* of R in $\mathfrak{J}(R)$ is a finite R-module. Then $[R^*/\mathfrak{C}(R) : R] = 2[R/\mathfrak{C}(R) : R]$.

As a corollary of (42.1) we have:

(42.2) COROLLARY. Theorems (29.2) and (30.2) remain true without assuming condition (**).

(42.3) REMARK. In (29.3) we remarked that the genesis of (29.2.2) is the elementary and ancient idea of parametrization of a conic, and in (29.4) we gave a version of it illustrated by a parabola and a hyperbola. Let us conclude with another familiar version illustrated by a circle (of radius 1):

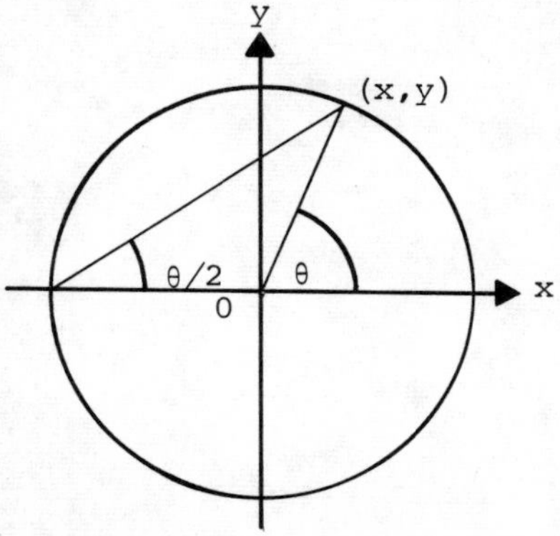

circle: $x^2 + y^2 = 1$.

$$\begin{cases} x = \cos\theta \\ y = \sin\theta \end{cases}$$

To get rid of trigometric functions, make the substitution which enables us to compute all trigometric integral by converting them into integrals of rational functions. To wit:

putting $t = \tan\theta/2$

$$\text{we get} \begin{cases} x = (1-t^2)/(1+t^2) \\ y = 2t/(1+t^2). \end{cases}$$

§43. GEOMETRIC LANGUAGE.

Geometrically speaking, varieties are thought of as sets of points in an ambient space, and subvarieties are certain subsets of such varieties. For algebraic varieties, all such sets should be "algebraically defined". If one starts to make rigorous, what "algebraically defined" exactly means, the "ideals" and "coordinate rings" of varieties naturally present themselves as a solution to the problem; and they can of course take care of all the terminology (although sometimes in a rather awkward manner) as illustrated in the previous chapters. Here we present the reader with algebraic definitions of various relavent geometric concepts that we have used. The general principal is, that; irreducible abstract varieties are thought of as their coordintate rings, embedded varieties are thought of as pairs [embedding variety, ideal] and we try to avoid treating reducible varieties as varieties unless they are a bunch of points or hypersurfaces.

(43.1) PROJECTIVE VARIETIES. By an <u>abstract projective variety</u> we mean a homogeneous domain A. The variety A is said to be <u>r-dimensional</u> if $\text{Dim } A = r$. The variety A is said to be a <u>projective space</u> (<u>projective</u> (Dim A)-<u>space</u>) if, $\text{Dim } A = \text{Emdim } A$. In particular, by an <u>abstract irreducible projective curve</u>, we mean a homogeneous domain A with $\text{Dim } A = 1$. Also an <u>abstract projective line</u> and an <u>abstract projective plane</u> are respectively a projective 1-space and a projective 2-space.

Now let A be an abstract irreducible projective variety. In the algebraic sense, any homogeneous ideal in A may be thought to be representing a subvariety of A. Geometrically one should restrict to ideals which are their own radicals. For irreducible, subvarieties, however, the two ideas agree and we say that any member of $\mathfrak{Q}(A)$ is an <u>irreducible subvariety</u> of A. Formally an irreducible subvariety C of A may be denoted by $[A,C]$, and this is how it appears in various formulas. Since any $C \in \mathfrak{Q}(A)$ is not the unique maximal homogeneous ideal of A, A/C is an abstract irreducible projective variety and we may say that A/C <u>is embedded in</u> A or that C defines the abstract variety A/C in A.

In the above sense, Emdim A gets an appropriate meaning, "embedding dimension of (the variety) A", namely, Emdim A = min{r : A can be embedded in a projective r-space.}.

A <u>hypersurface</u> in A is any member of $H^*(A)\setminus\{A\}$. The <u>degree of a hypersurface</u> Φ is equal to $\mathrm{Deg}_A\Phi$. In the special cases when A is a projective 3-space or a projective plane, a hypersurface is respectively called a <u>projective surface</u> or a <u>projective curve</u>. Note that a hypersurface is a subvariety in the algebraic sense, and we have to treat it as such, although formally we do not call it a subvariety. Hypersurfaces of degree 1 are said to be <u>hyperplanes</u>. A hypersurface Φ is irreducible if, and only if, $\Phi \in \mathfrak{Q}(A)$.

For an irreducible variety C of A, by $\mathfrak{R}(A,C)$ we denote the <u>local ring of</u> C <u>on</u> A, i.e., the ring of rational functions on A whose denominators do not belong to C. $\mathfrak{R}(A,I,C)$ denotes the ideal generated by hypersurfaces in $I \subset A$, in the ring $\mathfrak{R}(A,C)$. The above notations are also extended to cover the case when A itself is embedded i.e. is $[R,P]$, where P is an irreducible subvariety of a projective variety R. In particular, the above notation gives the function field $\mathfrak{K}(A)$ (or $\mathfrak{K}([R,P])$, as the case may be,) by taking $C = \{0\}$.

Ideals generated by a set of hyperplanes in A define (in the algebraic sense) linear subvarieties of A which are irreducible subvarieties of A, in case A is a projective space. In the general case we call such ideals "<u>flats</u>" or members of $\mathfrak{M}^*(A)$. The $H_0(A)$-vector-space of all the hyperplanes in a flat is also a useful concept (especially for studying affine pieces) and their collection is denoted by $\mathfrak{M}(A)$. Members of $\mathfrak{M}(A)$ may be called <u>linearities</u> in A. If A

is a projective space, then for all $N \in \mathfrak{M}^*(A)$, we have, $\mathrm{Emdim}[A,N] = \mathrm{Dim}(A/N)$. Hence $\mathrm{Emdim}[A,N]$ is thought of as the formal dimension of N in the general case; then $\mathfrak{M}_i^*(A)$ = the set of all i-dimensional flats in A. $\mathfrak{M}_i(A)$ is defined similarly.

If $N \in \mathfrak{M}(A)$ or $\mathfrak{M}^*(A)$ a <u>cone</u> with vertex N and of degree i is a hypersurface which is a member of $H_i^*(A,N)$.

To express the idea of "<u>the linear variety spanned by</u>" (like the line joining two points, plane containing two intersecting lines etc.) we use the notation $\Delta(A,-,-,\ldots)$; thus for homogeneous ideals $Q_1,\ldots,Q_s$, $\Delta(A,Q_1,\ldots,Q_s)$ denotes the flat spanned by $Q_1,\ldots,Q_s$.

Now let us point out the algebraic counterparts of some more <u>concepts about irreducible subvarieties</u> of A:

i-dimensional = member of $\mathfrak{Q}_i(A)$,
linear = member of $\mathfrak{Q}^1(A)$,
linear i-dimensional = member of $\mathfrak{Q}_i^1(A)$;

and hence, as particular cases, we have

point = member of $\mathfrak{Q}_0(A)$,
curve = member of $\mathfrak{Q}_1(A)$,
line = member of $\mathfrak{Q}_1^1(A)$,
plane = member of $\mathfrak{Q}_2^1(A)$, etc.

Now let C be an irreducible subvariety of A. Above concepts can be easily generalized to the embedded variety $[A,C]$. For example, irreducible subvarieties of C are members of

$$\mathfrak{Q}([A,C]) = \{P \in \mathfrak{Q}(A) : C \subset P\} .$$

A whole array of notations to describe (algebraically) subvarieties of several types can be located in §15. We illustrate the use of some of them. Thus let $Q_1,\ldots,Q_s$ be irreducible subvarieties of a projective space A. Then we have the following:

Q_1 lies on $Q_2 \Leftrightarrow Q_2$ passes through Q_1

$$\Leftrightarrow Q_2 \subset Q_1$$

$$\Leftrightarrow Q_1 \in \mathfrak{Q}(A,Q_2) ,$$

Q_1 lies outside $Q_2 \Leftrightarrow Q_2 \not\subset Q_1$

$$\Leftrightarrow Q_1 \in \mathfrak{Q}(A,\backslash Q_2) ,$$

Q_1,Q_2 are skew $\Leftrightarrow Q_1,Q_2$ do not meet

$$\Leftrightarrow \mathfrak{Q}(A,Q_1) \cap \mathfrak{Q}(A,Q_2) = \phi ,$$

and in case $\mathrm{Dim}\, A \geq 1$,

$Q_1,\ldots,Q_s$ are collinear $\Leftrightarrow$ some line Q passes through $Q_1,\ldots,Q_s$

$\Leftrightarrow Q \subset \mathfrak{Q}_1 \cap \mathfrak{Q}_2 \cap \ldots \cap Q_s$, for some $Q \in \mathfrak{Q}_1^1(A)$,

$\Leftrightarrow \mathrm{Emdim}[A,Q_1,\ldots,Q_s] \leq 1$, etc.

Also, when A is a projective space, osculating flats are osculating linear varieties and $T([A,\Phi],N)$ = the tangent cone of Φ at N where Φ is a hypersurface passing through a point N.

Finally we come to the concept of intersection multiplicity. There are two types of situations which we study.

For two hypersurfaces Φ,ψ in a projective space A such that Φ,ψ have no common factor, the interesection multiplicity of Φ and ψ at (along) a (Dim A-2)-dimensional (irreducible) subvariety P of A is defined to be $\mu([A,\Phi,\psi],P)$.

If C is an irreducible curve in an irreducible projective variety A, then the function field of C is $\mathfrak{R}([A,C]) = \mathfrak{R}(A/C)$, the set of points of C (in A) is $\mathfrak{Q}_0([A,C])$ and the set of all branches (places) at various points of C is $\mathfrak{Z}([A,C]) = \mathfrak{Z}(A/C)$. Note that in this, C may be $\{0\}$ and we may be talking about the abstract irreducible curve A, which may also be thought of as the curve

$[A,\{0\}]$ embedded in A.

For each branch $V \in \mathfrak{Z}([A,C])$ there is a unique point P of C in A at which V is centered and we denote P by $\mathfrak{Z}^*([A,C],V)$. Any set I of homogeneous elements represents, in the algebraic sense, a subvariety (divisor) of C in A which depends only on the ideal IA. We define the <u>intersection multiplicity of</u> C <u>and</u> IA <u>in</u> A <u>along a branch</u> V of C to be $\mu([A,C],I,V)$. To get the intersection multiplicity at points or various pointsets of C we extend the definition by linearity. Points of C which have a bigger residue field than $H_0(A)$ should be thought of as several points (each point as many times as its residue degree) in the "proper" counting for Bezout's theorem; and hence we introduce the "corrected" intersection multiplicity $\mu^*([A,C],I,-)$. Of special interest is the measure of singularity of C (or the intersection multiplicity of local conductor ideals) which we denote by the symbol $\mu_{\mathfrak{C}}$ (uncorrected) or by $\mu^*_{\mathfrak{C}}$ (corrected). Hypersurfaces, whose intersection multiplicity along each branch of C is bigger than or equal to the measure of singularity are called adjoints, while those for which equality holds are true adjoints. The concepts of adjoints or true adjoints are defined for individual points, pointsets or all points of C. (See §23,24 for details.)

The genus of the curve $[A,C]$ as a birational invariant of the function field $\mathfrak{K}([A,C])$ is defined as $\text{genus}(\mathfrak{K}([A,C]))$ and is studied in Chapter III (§30 and for $C = \{0\}$ case, in §29). We have included a proof, that (genus$(\mathfrak{K}([A,C])$ is zero and there is a residually rational place) $\Leftrightarrow$ ($\mathfrak{K}([A,C])$ is of the form $H_0(A/C)(t)$ for some t). (See (28.12) and (29.2.2) for the abstract $(C = 0)$ case, and (30.2.2) for the embedded case.)

(43.2) AFFINE VARIETIES. An affine irreducible variety over a ground field k is basically any variety whose coordinate ring is an affine domain over k; but we consider those varieties which also possess an affine coordinate system, or more intrinsically, an equivalence class of affine coordinate systems (under nonsingular affine transformations). Thus, an <u>abstract affine irreducible variety over a ground field</u> k is a filtered domain over the ground field k. An affine irreducible variety A is an <u>affine space</u> (affine (dim A)-space), if dim A = emdim A. An <u>affine line</u> and <u>an affine plane</u> are are respectively an affine 1-space and an affine 2-space.

Now let A be an affine irreducible variety. As before, any ideal in A may be thought of as representing a subvariety of A, in the algebraic sense, while we restrict to ideals which are their own radicals in a geometric set-up. Again we have that an <u>irreducible subvariety of</u> A is any member of $\mathfrak{P}(A)$. Formally an irreducible subvariety C of A may be denoted by [A,C], and this is how it appears in various formulas. We have that A/C is naturally an abstract affine irreducible variety and we may also say that A/C is embedded in A or that C defines the abstract variety A/C in A.

In this sense, emdim A gets an appropriate meaning, "embedding dimension of (the variety) A," namely, emdim A = min{r: A can be embedded in an affine r-space}.

The basic reason for attaching the affine structure (filtration) to A, is that then we are able to put A naturally "inside" a projective variety which may be termed as <u>a natural projective completion of</u> A, which is what we defined as a homogenization of A, i.e., $H_z(A)$ Note that any natural projective completion of an affine r-space is a projective r-space. Under the natural projective completion of A, all subvarieties of A (including those in the algebraic sense) also acquire their natural projective completion as subvarieties of $H_z(A)$ given by the same process of homogenization i.e. of "taking H_z of". Note that the hyperplane $zH_z(A)$ is usually termed as the hyperplane at infinity for A.

Conversely, starting with an abstract projective variety B, we may fix a hyperplane π and replace all subvarieties (including those in the algebraic sense) by their "parts outside π " or apply the process of dehomogenization, denoted by $F_z(-)$ where $\pi = zB$.

We introduce the appropriate notations for $H_z(-)$ and $F_z(-)$ in Chapter IV, and study various properties. The basic philosophy is, of course, that part of B outside π is "isomorphic" as an affine variety to $F_z(B)$ and B is in turn a natural projective completion of $F_z(B)$.

Again, let A be an abstract irreducible affine variety. A <u>hypersurface in</u> A is a member of $F^*(A)\setminus\{A\}$, and degree of a hypersurface Φ in A is $\deg_A\Phi$. If A is an affine 3-space or an affine plane, then a hypersurface is called an <u>affine surface</u> and an <u>affine curve</u> respectively. Hypersurfaces of degree 1 are called hyperplanes. A hypersurface in A is irreducible if, and only if, it belongs to $\mathfrak{P}(A)$.

$F_0(A)$-vector-spaces of hyperplanes of A serve the same function as linearities in an abstract irreducible projective variety. We are particularly interested in those vector-spaces which contain $F_0(A)$ or equivalently whose natural projective completion under a natural projective completion of A is a linearity in the hyperplane at infinity. These vector-spaces form $\mathfrak{L}^\infty(A)$. $\mathfrak{L}_i^\infty(A)$ consists of those vector-spaces which have dimension equal to (emdim A - i). In general $\mathfrak{L}(A)$ and $\mathfrak{L}_i(A)$ can be similarly defined, but we had no use for them.

Descriptions of i-dimensional, linear etc. irreducible subvarieties of A, run exactly parallel to those of projective varieties, except that $\mathfrak{Q}$ is replaced by $\mathfrak{P}$ throughout. (See §2.)

<u>Function field</u> of an affine irreducible variety A is its quotient field $\mathfrak{Z}(A)$ and in the embedded case when A is replaced by [R, P], we take $\mathfrak{Z}([R,P]) = \mathfrak{Z}(R/P)$.

The genus of an affine irreducible curve A (or [R, P] in the embedded case) is defined to be the genus of its natural projective completion, which in fact is simply genus($\mathfrak{Z}(A)$, $F_0(A)$)

(genus$(\mathfrak{J}(R/P), F_0(R/P))$).

In particular, let us note that if A is an irreducible affine curve with only one place at infinity (which, then, is residually rational), it follows that

A has genus zero $\Rightarrow \mathfrak{J}(A) = k(t)$ for some t where $k = F_0(A)$

$\Rightarrow A = k[t']$ for a suitable t', where the unique place at infinity for A is the $(1/t')$-adic valuation.

Finally we come to the concept of intersection multiplicity. This also runs entirely parallel to the projective case; in fact intersection multiplicities in both affine and projective cases at a point (or along an irreducible subvariety) are defined by extending all relavent ideals to the local ring of the point (or of the irreducible subvariety). What is different in the two cases is the process of constructing the local ring: in the affine case, we take the usual localization, in the projective case, we take the "homogeneous" localization (denoted by $\mathfrak{R}(-,-)$). To keep the distinction between the two cases, we use the letter λ in place of μ, throughout. Also $F_0(A)$ takes the place of $H_0(A)$ and λ^* takes the place of μ^*. The various definitions related to the local intersection multiplicity λ appear in §5,6. Note that the filtration on A is not involved in the definitions of intersection multiplicity except in the definition of λ^*.

The concepts of conductor, adjoint, true adjoint etc. are also similar to those in projective curves; the definitions are given in §5. Again the filtration is not involved in the definitions. Note that we have studied in Chapter I, those concepts of affine curves which do not depend on the filtration (affine structure); and the rest of them are studied in Chapter IV.

§44. Index to notations.

NOTE. The following list of notations is prepared in the order of appearance. Whenever a ( ) appears in the following list, in the actual notation it may be filled with any tuple (as defined) or sometimes dropped altogether. For example, when we list " $\mathfrak{Q}^{()}_{()}( )$ " we are refering to any of " $\mathfrak{Q}^1(A,I)$ " , " $\mathfrak{Q}^1_i([A,I])$ ", " $\mathfrak{Q}(A,x,P)$ ", ... etc.

Notation	Page Number
$\mathfrak{E}_{(\ )}^{(\ )}$	4.19
$\mathfrak{g}(\quad)$	5.13, 5.18

§45. Index to topics.

(45.1) INDEX TO TERMINOLOGY.

The following index provides listings for various terminology and certain theorems. We point out that §43 may be further useful in locating the different meanings of a word with several listings.

(45.2) BIBLIOGRAPHY

[1] Abhyankar, Shreeram S.: Resolution of singularities of embedded algebraic surfaces, Academic Press, New York and London (1966).

[2] ______________________: Local rings of high embedding dimension, Amer. J. Math., 89, No.4, 1073-1077 (1967).

[3] ______________________: Algebraic space curves, Les Presses de l'Université de montreal (1971).

[4] ______________________ and Moh Tzuong-tsieng: Embeddings of the line in the plane, submitted to J. Reine Angew. Math.

[5] ______________________: The adjoint condition revisited, to appear in Math. Student.

[6] Bass, Hyman: On the ubiquity of Gorenstein rings, Math. Z., 82, 8-28 (1963).

[7] Berger, Robert: Differentialmoduln eindimensionaler localer Ringe, Math. Z., 81, 326-354 (1963).

[8] Cohen, I.S.: On the structure and ideal theory of complete rings, Trans. Amer. Math. Soc., 59, No.1, 54-106 (1946).

[9] Chevalley, Claude: Introduction to the theory of algebraic functions of one variable, American Mathematical Society, Mathematical Surveys VI (1951).

[10] Kunz, Ernst: Differentialformen inseparabler algebraischer Funktionenköper, Math. Z., 76, 56-74 (1961).

[11] ___________: Volständige Durchschnitte und Differente, Arch. Math. (Basel), XIX , 47-58 (1968).

[12] Murthy, M.P. and Towber J.: Algebraic vector bundles over $\mathbb{A}^3$ are trivial, Invent. Math., 24, Fasc 3, 173-189 (1974).

[13] Nastold, Hans-Joachim: Zum Dualitätssatz inseparblen Funktionenkörper der Dimension 1, Math. Z., 76, 75-84 (1961).

[14] Nöther, Max: Zur Theorie des eindeutigen Entsprechens algebraischer Gebilde von beliebig vièlen Dimension, Math. Ann., 2, 293-316 (1870).

[15] ___________: Über einen Satz aus der Theorie der algebraischen Functionen, Math. Ann., 6, 351-359 (1873).

[16] Roquette, P.: Über den Singularitätsgrad endimensionaler Ringe II, J. Reine Angew. Math., 209, 12-16 (1962).

[17] Zariski, Oscar and Falb, Peter: On differentials in function fields, Amer. J. Math., 88, No.3, 542-566 (1961).

[18] Zariski, Oscar and Samuel Pierre: Commutative algebra (vol. I,II), Van Nonstrand, 1958.

Vol. 247: Lectures on Operator Algebras. Tulane University Ring and Operator Theory Year, 1970–1971. Volume II. XI, 786 pages. 1972. DM 40,–

Vol. 248: Lectures on the Applications of Sheaves to Ring Theory. Tulane University Ring and Operator Theory Year, 1970–1971. Volume III. VIII, 315 pages. 1971. DM 26,–

Vol. 249: Symposium on Algebraic Topology. Edited by P. J. Hilton. VII, 111 pages. 1971. DM 16,–

Vol. 250: B. Jónsson, Topics in Universal Algebra. VI, 220 pages. 1972. DM 20,–

Vol. 251: The Theory of Arithmetic Functions. Edited by A. A. Gioia and D. L. Goldsmith VI, 287 pages. 1972. DM 24,–

Vol. 252: D. A. Stone, Stratified Polyhedra. IX, 193 pages. 1972. DM 18,–

Vol. 253: V. Komkov, Optimal Control Theory for the Damping of Vibrations of Simple Elastic Systems. V, 240 pages. 1972. DM 20,–

Vol. 254: C. U. Jensen, Les Foncteurs Dérivés de $\varprojlim$ et leurs Applications en Théorie des Modules. V, 103 pages. 1972. DM 16,–

Vol. 255: Conference in Mathematical Logic – London '70. Edited by W. Hodges. VIII, 351 pages. 1972. DM 26,–

Vol. 256: C. A. Berenstein and M. A. Dostal, Analytically Uniform Spaces and their Applications to Convolution Equations. VII, 130 pages. 1972. DM 16,–

Vol. 257: R. B. Holmes, A Course on Optimization and Best Approximation. VIII, 233 pages. 1972. DM 20,–

Vol. 258: Séminaire de Probabilités VI. Edited by P. A. Meyer. VI, 253 pages. 1972. DM 22,–

Vol. 259: N. Moulis, Structures de Fredholm sur les Variétés Hilbertiennes. V, 123 pages. 1972. DM 16,–

Vol. 260: R. Godement and H. Jacquet, Zeta Functions of Simple Algebras. IX, 188 pages. 1972. DM 18,–

Vol. 261: A. Guichardet, Symmetric Hilbert Spaces and Related Topics. V, 197 pages. 1972. DM 18,–

Vol. 262: H. G. Zimmer, Computational Problems, Methods, and Results in Algebraic Number Theory. V, 103 pages. 1972. DM 16,–

Vol. 263: T. Parthasarathy, Selection Theorems and their Applications. VII, 101 pages. 1972. DM 16,–

Vol. 264: W. Messing, The Crystals Associated to Barsotti-Tate Groups: With Applications to Abelian Schemes. III, 190 pages. 1972. DM 18,–

Vol. 265: N. Saavedra Rivano, Catégories Tannakiennes. II, 418 pages. 1972. DM 26,–

Vol. 266: Conference on Harmonic Analysis. Edited by D. Gulick and R. L. Lipsman. VI, 323 pages. 1972. DM 24,–

Vol. 267: Numerische Lösung nichtlinearer partieller Differential- und Integro-Differentialgleichungen. Herausgegeben von R. Ansorge und W. Törnig, VI, 339 Seiten. 1972. DM 26,–

Vol. 268: C. G. Simader, On Dirichlet's Boundary Value Problem. IV, 238 pages. 1972. DM 20,–

Vol. 269: Théorie des Topos et Cohomologie Etale des Schémas. (SGA 4). Dirigé par M. Artin, A. Grothendieck et J. L. Verdier. XIX, 525 pages. 1972. DM 50,–

Vol. 270: Théorie des Topos et Cohomologie Etale des Schémas. Tome 2. (SGA 4). Dirigé par M. Artin, A. Grothendieck et J. L. Verdier. V, 418 pages. 1972. DM 50,–

Vol. 271: J. P. May, The Geometry of Iterated Loop Spaces. IX, 175 pages. 1972. DM 18,–

Vol. 272: K. R. Parthasarathy and K. Schmidt, Positive Definite Kernels, Continuous Tensor Products, and Central Limit Theorems of Probability Theory. VI, 107 pages. 1972. DM 16,–

Vol. 273: U. Seip, Kompakt erzeugte Vektorräume und Analysis. IX, 119 Seiten. 1972. DM 16,–

Vol. 274: Toposes, Algebraic Geometry and Logic. Edited by. F. W. Lawvere. VI, 189 pages. 1972. DM 18,–

Vol. 275: Séminaire Pierre Lelong (Analyse) Année 1970–1971. VI, 181 pages. 1972. DM 18,–

Vol. 276: A. Borel, Représentations de Groupes Localement Compacts. V, 98 pages. 1972. DM 16,–

Vol. 277: Séminaire Banach. Edité par C. Houzel. VII, 229 pages. 1972. DM 20,–

Vol. 278: H. Jacquet, Automorphic Forms on GL(2). Part II. XIII, 142 pages. 1972. DM 16,–

Vol. 279: R. Bott, S. Gitler and I. M. James, Lectures on Algebraic and Differential Topology. V, 174 pages. 1972. DM 18,–

Vol. 280: Conference on the Theory of Ordinary and Partial Differential Equations. Edited by W. N. Everitt and B. D. Sleeman. XV, 367 pages. 1972. DM 26,–

Vol. 281: Coherence in Categories. Edited by S. Mac Lane. VII, 235 pages. 1972. DM 20,–

Vol. 282: W. Klingenberg und P. Flaschel, Riemannsche Hilbertmannigfaltigkeiten. Periodische Geodätische. VII, 211 Seiten. 1972. DM 20,–

Vol. 283: L. Illusie, Complexe Cotangent et Déformations II. VII, 304 pages. 1972. DM 24,–

Vol. 284: P. A. Meyer, Martingales and Stochastic Integrals I. VI, 89 pages. 1972. DM 16,–

Vol. 285: P. de la Harpe, Classical Banach-Lie Algebras and Banach-Lie Groups of Operators in Hilbert Space. III, 160 pages. 1972. DM 16,–

Vol. 286: S. Murakami, On Automorphisms of Siegel Domains. V, 95 pages. 1972. DM 16,–

Vol. 287: Hyperfunctions and Pseudo-Differential Equations. Edited by H. Komatsu. VII, 529 pages. 1973. DM 36,–

Vol. 288: Groupes de Monodromie en Géométrie Algébrique. (SGA 7 I). Dirigé par A. Grothendieck. IX, 523 pages. 1972. DM 50,–

Vol. 289: B. Fuglede, Finely Harmonic Functions. III, 188. 1972. DM 18,–

Vol. 290: D. B. Zagier, Equivariant Pontrjagin Classes and Applications to Orbit Spaces. IX, 130 pages. 1972. DM 16,–

Vol. 291: P. Orlik, Seifert Manifolds. VIII, 155 pages. 1972. DM 16,–

Vol. 292: W. D. Wallis, A. P. Street and J. S. Wallis, Combinatorics: Room Squares, Sum-Free Sets, Hadamard Matrices. V, 508 pages. 1972. DM 50,–

Vol. 293: R. A. DeVore, The Approximation of Continuous Functions by Positive Linear Operators. VIII, 289 pages. 1972. DM 24,–

Vol. 294: Stability of Stochastic Dynamical Systems. Edited by R. F. Curtain. IX, 332 pages. 1972. DM 26,–

Vol. 295: C. Dellacherie, Ensembles Analytiques, Capacités, Mesures de Hausdorff. XII, 123 pages. 1972. DM 16,–

Vol. 296: Probability and Information Theory II. Edited by M. Behara, K. Krickeberg and J. Wolfowitz. V, 223 pages. 1973. DM 20,–

Vol. 297: J. Garnett, Analytic Capacity and Measure. IV, 138 pages. 1972. DM 16,–

Vol. 298: Proceedings of the Second Conference on Compact Transformation Groups. Part 1. XIII, 453 pages. 1972. DM 32,–

Vol. 299: Proceedings of the Second Conference on Compact Transformation Groups. Part 2. XIV, 327 pages. 1972. DM 26,–

Vol. 300: P. Eymard, Moyennes Invariantes et Représentations Unitaires. II. 113 pages. 1972. DM 16,–

Vol. 301: F. Pittnauer, Vorlesungen über asymptotische Reihen. VI, 186 Seiten. 1972. DM 18,–

Vol. 302: M. Demazure, Lectures on p-Divisible Groups. V, 98 pages. 1972. DM 16,–

Vol. 303: Graph Theory and Applications. Edited by Y. Alavi, D. R. Lick and A. T. White. IX, 329 pages. 1972. DM 26,–

Vol. 304: A. K. Bousfield and D. M. Kan, Homotopy Limits, Completions and Localizations. V, 348 pages. 1972. DM 26,–

Vol. 305: Théorie des Topos et Cohomologie Etale des Schémas. Tome 3. (SGA 4). Dirigé par M. Artin, A. Grothendieck et J. L. Verdier. VI, 640 pages. 1973. DM 50,–

Vol. 306: H. Luckhardt, Extensional Gödel Functional Interpretation. VI, 161 pages. 1973. DM 18,–

Vol. 307: J. L. Bretagnolle, S. D. Chatterji et P.-A. Meyer, Ecole d'été de Probabilités: Processus Stochastiques. VI, 198 pages. 1973. DM 20,–

Vol. 308: D. Knutson, λ-Rings and the Representation Theory of the Symmetric Group. IV, 203 pages. 1973. DM 20,–

Vol. 309: D. H. Sattinger, Topics in Stability and Bifurcation Theory. VI, 190 pages. 1973. DM 18,–

Vol. 310: B. Iversen, Generic Local Structure of the Morphisms in Commutative Algebra. IV, 108 pages. 1973. DM 16,–

Vol. 311: Conference on Commutative Algebra. Edited by J. W. Brewer and E. A. Rutter. VII, 251 pages. 1973. DM 22,–

Vol. 312: Symposium on Ordinary Differential Equations. Edited by W. A. Harris, Jr. and Y. Sibuya. VIII, 204 pages. 1973. DM 22,–

Vol. 313: K. Jörgens and J. Weidmann, Spectral Properties of Hamiltonian Operators. III, 140 pages. 1973. DM 16,–

Vol. 314: M. Deuring, Lectures on the Theory of Algebraic Functions of One Variable. VI, 151 pages. 1973. DM 16,–

Vol. 315: K. Bichteler, Integration Theory (with Special Attention to Vector Measures). VI, 357 pages. 1973. DM 26,–

Vol. 316: Symposium on Non-Well-Posed Problems and Logarithmic Convexity. Edited by R. J. Knops. V, 176 pages. 1973. DM 18,–

Vol. 317: Séminaire Bourbaki – vol. 1971/72. Exposés 400–417. IV, 361 pages. 1973. DM 26,–

Vol. 318: Recent Advances in Topological Dynamics. Edited by A. Beck, VIII, 285 pages. 1973. DM 24,–

Vol. 319: Conference on Group Theory. Edited by R. W. Gatterdam and K. W. Weston. V, 188 pages. 1973. DM 18,–

Vol. 320: Modular Functions of One Variable I. Edited by W. Kuyk. V, 195 pages. 1973. DM 18,–

Vol. 321: Séminaire de Probabilités VII. Edité par P. A. Meyer. VI, 322 pages. 1973. DM 26,–

Vol. 322: Nonlinear Problems in the Physical Sciences and Biology. Edited by I. Stakgold, D. D. Joseph and D. H. Sattinger. VIII, 357 pages. 1973. DM 26,–

Vol. 323: J. L. Lions, Perturbations Singulières dans les Problèmes aux Limites et en Contrôle Optimal. XII, 645 pages. 1973. DM 42,–

Vol. 324: K. Kreith, Oscillation Theory. VI, 109 pages. 1973. DM 16,–

Vol. 325: Ch.-Ch. Chou, La Transformation de Fourier Complexe et L'Equation de Convolution. IX, 137 pages. 1973. DM 16,–

Vol. 326: A. Robert, Elliptic Curves. VIII, 264 pages. 1973. DM 22,–

Vol. 327: E. Matlis, 1-Dimensional Cohen-Macaulay Rings. XII, 157 pages. 1973. DM 18,–

Vol. 328: J. R. Büchi and D. Siefkes, The Monadic Second Order Theory of All Countable Ordinals. VI, 217 pages. 1973. DM 20,–

Vol. 329: W. Trebels, Multipliers for (C, α)-Bounded Fourier Expansions in Banach Spaces and Approximation Theory. VII, 103 pages. 1973. DM 16,–

Vol. 330: Proceedings of the Second Japan-USSR Symposium on Probability Theory. Edited by G. Maruyama and Yu. V. Prokhorov. VI, 550 pages. 1973. DM 36,–

Vol. 331: Summer School on Topological Vector Spaces. Edited by L. Waelbroeck. VI, 226 pages. 1973. DM 20,–

Vol. 332: Séminaire Pierre Lelong (Analyse) Année 1971-1972. V, 131 pages. 1973. DM 16,–

Vol. 333: Numerische, insbesondere approximationstheoretische Behandlung von Funktionalgleichungen. Herausgegeben von R. Ansorge und W. Törnig. VI, 296 Seiten. 1973. DM 24,–

Vol. 334: F. Schweiger, The Metrical Theory of Jacobi-Perron Algorithm. V, 111 pages. 1973. DM 16,–

Vol. 335: H. Huck, R. Roitzsch, U. Simon, W. Vortisch, R. Walden, B. Wegner und W. Wendland, Beweismethoden der Differentialgeometrie im Großen. IX, 159 Seiten. 1973. DM 18,–

Vol. 336: L'Analyse Harmonique dans le Domaine Complexe. Edité par E. J. Akutowicz. VIII, 169 pages. 1973. DM 18,–

Vol. 337: Cambridge Summer School in Mathematical Logic. Edited by A. R. D. Mathias and H. Rogers. IX, 660 pages. 1973. DM 42,–

Vol: 338: J. Lindenstrauss and L. Tzafriri, Classical Banach Spaces. IX, 243 pages. 1973. DM 22,–

Vol. 339: G. Kempf, F. Knudsen, D. Mumford and B. Saint-Donat, Toroidal Embeddings I. VIII, 209 pages. 1973. DM 20,–

Vol. 340: Groupes de Monodromie en Géométrie Algébrique. (SGA 7 II). Par P. Deligne et N. Katz. X, 438 pages. 1973. DM 40,–

Vol. 341: Algebraic K-Theory I, Higher K-Theories. Edited by H. Bass. XV, 335 pages. 1973. DM 26,–

Vol. 342: Algebraic K-Theory II, "Classical" Algebraic K-Theory, and Connections with Arithmetic. Edited by H. Bass. XV, 527 pages. 1973. DM 36,–

Vol. 343: Algebraic K-Theory III, Hermitian K-Theory and Geometric Applications. Edited by H. Bass. XV, 572 pages. 1973. DM 38,–

Vol. 344: A. S. Troelstra (Editor), Metamathematical Investigation of Intuitionistic Arithmetic and Analysis. XVII, 485 pages. 1973. DM 34,–

Vol. 345: Proceedings of a Conference on Operator Theory. Edited by P. A. Fillmore. VI, 228 pages. 1973. DM 20,–

Vol. 346: Fučík et al., Spectral Analysis of Nonlinear Operators. II, 287 pages. 1973. DM 26,–

Vol. 347: J. M. Boardman and R. M. Vogt, Homotopy Invariant Algebraic Structures on Topological Spaces. X, 257 pages. 1973. DM 22,–

Vol. 348: A. M. Mathai and R. K. Saxena, Generalized Hypergeometric Functions with Applications in Statistics and Physical Sciences. VII, 314 pages. 1973. DM 26,–

Vol. 349: Modular Functions of One Variable II. Edited by W. Kuyk and P. Deligne. V, 598 pages. 1973. DM 38,–

Vol. 350: Modular Functions of One Variable III. Edited by W. Kuyk and J.-P. Serre. V, 350 pages. 1973. DM 26,–

Vol. 351: H. Tachikawa, Quasi-Frobenius Rings and Generalizations. XI, 172 pages. 1973. DM 18,–

Vol. 352: J. D. Fay, Theta Functions on Riemann Surfaces. V, 137 pages. 1973. DM 16,–

Voi. 353: Proceedings of the Conference. on Orders, Group Rings and Related Topics. Organized by J. S. Hsia, M. L. Madan and T. G. Ralley. X, 224 pages. 1973. DM 20,–

Vol. 354: K. J. Devlin, Aspects of Constructibility. XII, 240 pages. 1973. DM 22,–

Vol. 355: M. Sion, A Theory of Semigroup Valued Measures. V, 140 pages. 1973. DM 16,–

Vol. 356: W. L. J. van der Kallen, Infinitesimally Central-Extensions of Chevalley Groups. VII, 147 pages. 1973. DM 16,–

Vol. 357: W. Borho, P. Gabriel und R. Rentschler, Primideale in Einhüllenden auflösbarer Lie-Algebren. V, 182 Seiten. 1973. DM 18,–

Vol. 358: F. L. Williams, Tensor Products of Principal Series Representations. VI, 132 pages. 1973. DM 16,–

Vol. 359: U. Stammbach, Homology in Group Theory. VIII, 183 pages. 1973. DM 18,–

Vol. 360: W. J. Padgett and R. L. Taylor, Laws of Large Numbers for Normed Linear Spaces and Certain Fréchet Spaces. VI, 111 pages. 1973. DM 16,–

Vol. 361: J. W. Schutz, Foundations of Special Relativity: Kinematic Axioms for Minkowski Space Time. XX, 314 pages. 1973. DM 26,–

Vol. 362: Proceedings of the Conference on Numerical Solution of Ordinary Differential Equations. Edited by D. Bettis. VIII, 490 pages. 1974. DM 34,–

Vol. 363: Conference on the Numerical Solution of Differential Equations. Edited by G. A. Watson. IX, 221 pages. 1974. DM 20,–

Vol. 364: Proceedings on Infinite Dimensional Holomorphy. Edited by T. L. Hayden and T. J. Suffridge. VII, 212 pages. 1974. DM 20,–

Vol. 365: R. P. Gilbert, Constructive Methods for Elliptic Equations. VII, 397 pages. 1974. DM 26,–

Vol. 366: R. Steinberg, Conjugacy Classes in Algebraic Groups (Notes by V. V. Deodhar). VI, 159 pages. 1974. DM 18,–

Vol. 367: K. Langmann und W. Lütkebohmert, Cousinverteilungen und Fortsetzungssätze. VI, 151 Seiten. 1974. DM 16,–

Vol. 368: R. J. Milgram, Unstable Homotopy from the Stable Point of View. V, 109 pages. 1974. DM 16,–

Vol. 369: Victoria Symposium on Nonstandard Analysis. Edited by A. Hurd and P. Loeb. XVIII, 339 pages. 1974. DM 26,–

Vol. 370: B. Mazur and W. Messing, Universal Extensions and One Dimensional Crystalline Cohomology. VII, 134 pages. 1974. DM 16,–